T5-AVM-703

Glutamine Repeats and Neurodegenerative Diseases: Molecular Aspects

Glutamine Repeats and Neurodegenerative Diseases: Molecular Aspects

Edited by

PETER HARPER
Professor of Medical Genetics, Institute of Medical Genetics, University of Wales College of Medicine, Cardiff

and

MAX PERUTZ
MRC Laboratory of Molecular Biology, Cambridge

Oxford University Press, Great Clarendon Street, Oxford OX2 6DP

Oxford University Press is a department of the University of Oxford.
It furthers the University's objective of excellence in research, scholarship,
and education by publishing worldwide in
Oxford New York
Athens Auckland Bangkok Bogotá Buenos Aires Calcutta
Cape Town Chennai Dar es Salaam Delhi Florence Hong Kong Istanbul
Karachi Kuala Lumpur Madrid Melbourne Mexico City Mumbai
Nairobi Paris São Paulo Singapore Taipei Tokyo Toronto Warsaw
with associated companies in
Berlin Ibadan

Oxford is a registered trade mark of Oxford University Press
in the UK and in certain other countries

Published in the United States
by Oxford University Press Inc., New York

First published 2001

British Library Cataloguing in Publication Data available

Library of Congress Cataloging in Publication Data

1 3 5 7 9 10 8 6 4 2

ISBN 0 19 8506856

Typeset by
Florence Production Ltd, Stoodleigh, Devon

Printed in Great Britain on acid-free paper by

TJ International Ltd, Padstow, Cornwall

Contents

Preface

The recognition that a series of important human inherited diseases is caused by the expansion of unstable mutations in trinucleotide repeat sequences is still less than ten years old; the hypothesis that for a sub-group of these with CAG repeats, notably Huntington's disease, the expanded polyglutamine sequence corresponding to this might itself be directly responsible for the progressive central nervous system pathology characteristic of this group is still more recent.

The contributions in this volume, originally submitted to the Royal Society no more than five and a half years after the discovery of the gene for Huntington's disease, testify to that discovery's tremendous stimulus to new research and to the vigour with which the scientific community responded to the challenge presented by their new type of genetic disease, caused by the expansion of a repeat of a single amino acid, glutamine. At the same time, the papers show how far we still are from a full understanding of the biochemical mechanism of the disease.

Themes for future discussion meetings must be submitted to the Royal Society two years in advance. In the autumn of 1996, there was as yet little work to discuss, but we decided to take a chance that the next two years would bring important new developments. Once the Royal Society had agreed to sponsor a two-day discussion, additional support by the Hereditary Disease Foundation and by Merck Research Laboratories made it possible to bring together in London most workers in the field from all over the world, and support from the Novartis Foundation allowed us to extend the discussion to a third day, so as to cover the several other diseases due to expansion of glutamine diseases which recently have come to light.

Our gamble has paid off: 1997 saw the beginning of discoveries which have transformed the field and greatly expanded the scope of research in it. Gillian Bates and her collaborators in London introduced a fragment of the Huntington gene, including the codons for the glutamine repeat, into mice and showed that mice transgenic for the fragment with a repeat of 18 glutamines remained healthy, while those with repeats of *ca.*150 glutamines developed symptoms similar to some of those of the human disease. Stephen Davies, also in London, found in the cortex and striatum of the affected mice neural intranuclear inclusions consisting of fibrous and granular aggregates of Bates' Huntington fragment. Shortly afterwards, Marian DiFiglia at the Harvard Medical School, in collaboration with the London workers, announced the discovery of similar aggregates in post-mortem sections of human brains.

At the same time, Erich Wanker and his colleagues in Berlin introduced Bates' fragment into an *Escherichia coli* expression system and synthesized the fragment with an entire series of glutamine repeats of different length.

Wanker's work has led to a striking parallel between the behaviour of the Huntington fragments *in vitro* and the pathology of the diseases due to expanded glutamine repeats *in vivo*. *In vitro*, fragments with fewer than 37 repeats are soluble, while those with more than 40 repeats form insoluble aggregates. *In vivo*, humans (and mice) with fewer than 37 repeats remain healthy, while those with more than 40 sooner or later suffer from neurodegeneration. This correlation can hardly be coincidental.

Nevertheless, this volume includes reports of the development of neurodegeneration without aggregates and of aggregates without neurodegeneration. These findings led some of the workers in the field to the hypothesis that the aggregates are merely an epiphenomenon and that the true cause of the disease lies elsewhere. Further research will be needed to clarify the reasons for these contradictory observations.

The arrangement of the chapters in this book reflects the main themes of the work presented at the Royal Society meeting. A clinically oriented introduction precedes accounts of basic work on Huntington's disease and other ataxias caused by expanded glutamine repeats. Studies of transgenic mice, *Drosophila* and unicellular systems make up most of the first section of the book, while later sections deal with studies of the huntingtin protein and the general mechanism of glutamine toxicity. The final section is devoted to work on other glutamine repeat disorders; in all of these, aggregates of the proteins with expanded glutamine repeats have now been discovered in the affected neurons.

The final chapter shows that the discovery of the neural aggregates in Huntington's disease and related ataxias has bought a new unity to the molecular pathology of neurodegenerative diseases. Alzheimer's disease is associated with extracellular deposits of the β-amyloid protein Aβ and intracellular deposits of the microtubule-associated protein tau. Parkinson's disease is associated with Lewy bodies made of filaments of the protein α-synuclein. Prion diseases are associated with aggregation of prion proteins, and a host of amyloid diseases with aggregation of other proteins. The aggregates seem to poison neurons—no matter whether they are deposited in the cell cytoplasm or in the cell nucleus. We do not yet understand why, but it might be because living cells are highly organized structures and cannot tolerate disruption by protein precipitates or because these precipitates adsorb and inactivate other essential proteins.

Most of the work described in this book represents fundamental research, but its potential applications to the understanding of pathology and to future therapy are striking. The transgenic models provide both an experimental approach to the disease process and an opportunity for testing new therapeutic agents which is already being taken up by more clinically orientated workers. The research on how the physical properties of huntingtin are altered by the length of the polyglutamine repeat provide a remarkable parallel with the clinical and genetic observations and an alternative approach to the testing of possible drugs.

We hope that bringing together this important body of work will help to stimulate further the already very active field, and to make it more accessible to clinicians and workers in allied areas of neuroscience. Other excellent books on trinucleotide repeat disorders have appeared, but they have concentrated on the mechanisms and consequences of genome instability, whereas the present book deals largely with the properties and effects of the polyglutamine repeat expansion at the protein level and its molecular and cellular consequences. We hope that readers will find this work as exciting and relevant as we do.

Cardiff and Cambridge

Peter Harper
Max Perutz

List of Contributors

A. Abel Neurogenetics Branch, National Institue of Neurological Diseases and Stroke, National Institutes of Health, Bethesda, MD 20892, USA

B. Amos Department of Zoology, Downing Street, Cambridge CB2 3EJ, UK

G. Annesi Istituto di Medicina Sperimentale e Biotecnologie CNR, Contzada Burga, 87050 Cosenza, Italy

N. Aronin Neuroendocrinology Laboratory, Department of Medicine, University of Massachusetts Medical School, 55 Lake Avenue North, Worcester, MA 01655, USA

C. K. Bailey Neurology Department, University of Pennsylvania School of Medicine, Philadelphia, PA 19104, USA

G. P. Bates Medical and Molecular Genetics, GKT Medical and Dental School, King's College, 8th Floor Guy's Tower, Guy's Hospital, London SE1 9RT, UK

N. M. Bonini Department of Biology, University of Pennsylvania, Philadelphia, PA 19104-6018, USA

D. R. Borchelt Department of Pathology, The Johns Hopkins University School of Medicine, Baltimore, MD 21205, USA

M. Bout MGC-Department of Human Genetics/Section of Molecular Carcinogenesis, Sylvius Laboratory, Leiden University Medical Center, Wassenaarseweg 72, 2333AL Leiden, The Netherlands

A. Brice INSERM U289, Hôpital de la Salpêtrière, 47 Bd de l'Hôpital 75651 Paris Cedex 13, France

J. A. Cearley Department of Biochemistry and Molecular Genetics, University of Alabama at Birmingham, Birmingham, AL 35294, USA

V. Charles Genetics and Molecular Biology Branch; National Human Genome Research Institute, National Institutes of Health, Building 49, Room 3A26, 49 Convent Drive MSC 4442, Bethesda, MD 20892, USA

G. Cooper Department of Zoology, Downing Street, Cambridge CB2 3EJ, UK

J. K. Cooper Department of Psychiatry and Behavioral Sciences, The Johns Hopkins University School of Medicine, Baltimore, MD 21205, USA

B. A. Cozens Department of Anatomy and Developmental Biology, University College London, Gower Street, London WC1E 6BT, UK

C. J. Cummings Department of Pediatrics and Program in Cell and Molecular Biology, Baylor College of Medicine, Houston, TX 77030, USA

S. W. Davies Department of Anatomy and Developmental Biology, University College London, Gower Street, London WC1E 6BT, UK

J. T. den Dunnen MGC-Department of Human Genetics/Section of Molecular Carcinogenesis, Sylvius Laboratory, Leiden University Medical Center, Wassenaarseweg 72, 2333AL Leiden, The Netherlands

P. J. Detloff Department of Biochemistry and Molecular Genetics, University of Alabama at Birmingham, Birmingham, AL 35294, USA

M. DiFiglia Laboratory of Cellular Neurobiology, Department of Neurology, Massachusetts General Hospital East, Charlestown, MA 02129, USA

P. Doherty Experimental Pathology, GKT Medical and Dental School, King's College, 4th Floor Hodgkin Building, Guy's Hospital, London SE1 9RT, UK

J. C. Dorsman MGC-Department of Human Genetics/Section of Molecular Carcinogenesis, Sylvius Laboratory, Leiden University Medical Center, Wassenaarseweg 72, 2333AL Leiden, The Netherlands

J. Fagart Institut de Génétique et de Biologie Moléculaire et Cellulaire (IGBMC), CNRS/INSERM/ULP, B.P. 163, 67404 Illkirch Cédex, C.U. de Strasbourg, France

K. H. Fischbeck Neurogenetics Branch, National Institue of Neurological Diseases and Stroke, National Institutes of Health, Bethesda, MD 20892, USA

M. Frontali Istituto di Medicina Sperimentale, CNR, Via Fosso del Cavaliere, 00133 Roma, Italy

L. Gan Centre for Molecular Medicine and Therapeutics, 980 West 28th Avenue and Department of Medical Genetics, University of British Columbia, Vancouver, British Columbia, Canada V5Z 4H4

M. Goedert Medical Research Council Laboratory of Molecular Biology, Hills Road, Cambridge CB2 2QH, UK

C.-A. Gutekunst Emory University School of Medicine, Atlanta, GA 30322, USA

A. S. Hackam Department of Ophthalmology, Maumenee 815, Wilmer Eye Institute, John Hopkins Hospital, 600 North Wolfe St., Baltimore, MD 21287, USA

P. S. Harper Institute of Medical Genetics, University of Wales College of Medicine, Heath Park, Cardiff CF4 4XN, UK

M. R. Hayden Centre for Molecular Medicine and Therapeutics, 980 West 28th Avenue and Department of Medical Genetics, University of British Columbia, Vancouver, British Columbia, Canada V5Z 4H4

S. M. Hersch Emory University School of Medicine, Atlanta, GA 30322, USA

J. G. Hodgson Centre for Molecular Medicine and Therapeutics, 980 West 28th Avenue and Department of Medical Genetics, University of British Columbia, Vancouver, British Columbia, Canada V5Z 4H4

B. Hollenbach Max-Planck-Institut für Molekulare Genetik, Ihnestrasse 73, D-14195 Berlin, Germany

D. Housman Center for Cancer Research, Massachusetts Institute of Technology, Cambridge, MA 02139, USA

C. Jodice Dipartimento di Biologia, Università Tor Vergata, Via Ricerca Scientifica, 00133 Roma, Italy

B. M. Jordan Center for Cancer Research, Massachusetts Institute of Technology, Cambridge, MA 02139, USA

A. L. Jones Institute of Medical Genetics, University of Wales College of Medicine, Cardiff CF4 4XN, UK

I. Kanazawa Department of Neurology, Graduate School of Medicine, University of Tokyo, Tokyo 113–8655, Japan

A. Kazantsev Center for Cancer Research, Massachusetts Institute of Technology, Cambridge, MA 02139, USA

M. Kim Laboratory of Cellular Neurobiology, Department of Neurology, Massachusetts General Hospital East, Charlestown, MA 02129, USA

G. Laforet Neuroendocrinology Laboratory, Department of Medicine, University of Massachusetts Medical School, 55 Lake Avenue North, Worcester, MA 01655, USA

H. Lehrach Max-Planck-Institut für Molekulare Genetik, Ihnestrasse 73, D-14195 Berlin, Germany

A. Lieberman Neurogenetics Branch, National Institue of Neurological Diseases and Stroke, National Institutes of Health, Bethesda, MD 20892, USA

A. Lunkes Institut de Génétique et de Biologie Moléculaire et Cellulaire (IGBMC), CNRS/INSERM/ULP, B.P. 163, 67404 Illkirch Cédex, C.U. de Strasbourg, France

R. Lurz Max-Planck-Institut für Molekulare Genetik, Ihnestrasse 73, D-14195 Berlin, Germany

M. L. C. Maat-Schieman Department of Neurology, Leiden University Medical Center, Leiden, The Netherlands

A. Mahal Medical and Molecular Genetics, GKT Medical and Dental School, King's College, 8th Floor Guy's Tower, Guy's Hospital, London SE1 9RT, UK

J.-L. Mandel Institut de Génétique et de Biologie Moléculaire et Cellulaire (IGBMC), CNRS/INSERM/ULP, B.P. 163, 67404 Illkirch Cédex, C.U. de Strasbourg, France

L. Mangiarini Medical and Molecular Genetics, GKT Medical and Dental School, King's College, 8th Floor Guy's Tower, Guy's Hospital, London SE1 9RT, UK

R. L. Margolis Department of Psychiatry and Behavioral Sciences, The Johns Hopkins University School of Medicine, Baltimore, MD 21205, USA

D. E. Merry Neurology Department, University of Pennsylvania School of Medicine, Philadelphia, PA 19104, USA

D. Moras Institut de Génétique et de Biologie Moléculaire et Cellulaire (IGBMC), CNRS/INSERM/ULP, B.P. 163, 67404 Illkirch Cédex, C.U. de Strasbourg, France

A. Novelletto Dipartimento di Biologia, Università Tor Vergata, Via Ricerca Scientifica, 00133 Roma, Italy

F. C. Nucifora Jr Department of Psychiatry and Behavioral Sciences, The Johns Hopkins University School of Medicine, Baltimore, MD 21205, USA

J. M. Ordway Department of Biochemistry and Molecular Genetics, University of Alabama at Birmingham, Birmingham, AL 35294, USA

H. T. Orr Departments of Laboratory Medicine and Pathology, and Biochemistry and Institute of Human Genetics, University of Minnesota, Minneapolis, MN 55455, USA

M. F. Peters Department of Psychiatry and Behavioral Sciences, The Johns Hopkins University School of Medicine, Baltimore, MD 21205, USA

E. Preisinger Center for Cancer Research, Massachusetts Institute of Technology, Cambridge, MA 02139, USA

A. S. Raza Department of Anatomy and Developmental Biology, University College London, Gower Street, London WC1E 6BT, UK

P. H. Reddy Genetics and Molecular Biology Branch; National Human Genome Research Institute, National Institutes of Health, Building 49, Room 3A26, 49 Convent Drive MSC 4442, Bethesda, MD 20892, USA

R. A. C. Roos Department of Neurology, Leiden University Medical Center, Leiden, The Netherlands

C. A. Ross Department of Psychiatry and Behavioral Sciences, The Johns Hopkins University School of Medicine, Baltimore, MD 21205, USA

D. C. Rubinsztein Department of Medical Genetics, Cambridge Institute for Medical Research, Wellcome/MRC Building, Addenbrooke's Hospital, Hills Road, Cambridge CB2 2XY, UK

K. Sathasivam Medical and Molecular Genetics, GKT Medical and Dental School, King's College, 8th Floor Guy's Tower, Guy's Hospital, London SE1 9RT, UK

A. Sawa Department of Psychiatry and Behavioral Sciences, The Johns Hopkins University School of Medicine, Baltimore, MD 21205, USA

G. Schilling Department of Psychiatry and Behavioral Sciences, The Johns Hopkins University School of Medicine, Baltimore, MD 21205, USA

P. Schultz École Supérieure de Biotechnologie de Strasbourg, Boullevard Sébastien Brant, 67400 Illkirch, C.U. de Strasbourg, France

E. Scherzinger Max-Planck-Institut für Molekulare Genetik, Ihnestrasse 73, D-14195 Berlin, Germany

K. Schweiger Max-Planck-Institut für Molekulare Genetik, Ihnestrasse 73, D-14195 Berlin, Germany

A. H. Sharp Department of Psychiatry and Behavioral Sciences, The Johns Hopkins University School of Medicine, Baltimore, MD 21205, USA

S. Siesling Department of Neurology, Leiden University Medical Center, Leiden, The Netherlands

R. Singaraja Centre for Molecular Medicine and Therapeutics, 980 West 28th Avenue and Department of Medical Genetics, University of British Columbia, Vancouver, British Columbia, Canada V5Z 4H4

M. A. Smoor MGC-Department of Human Genetics/Section of Molecular Carcinogenesis, Sylvius Laboratory, Leiden University Medical Center, Wassenaarseweg 72, 2333AL Leiden, The Netherlands

D. A. Tagle Genetics and Molecular Biology Branch; National Human Genome Research Institute, National Institutes of Health, Building 49, Room 3A26, 49 Convent Drive MSC 4442, Bethesda, MD 20892, USA

Y. Trottier Institut de Génétique et de Biologie Moléculaire et Cellulaire (IGBMC), CNRS/INSERM/ULP, B.P. 163, 67404 Illkirch Cédex, C.U. de Strasbourg, France

M. Turmaine Department of Anatomy and Developmental Biology, University College London, Gower Street, London WC1E 6BT, UK

S. G. van Duinen Department of Neurology, Leiden University Medical Center, Leiden, The Netherlands

G. J. B. van Ommen MGC-Department of Human Genetics/Section of Molecular Carcinogenesis, Sylvius Laboratory, Leiden University Medical Center, Wassenaarseweg 72, 2333AL Leiden, The Netherlands

J. J. G. M. Verschuuren Department of Neurology, Leiden University Medical Center, Leiden, The Netherlands

E. E. Wanker Max-Planck-Institut für Molekulare Genetik, Ihnestrasse 73, D-14195 Berlin, Germany

W. O. Whetsell, Jr Department of Pathology, Vanderbilt University Medical Center, Nashville, TN 37232, USA.

J. D. Wood Department of Psychiatry and Behavioral Sciences, The Johns Hopkins University School of Medicine, Baltimore, MD 21205, USA

N. W. Wood Department of Clinical Neurology, Institute of Neurology, Queen Square, London WC1N 3BG, UK

B. Woodman Medical and Molecular Genetics, GKT Medical and Dental School, King's College, 8th Floor Guy's Tower, Guy's Hospital, London SE1 9RT, UK

P. F. Worth Department of Clinical Neurology, Institute of Neurology, Queen Square, London WC1N 3BG, UK

G. Zeder-Lutz Institut de Biologie Moléculaire et Cellulaire (IBMC), CNRS/ULP-9021, 67000 Strasbourg, France

T. Zhang Centre for Molecular Medicine and Therapeutics, 980 West 28th Avenue and Department of Medical Genetics, University of British Columbia, Vancouver, British Columbia, Canada V5Z 4H4

H. Y. Zoghbi Department of Molecular and Human Genetics, Baylor College of Medicine and Howard Hughes Medical Institute, Houston, TX 77030, USA

1

Huntington's disease: a clinical, genetic and molecular model for polyglutamine repeat disorders

Peter S. Harper

It is now well over a century since George Huntington, in a brief description of admirable clarity, delineated the main features of the neurodegenerative disorder that has since been known as Huntington's disease (HD) (Huntington 1872). Working as a family practitioner on Long Island, New York State, he drew on not only his own observations of the affected families under his care, but also on those of his father and grandfather in the same practice before him, covering a total period of 60 years.

Although limited to less than two pages, Huntington's 1872 description, later cited by William Osler as a model of clinical observation (Osler, 1894), covered almost all of the key features of the disease. He noted the involuntary movements or 'chorea', the adult onset, the relentlessly progressive nature and fatal course of the condition, the progressive motor disability, as well as the loss of mental function and behavioural problems, these last often occurring from an early stage of the disease. He also recognized its familial nature, correctly stating that it did not normally skip generations.

It is worth noting that George Huntington's classical description was not only brief, but that it was the only scientific paper he ever wrote. Despite this, his name is rightly remembered as having provided the foundations for the later detailed studies that have made, and continue to make, HD a model not only for the study of other neurodegenerative disorders, but for a wide range of aspects involving research and practice relating to genetic disorders, and now as the principal model for the study of the newly recognized group of polyglutamine repeat disorders.

During the century following Huntington's description, a wealth of details was recorded on the clinical and neuropathological aspects of HD (Harper 1996). Table 1.1 summarizes the main clinical features. The variation in severity, in relative prominence of different problems and in age at onset is considerably greater than was originally recognized. Neuropathological studies have shown clearly that it is a primary brain degeneration, with a rather characteristic distribution involving cell loss in parts of the

Figure 1.1 George Huntington in later life. (From Harper (1996), by permission of W. B. Saunders.)

basal ganglia and cerebral cortex, notably the caudate nucleus. However, although these early descriptions provided a firm foundation on which later research could be based, no clear indications could be obtained from pathological, clinical or experimental studies as to what might be the specific underlying cause of the disorder; increasingly, workers involved in HD research looked towards the developing field of genetics to provide the answers.

The inheritance of HD was recognized as following an autosomal dominant pattern from the time that Mendel's laws had been rediscovered. Affecting both sexes equally, transmitted only by affected individuals (apart from those dying young) and with 50% of offspring of an affected parent developing the disorder, HD provided a striking example of this mode of inheritance, while its high frequency in many rapidly expanding

Table 1.1 *Clinical features of Huntington's disease*

Neurological
Progressive involuntary movements (chorea)
Inco-ordination
General motor disability (in late stages)
Rigidity (particularly in early-onset cases)
Mental
Depression
Irritability, mood changes
Behavioural disorder
Progressive cognitive decline
Psychotic episodes (infrequent)

immigrant populations of European origin made it a major problem in relation to genetic counselling and to prevention of the condition. However, there were puzzling genetic features noted which did not fit with conventional Mendelian inheritance, notably the paternal transmission of the rare juvenile form of the disease (Merrit *et al.* 1969) and the recognition that 'anticipation', long recognized and debated in relation to another dominantly inherited disorder, myotonic dystrophy (Penrose 1948), applied also to HD, at least in the male line (Ridley *et al.* 1991). Until the gene was isolated, solutions to these questions remained the subject of speculation only. Table 1.2 summarizes the particular genetic aspects that posed problems and which subsequently have proved highly relevant to the mutational mechanism.

Gene mapping was first attempted for HD in the 1970s, using blood groups and other protein markers, but was not then successful owing to the lack of power of these markers and the scarcity of large families with sufficient living affected members. It was the advent of DNA polymorphisms, along with recognition of the value of the large and extended Venezuela kindred with HD, that made localization of the gene possible, added to which should be mentioned the foresight of the Hereditary Disease Foundation (HDF) in supporting the long-term effort needed and in attracting high-quality

Table 1.2 *Huntington's disease: genetic aspects*

Autosomal dominant inheritance
Homozygotes no more severe than heterozygotes
Close phenotypic similarity between identical twins
Marked variation within a kindred
Absence of clear *de novo* mutations
Anticipation (mainly in male line of transmission)
Juvenile-onset cases mainly paternally transmitted

scientific groups to work on the topic. In fact, the initial localization, to everybody's surprise, came much earlier than expected with the finding in 1983 (Gusella *et al.* 1983) that one of the first DNA polymorphisms available clearly localized the gene to the short arm of chromosome 4.

From this point on, the eventual isolation and identification of the gene was never in doubt, even though it took ten years to achieve this goal (Huntington's Disease Collaborative Research Group 1993). It should be remembered that at the outset of the work in 1983, the actual isolation of a gene by positional cloning was entirely a conjecture rather than an established fact. Again, the formation of the Huntington's Disease Collaborative Research Group, its funding and its co-ordination and general nurturing by the HDF through years of apparently slow progress provide an object lesson on how a truly collaborative major scientific project can be sustained and ultimately succeed.

It should not be thought that the work of the Collaborative Group proceeded on a fixed plan; on the contrary, new techniques were introduced as they became feasible, while new ideas from other fields of work were also brought in. It was during the final but time-consuming stages of the work, when a series of possible candidate genes in the critical region were being examined for mutations, that the concept of trinucleotide repeat disorders was brought onto the scene, helping greatly in the recognition of the gene and then transforming the course of all subsequent HD molecular research.

Although two disorders, Kennedy's disease (LaSpada *et al.* 1991) and fragile X mental retardation (Oberlé *et al.* 1991), had been shown in 1991 to result from trinucleotide repeat mutations, this did not have an immediate impact on the search for the HD gene, in part because of the X-linked nature of these disorders. Rather, it was the example of myotonic dystrophy, recognized as a trinucleotide repeat disorder in 1992 (Brook *et al.* 1992), which had a profound effect, largely because two of the Collaborative Group's member teams had also been directly involved with this disease. Despite the clinical differences between the conditions, the shared genetic features of anticipation and parent of origin effects provided compelling reasons for HD also being a trinucleotide repeat disorder, so that it was the search for and eventual finding of such a repeat, and its expansion in the HD patients, that provided the conclusive proof that the gene and mutation responsible for HD had finally been discovered (Huntington's Disease Collaborative Research Group 1993).

In the immediate aftermath of this discovery, attention was focused not so much on the function of the gene or the possible role of the mutation in pathogenesis, but on the relationship of the variability of the CAG repeat to the clinical and genetic aspects of the disease. As with myotonic dystrophy, so for HD most of the puzzling issues could now be explained (Snell *et al.* 1993): the anticipation was clearly related to genetic instability and intergenerational change in repeat length, while severity of disease, in particular juvenile HD, could also be closely related to this. Parent of origin effects could be related to differences in expansion at male and female meiosis, while the immediate origins of the disease could be explained in terms of healthy individuals carrying an expanded allele below the repeat length of *ca.* 38 that is now recognized as critical in terms of clinical disease (Rubinsztein *et al.* 1996). Table 1.3 lists some of the ways in which the recognition of an unstable trinucleotide repeat sequences as the mutational basis for HD has clarified the genetic problems mentioned earlier. In many respects, the germline and somatic cell genetic instability is less striking in HD and the other CAG repeat disorders, where the expanded sequence rarely exceeds 100 repeats, than in myotonic dystrophy and fragile X, where the sequences are not translated and may exceed 1000 repeats (Harper, 1998).

While these studies of mutational instability were producing rapid increases in our understanding of the genetic aspects of HD, research on the nature and function of the protein produced by the gene inevitably proceeded at a slower pace. This in part resulted from the lack of clues from the gene sequence, but it also reflected the fact that the skills and techniques now needed for this work were very different from those possessed by most of the genetic groups who had successfully found and isolated the gene itself. The necessary restructuring of the groups took time, as did adjustment to the fact that the field was now open to many experts who previously had had no involvement with HD. Again it is a tribute to the loyalty that HD research has produced that so many workers chose to stay with HD, while radically altering the nature of their work, rather than simply move on to another genetic disorder. Equally, the formation of new collaborations involving those with completely new relevant approaches has allowed the field of HD molecular research to remain extremely cohesive.

Perhaps the most striking example of this attraction into HD research of people with different skills has been the one that has given rise to the present book and the discussion meeting that generated it: the involvement of Dr Max Perutz. It has already been noted that the initial focus of research after isolation of the HD gene was on relating the mutation to the genetic features of the disorder, but it was not long before it became clear that within the

Table 1.3 *Consequences of genetic instability in Huntington's disease*

Intergenerational increase in repeat number—anticipation
Repeat number correlated with onset—juvenile cases show largest repeats
Greater expansion in male than female meiosis—juvenile cases mainly male-transmitted
Critical range of repeat number for causing clinical disease—rarely occurs with < 37 repeats; normally develops with > 38
Instability already present below threshold for clinical manifestation—origin from asymptomatic carriers of 'intermediate alleles'
Somatic instability not marked—no significant change in repeat number during embryonic development or later life

broader grouping of trinucleotide repeat disorders, a sub-group could be defined in which the expanded repeat sequence was CAG, where in all cases the repeat appeared in the protein sequence, and where the clinical features represented progressive CNS degenerations of a closely similar nature. Kennedy's disease (spinobulbar muscular atrophy) has already been mentioned, but HD was closely followed by other disorders including dentatorubral–pallidoluysian atrophy (DRPLA) and a series of dominantly inherited spinocerebellar ataxias (see table 1.4) (Orr *et al.* 1993; Koide *et al.* 1994). The fact that all these disorders, despite widely different genes and protein sequences, shared a CAG repeat expansion suggested that the expanded polyglutamine sequence in the various proteins might be of direct relevance in the pathogenesis of the neurodegeneration of this group of disorders, something which greatly increased the interaction of people working primarily on different disorders in this group, whose research had often previously had little contact with the ideas and experimental work being done on the other polyglutamine repeat disorders.

That this concept of a common pathogenetic process turned from being an unfocused idea into a hypothesis with a clear molecular basis that could be experimentally tested can undoubtedly be attributed to Max Perutz's entry into the HD field, (Perutz *et al.* 1994; Perutz 1996) and can best be illustrated by quoting from his 1996 review of the topic (Perutz 1996)

"These remarkable discoveries posed a great challenge to biomedical research. What is the molecular mechanism of CAG expansion? Can it be prevented? What is the structure and function of the normal glutamine repeats? How does expansion affect them? Why is it toxic to specific neurons in the central nervous system? Can the toxic effects of expanded glutamine repeats be prevented or at least alleviated?"

This clearly set out the challenges to be met, some of which are beginning to be answered by work such as that reported in the various chapters of this book. However, Perutz did not stop there, but produced a model that could explain the possible role of glutamine repeats in terms of structural biology. This 'polar zipper' model (figure 1.2) has provided the starting point for much of current HD research and its origin is again best described in his own words from the same review.

"By a strange accident, my attention was drawn to glutamine repeats before the publication of the gene for Huntington's disease. Such repeats have been found, for instance, in some homeodomain proteins of *Drosophila*. A survey of the Swiss Prot data bank showed that 33 out of 40 proteins with 20 or more glutamines in a row are transcription factors, many of them in *Drosophila*, involved mostly in the developmental regulation especially of the nervous system. Wondering what the structure of glutamine repeats might be, I built an atomic model which showed that β-strands of poly-L-glutamine could be linked together by hydrogen bonds between both their main chain and side chain amides. In other words, they acted as polar zippers, which made me wonder if they attached β-strands of proteins to each other, while leucine

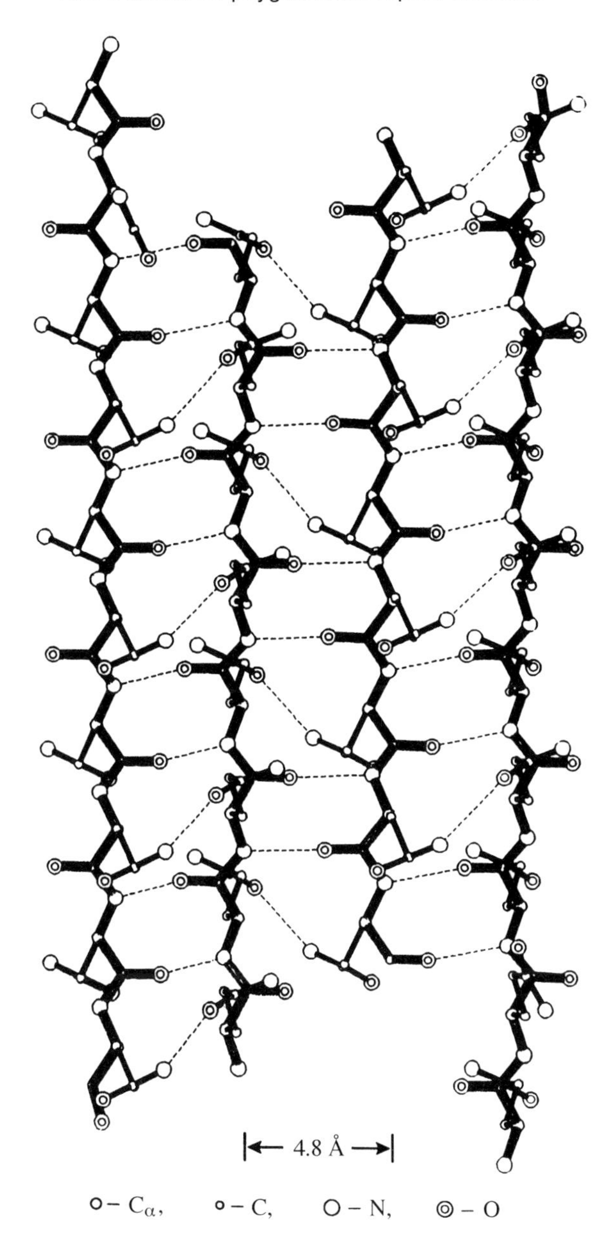

Figure 1.2 The 'polar zipper' model for polyglutamine repeats. (From Perutz (1996), with kind permission of the author and *Current Opinion in Structural Biology*.)

Table 1.4 *CAG repeat expansions and inherited neurological degenerations*

disorder	*protein*	*chromosome location*	*areas of brain predominantly involved*
Spinobulbar muscular atrophy (Kennedy's disease)	androgen receptor	Xq13	spinal and bulbar motor neurons
Huntington's disease	huntingtin	4p16	caudate nucleus, putamen; also cerebral cortex
Dentatorubral–pallidoluysian atrophy (DRPLA)	atrophin	12p13	dentate and red nuclei, cerebellum, brain stem
Spinocerebellar ataxia (SCA)			
type 1	ataxin-1	6p23	cerebellum and brain stem, especially Purkinje cells
type 2	ataxin-2	12q24	comparable to SCA1
type 3 (Machado–Joseph disease)	ataxin-3	14q32	cerebellum; striato-nigral pathways
type 6	Λ1A voltage-dependent calcium channel subunit	19p13	cerebellum, brain stem
type 7	ataxin-7	13p12	cerebellum; additional retinal degeneration

zippers had evolved to make α-helices stick together. When I read the astonishing paper in *Cell* on the gene for Huntington's disease, it occurred to me that the polar zipper action of glutamine repeats might furnish a possible clue to the molecular mechanism of the disease, but in view of the medical importance of the problem it seemed essential to test that idea experimentally."

Testing that idea experimentally has, as the work presented in this book shows, already born fruit in terms of our increased understanding of HD and other glutamine repeat disorders, to the extent that possibilities for modifying their course look feasible in a way that would have seemed unrealistic even five years ago. Such a prospect has already further increased the impetus of the research, both basic and clinical, and now allows hope for affected patients and relatives at risk that therapy of real value will become available in the foreseeable future.

References

Brook, J. D., McCurrach, M. R., Harley, H. G. *et al.* 1992 Molecular basis of myotonic dystrophy: expansion of a trinucleotide (CTG) repeat at the 3' end of a transcript encoding a protein kinase family member. *Cell* **68**, 799–808.

Gusella, J. F., Wexler, N. S., Conneally, P. M. *et al.* 1983 A polymorphic DNA marker genetically linked to Huntington's disease. *Nature* **306**, 234–238.

Harper, P. S. (ed.) 1996 *Huntington's disease*, 2nd edn. London: Saunders.

Harper, P. S. 1998 Myotonic dystrophy as a trinucleotide repeat disorder—a clinical perspective. In *Genetic instabilities and hereditary neurological diseases* (ed R. D. Wells and S. T. Warren) pp. 115–130. New York. Academic Press.

Huntington, G. 1872 On chorea. *Med. Surg. Reporter* **26**, 317–321.

Huntington's Disease Collaborative Research Group 1993 A novel gene containing a trinucleotide repeat that is expanded and unstable on Huntington's disease chromosomes. *Cell* **72**, 971–983.

Koide, R., Ikeuchi, T., Onodera, O. *et al.* 1994 Unstable expansion of CAG repeat in hereditary dentatorubral–pallidoluysian atrophy (DRPLA). *Nature Genet.* **6**, 9–13.

LaSpada, A. R., Wilson, E. M., Lubahn, D. B. *et al.* 1991 Androgen receptor gene mutations in X-linked spinal and bulbar muscular atrophy. *Nature* **352**, 77–79.

Merrit, A. D., Conneally, P. M., Rahman, N. F. & Drew, A. L. 1969 Juvenile Huntington's chorea. In *Progress in neurogenetics*, vol. 1 (ed. A. Barbeau & J. R. Brunette), pp. 645–650. Amsterdam: Excerpta Medica.

Oberlé, I., Rousseau, F., Heitz D. *et al.* 1991 Instability of a 550 base pair DNA segment and abnormal methylation in fragile X syndrome. *Science* **252**,1097–1102.

Orr, N. T., Chung, M., Banfi, S. *et al.* 1993 Expansion of an unstable trinucleotide CAG repeat in spinocerebellar ataxia type I. *Nature Genet.* **4**, 221–226.

Osler, W. 1894 Case of hereditary chorea. *Johns Hopkins Hosp. Bull.* **5**, 119–129.

Penrose, L. S. 1948 The problem of anticipation in pedigrees of dystrophia myotonica. *Ann. Eugen.* **14**, 125–132.

Perutz, M. F., Johnson, T., Suzuki, M. and Finch, J. T. 1994 Glutamine repeats as polar zippers: their possible role in inherited neurodegenerative diseases. *Proc. Natl. Acad. Sci. USA* **9**, 3555–3787.

Perutz, M. F. 1996 Glutamine repeats and inherited neurodegenerative diseases. Molecular aspects. *Curr. Opin. Struct. Biol.* **6**, 658–848

Ridley, R. M., Frith, C. D., Farrer, L. A. & Conneally, P. M. 1991 Patterns of inheritance of the symptoms of Huntington's disease suggestive of an effect of genomic imprinting. *J. Med. Genet.* **28**, 224–231.

Rubinsztein, D. C. (and 36 others) 1996 Phenotypic characterisation of individuals with 30–40 CAG repeats in the Huntington's disease (HD) gene reveals HD cases with 36 repeats and apparently normal elderly individuals with 36–39 repeats. *Am. J. Hum. Genet.* **59**, 16–22.

Snell, R. G., MacMillan, J. C., Cheadle, J. P., Fenton, I., Lazarou, L. P., Davies, P., MacDonald, M., Gusella, J. F., Harper, P. S. & Shaw, D. J. 1993 Relationship between trinucleotide repeat expansions and phenotypic variation in Huntington's disease. *Nature Genet.* **4**, 393–397.

Animal models of Huntington's disease

2

A transgenic mouse model of Huntington's disease

Laura Mangiarini, Kirupa Sathasivam, Amarbirpal Mahal, Ben Woodman, Mark Turmaine, Stephen W. Davies and Gillian P. Bates

The polyglutamine neurodegenerative diseases

Huntington's disease (HD) is an autosomal dominant neurodegenerative disease. Whilst onset is generally within the 4th or 5th decade, the disease can start at any time from early childhood until very old age, with a mean duration of 15–20 years (Harper 1996) . The inheritance pattern of HD shows anticipation when transmitted through the male line in that the majority of early-onset cases inherit the mutation from their fathers (Farrer *et al.* 1992). Patients exhibit a diverse set of symptoms, with well-recognized emotional, cognitive and motor components. The movement disorder can appear drastically different between the adult- and early-onset forms of the disease in that the juvenile patients can exhibit Parkinsonian-like features and never express chorea. The dominant neuropathological finding is widespread neuronal loss occurring predominantly in the cortex, putamen and frontal lobes (Vonsattel *et al.* 1985; Robitaille *et al.* 1997; Vonsattel and DiFiglia 1998). However, degeneration can eventually appear throughout the brain (Vonsattel and DiFiglia 1998). In cases with juvenile onset, an additional focus for neurodegeneration are the Purkinje cells of the cerebellum (Robitaille *et al.* 1997; Young 1998). In addition, HD patients need a high calorific intake and usually find difficulty maintaining their body weight, with a marked loss in muscle bulk (Sanberg *et al.* 1981; Harper 1996).

The HD gene contains 67 exons and extends across 170 kb DNA (Ambrose *et al.* 1993; Baxendale *et al.* 1995). The CAG repeat that is expanded on HD chromosomes lies within exon 1 and is translated into a stretch of polyglutamine (polyQ) residues. The normal and expanded ranges are $(CAG)_{6-39}$ and $(CAG)_{36-250}$, respectively (Stine *et al.* 1993; Rubinsztein *et al.* 1996; Nance *et al.* 1999). The majority of adult-onset cases have expansions ranging from 40 to 55 units; expansions of 70 and above invariably cause the juvenile form of the disease, and expansions in excess of 100 repeats are rare. The normal and mutant forms of huntingtin have been shown to be expressed at similar levels in the central nervous system and in peripheral tissues (Trottier *et al.* 1995). Within the brain, huntingtin was found predominantly in neurons and was present in cell bodies, dendrites and also in the nerve terminals. Immunohistochemistry, electron microscopy and subcellular fractionations have shown that huntingtin is primarily a cytosolic protein associated with vesicles and/or microtubules, suggesting that it plays a functional role in cytoskeletal anchoring or transport of vesicles (DiFiglia *et al.* 1995; Gutekunst *et al.* 1995; Sharp *et al.* 1995). Huntingtin has also been detected in the nucleus (de Rooij *et al.* 1996) in mouse embryonic fibroblast, adult fibroblast and neuroblastoma cell lines.

In addition to HD, a polyQ expansion has been found to cause seven other late-onset, inherited neurodegenerative diseases, namely: spinal and bulbar muscular atrophy (SBMA), dentatorubral–pallidoluysian atrophy (DRPLA) and the spinocerebellar ataxias (SCA) 1, 2, 3, 6 and 7 (referenced in Bates *et al.* 1998). These diseases bear many similarities: they are autosomal dominant (with the exception of X-linked SBMA), often show anticipation and have broadly comparable normal and expanded repeat ranges. The proteins that harbour the polyQ tracts are otherwise unrelated and, with the exception of the androgen receptor (SBMA) and the α_{1A} voltage-dependent calcium channel (SCA6), are novel proteins of unknown function identified by positional cloning strategies. Although these proteins have wide and extensively overlapping expression profiles, the patterns of neurodegeneration are comparatively distinct. However, this selective neuronal vulnerability is not absolute and the earlier onset forms of these disease, associated with longer polyQ expansions, tend to result in much wider and more overlapping neuropathologies (Young 1998). The molecular event that triggers polyQ pathogenesis must correlate with the polyQ pathogenic size threshold and account for the late onset of these disorders.

The R6 transgenic lines

The R6 transgenic lines contain a 2 kb genomic fragment that spans the 5' end of the human gene, encompasses 1 kb of control elements and generates an N-terminal exon 1 protein corresponding to approximately 3% of huntingtin (Mangiarini *et al.* 1996). Six lines have been established, the main features of which are summarized in table 2.1. Four lines contain expanded CAG repeats of a size larger than that generally associated with the juvenile form of the disease: R6/1 $(CAG)_{115}$; R6/2 $(CAG)_{145}$; R6/5 $(CAG)_{135-156}$; and R6/0 $(CAG)_{142}$ (Mangiarini *et al.* 1997), and a further two lines, HDex6 and HDex27, contained $(CAG)_{18}$ as normal repeat controls. In all cases other than line R6/0, in which the transgene is probably silenced by the site of integration, expression of the transgene protein showed a ubiquitous tissue expression profile (Mangiarini *et al.* 1996).

A progressive neurological phenotype develops in lines R6/1, R6/2 and R6/5 in which the CAG repeats are expanded and the transgene is expressed.

Table 2.1 *Summary of the exon 1 huntingtin transgenic mouse lines*

transgenic line	*integration site*	*CAG repeat size*	*transgene expression*	*phenotype*
R6/1	single copy	113	+	+
R6/2	one intact copy	144	+	+
R6/5	four intact copies	128–156	+	+
R6/0	single copy	142	–	–
HDex6	~20 copies	18	+	–
HDex27	~7 copies	18	+	–

On the basis of home cage behaviour, the onset ages are approximately two months in line R6/2 and four to five months in line R6/1. The movement disorder includes an irregular gait, stereotypic grooming movements, rapid shudders (similar to a wet dog shake), a tremor (or very rapid myoclonus) and a tendency to clasp the hind- and forelimbs together when suspended by the tail. These movements are not associated with seizure activity (J. Noebels, unpublished data). In addition, a proportion of the mice develop tonic–clonic seizures. A detailed analysis of motor function as measured by performance on the rotarod, beam walking, swimming and footprint analysis shows that the age of onset differs dependent on the nature and difficulty of the task being measured (Carter *et al.* 1999) and that significant differences between the R6/2 transgenes and their littermate controls can be detected as early as five weeks. In addition to the movement disorder, the mice exhibit a progressive weight loss. The phenotype progresses rapidly and mice are rarely kept beyond 12 weeks of age, by which time the movement disorder is pronounced and the transgenetic mice weigh between 60% and 70% of their littermate controls. A small number of mice have been studied to four months of age.

PolyQ aggregation and the pathogenic repeat threshold

To gain insight into the molecular interactions that underlie the pathogenesis in the R6 lines, huntingtin exon 1 proteins containing polyQ repeats ranging from 20Q to 122Q have been expressed as a glutathione *S*-transferase (GST) fusion protein in *Escherichia coli* (Scherzinger *et al.* 1997). After purification via the GST tag, the proteins with 83Q and 122Q spontaneously formed ordered fibrils by self-aggregation, whilst those with 20Q and 30Q expansions in the non-pathogenic range remained soluble. In contrast, the fusion protein containing 51Q remained soluble until the removal of GST with the proteases factor Xa or trypsin, after which, it too formed ordered fibrils. When stained with Congo red and viewed under polarized light, the aggregates showed a green colour and birefringence indicative of amyloid

(Scherzinger *et al.* 1997). This is consistent with the prediction of Max Perutz that polyQ tracts are capable of interacting via hydrogen bonding between main-chain and side-chain amides in a cross-β-sheet structure, an interaction that he termed a polar zipper (Perutz *et al.* 1994). This solubility of exon 1 fusion proteins, carrying comparatively modest pathogenic polyQ expansions, prior to GST tag removal has made it possible to control the initiation of aggregation and study its kinetics. Analysis of exon 1 proteins containing polyQ tracts of 20, 27, 32, 37, 39, 40, 42, 45, 51 and 93 repeats has shown that the kinetics of the self-aggregation process increase markedly as the polyQ expansion increases from 32 to 37 repeats. This correlation between aggregation potential and the pathogenic repeat threshold suggests that the initiation of aggregation is the molecular event that triggers polyQ disease (Scherzinger *et al.* 1999). The large intracellular inclusions that can be identified by light microscopy mark the end point of the aggregation process. This phenomenon has been proposed to occur via a nucleation and aggregation pathway (Lansbury 1997; Scherzinger *et al.* 1999), the structural intermediates of which remain to be defined.

Polyglutamine aggregation in the R6 lines

Polyglutamine aggregates were first identified in the R6 lines in the form of neuronal intranuclear inclusions (Davies *et al.* 1997). At the light microscope level, they could be seen by immunohistochemistry with antibodies raised against exon 1 huntingtin and against ubiquitin, and appeared as an intense focus of staining in the neuronal nucleus. These inclusions could be identified by ultrastructure in the absence of immunostaining as a granular and fibrillar structure devoid of a membrane and slightly larger than the nucleolus. In line R6/2, they are apparent prior to four weeks in cortex and hippocampus, and by end stage disease are conspicuous throughout the grey matter. Neuronal inclusions form outside of the nucleus only in isolated cases in line R6/2 (termed dystrophic neurites in HD post-mortem brains (DiFiglia *et al.* 1997)), although this is a more frequent occurrence in the more slowly progressing R6/1 and R6/5 lines (S. W. Davies, unpublished data).

More recently, antibody EM48 has been shown to recognize a much greater extent of polyQ aggregation in the brains of the R6/2 mice, much of it localized to the neuronal processes (Li *et al.* 1999). These neuropil aggregates predominate in the cortex and striatum, are smaller than nuclear inclusions and do not have an obvious filamentous structure. Only a subset of the neuropil aggregates could be detected with antibodies to ubiquitin, and those that stained positive tended to be in the larger size class. Electron microscopy immunogold examination showed the nucleus to contain nuclear inclusions in the form of a single large aggregate and diffuse immunogold particles. Small aggregates were only present outside of the nucleus, frequently in axons and clustered in axon terminals in which synaptic vesicles and presynaptic junctions could be seen. No immunogold particles were seen in the post-synaptic densities. Synaptic morphology suggested that the transgene protein appears to form aggregates preferentially in excitatory axons. Both nuclear inclusions and neuropil aggregates are present before four weeks, but the formation of neuropil aggregates is more progressive such that by 12 weeks, neuropil aggregates were noticeably more abundant than nuclear inclusions (Li *et al.* 1999).

PolyQ aggregation has also been detected outside the central nervous system (CNS) in the R6 lines. Immunohistochemistry using anti-exon 1 huntingtin and ubiquitin antibodies identified nuclear inclusions in a variety of post-mitotic cells including skeletal and cardiac muscle, hepatocytes, medullary and reticular cells of the adrenal glands, the pancreatic islets of Langerhans and, at a low frequency, in the tubular, interstitial and glomerular cells of the kidney (Sathasivam *et al.* 1999). In all cases, the inclusions were only present in the nucleus and were absent at four weeks but could be detected from six weeks of age. The ultrastructure of nuclear inclusions was studied in skeletal muscle (figure 2.1) and found to be identical to that previously described in neurons (Sathasivam *et al.* 1999).

PolyQ aggregation in HD postmortem brains

PolyQ aggregation has been detected in HD post-mortem brains in the form of intranuclear inclusions (DiFiglia *et al.* 1997; Becher *et al.* 1998; Gutekunst

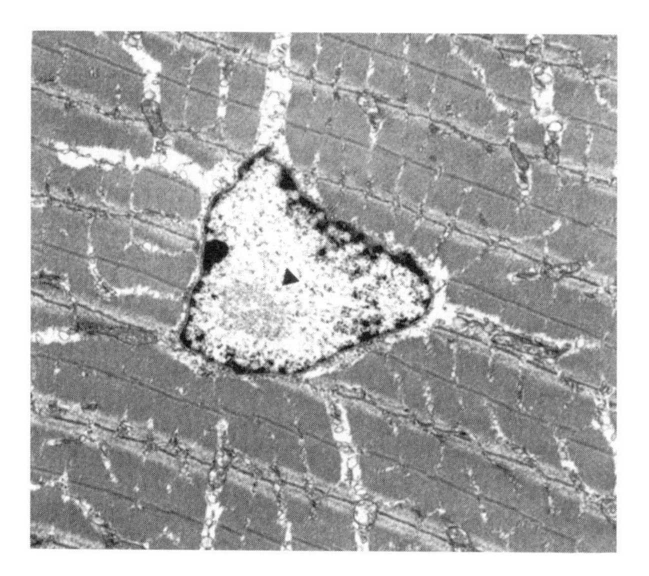

Figure 2.1 Ultrastructure of a nuclear inclusion (arrow) in the quadricep muscle of an R6/2 mouse at 12 weeks of age.

et al. 1999), dystrophic neurites (DiFiglia *et al.* 1997) and neuropil aggregates (Gutekunst *et al.* 1999). Nuclear inclusions and dystrophic neurites are detected by anti-huntingtin antibodies that recognize the N-terminus of the protein (DiFiglia *et al.* 1997; Becher *et al.* 1998). This has raised the possibility that full-length huntingtin must be cleaved prior to aggregate formation, which is supported by work with GST fusion proteins (Scherzinger *et al.* 1997), cell culture models (Lunkes and Mandel 1998; Martindale *et al.* 1998) and the identification of N-terminal 40 kDa fragments on Western blots from juvenile brains that are absent in controls (DiFiglia *et al.* 1997). If this is the case, the first step in the aggregation pathway has been by-passed in the R6 transgenic lines which only express the N-terminus of the huntingtin protein. The neuropil aggregates were detected with the N-terminal EM48 antibody which preferentially recognizes aggregated huntingtin and labels many more aggregates in neuronal nuclei, perikarya and processes in human brain than had been described previously (Gutekunst *et al.* 1999). Nuclear inclusions were found to predominate in juvenile HD brains (38–52% of all cortical neurons) (DiFiglia *et al.* 1997), whereas dystrophic neurites and neuropil aggregates were more frequent in those from patients with adult onset (DiFiglia *et al.* 1997; Gutekunst *et al.* 1999). The majority of polyglutamine aggregation is located in the cortex (DiFiglia *et al.* 1997; Becher *et al.* 1998; Gutekunst *et al.* 1999), and both dystrophic neurites and neuropil aggregates have been described in the cortex of presymptomatic HD brains (DiFiglia *et al.* 1997; Gutekunst *et al.* 1999). Therefore, polyglutamine aggregation is present in HD brains prior to the onset of symptoms and is more conspicuous in the cerebral cortex than in the striatum.

In line R6/2, neuronal cell death occurs late and is very selective

A neuropathological analysis of serial sections throughout the entire brain and spinal cord from R6/2 mice at 12 weeks found no evidence of neuronal cell death (Mangiarini *et al.* 1996). However, from 14 weeks, these mice do develop a late stage neurodegeneration within the anterior cingulate cortex, dorsal striatum and in the Purkinje cells of the cerebellum (Davies *et al.*, 1999. Turmaine *et al.*, 2000). Degenerating neurons have an enhanced affinity for either toluidine blue or osmium. They characteristically contain nuclear inclusions and exhibit condensation of both cytoplasm and nucleus and ruffling of the plasma membrane, but maintain an ultrastructural preservation of cellular organelles. Whilst they demonstrate certain features of apoptosis, they do not develop blebbing of the nucleus or cytoplasm, apoptotic bodies or fragmentation of DNA, and cell death appears to occur over a relative protracted time-course (Turmaine *et al.*, 2000). Similar dark degenerating cells have also been seen in HD post-mortem brains (Vonsattel and DiFiglia 1998; Turmaine *et al.*, 2000) and in a *Drosophila melanogaster* model of HD (Jackson *et al.* 1998) in which the cell death could not be rescued by crossing to flies expressing the anti-apoptotic gene, p35 (Jackson *et al.* 1998).

Outside of the CNS, no evidence of cell death can be detected in line R6/2 in the tissues containing nuclear inclusions at 14 weeks. A particularly extensive analysis was performed on skeletal muscle because the atrophy of this tissue is so pronounced in the R6 mice and it was hoped that this study may shed some light on the extensive loss of muscle bulk observed in HD patients. Transverse sections were cut from the quadricep muscle of R6/2 mice at 14 weeks and of R6/1 mice at 15 months. These were stained using haematoxylin and eosin (H&E), haematoxylin van Gieson (HVG) and the periodic acid–Schiff (PAS) assay. Acid phosphatase and NADH–tetrazolium reductase (NADH–TR) enzyme activities were measured and immunohistochemistry was performed using an anti-N-cam antibody and antibodies that detect specific fibre types (Sathasivam *et al.* 1999). Finally, an ultrastructural analysis was performed. No evidence of a myopathy or neuropathy could be detected. The only difference between the transgenic mice and their non-transgenic littermate controls was that the muscle fibres were uniformly smaller across all muscle types (Sathasivam *et al.* 1999). The mechanism underlying this uniform shrinkage is not known.

The progressive neurological phenotype in line R6/2 mice is caused by neuronal dysfunction and not neurodegeneration

In line R6/2, motor dysfunction is progressive and can be detected by five weeks of age. PolyQ aggregation is present within neurons before four weeks in the form of both nuclear inclusions and neuropil aggregates. However, there is no evidence of selective neurodegeneration until 14 weeks (figure 2.2). Therefore, the R6/2 symptoms must be caused by a neuronal dysfunction rather than by neurodegeneration. Insights into the underlying basis of neurodysfunction are beginning to arise from a number of specific studies.

Alterations in neurotransmitter receptors are a pathological hallmark of the neurodegeneration seen in HD. To explore the relationship between the glutamate and other receptors known to be affected in HD and the symptoms in R6/2 mice, receptors were examined in the brains of four-, eight- and twelve-week-old mice using receptor binding autoradiography, immunoblotting for receptor proteins and *in situ* hybridization. Neurotransmitter receptors that are altered in HD, i.e. receptors for glutamate, dopamine, acetylcholine and adenosine, are decreased in the brain of the R6/2 mice. The receptor alterations are selective in that other receptors, namely NMDA and GABA, are unaltered. By eight weeks, there are already major alterations in the glutamate, dopamine and adenosine neurotransmitter systems which have major importance in striatal function (Cha *et al.* 1998, 1999). The

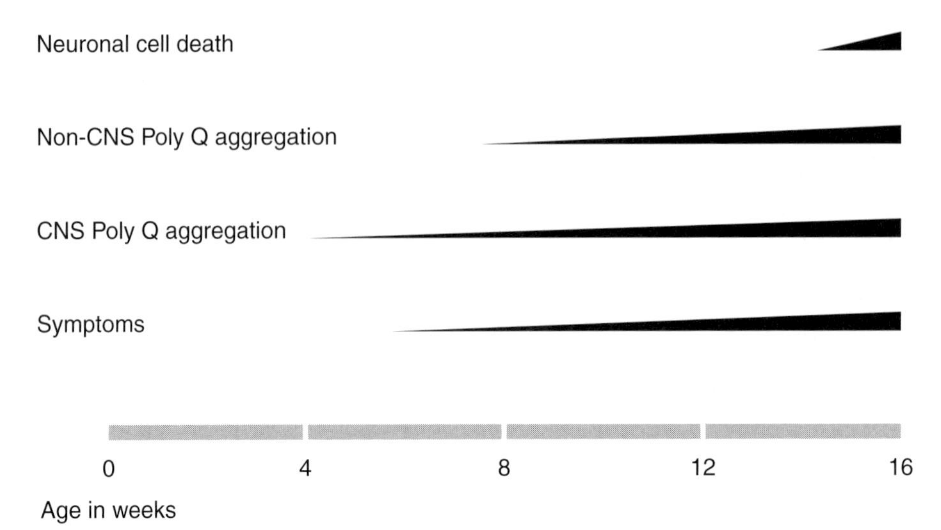

Figure 2.2 Schematic representation of the relative onset and progression of the formation of polyglutamine aggregates, the phenotype and neurodegeneration in line R6/2. Polyglutamine aggregation in the form of nuclear inclusions and neuropil aggregates is present within the brain before four weeks. An impairment in motor dysfunction can be detected by five weeks, whereas the onset of the phenotype is obvious by two months from home cage observations. Neurodegeneration can be seen from 14 weeks.

neurotransmitter receptor decreases are themselves preceded by selective decreases in the corresponding mRNA species. The decreases in mGluR1, A2a adenosine and D_1 dopamine receptor mRNA signals in the striatum were statistically significant by four weeks of age, and the decrease in D_2 dopamine receptor mRNA in the striatum and mGluR2 receptor mRNA expression in the cortex was significant by eight weeks of age (Cha *et al.* 1998, 1999). The earliest changes occur after, or are coincident with, the appearance of polyQ aggregates as detectable by light microscopy. They precede the onset of the neurological symptoms, and the striatal receptor changes are likely to contribute to the movement disorder seen in these mice. As they occur at a time at which there is no evidence of neuronal loss, they are unlikely to reflect damage to a particular set of neurons but rather occur at the level of specific receptor changes. It is possible that the mutant HD protein in the nucleus, perhaps in the form of a conformational intermediate or in the early stages of aggregation, may alter the expression of multiple genes including those encoding specific neurotransmitter receptors (Cha *et al.* 1998, 1999).

Synaptic plasticity and the development of nuclear inclusions in the hippocampal formation of R6/2 mice have been studied (Murphy *et al.* 2000). In the CA1 region, inclusions can be detected in the pyramidal cells from three weeks of age, and this precedes a reduction in the magnitude of long-term potentiation (LTP) expressed at CA1 synapses which occurs at five weeks. In addition, transgenic mice, unlike control mice, also exhibit marked long-term depression (LDP). Both the changes in synaptic plasticity and the appearance of nuclear inclusions occur well before the onset of the neurological phenotype. Interestingly, the reduction in LTP and the expression of LDP were independent of phenotype severity. Whilst inclusions were apparent in the CA1 subfield at three weeks, they were not detected throughout the stratum granulosum of the dentate gyrus until *ca.* 10 weeks. LTP at granule cell–perforant path synapses was severely impaired in animals aged more than 12 weeks. Therefore, the appearance of large polyQ aggregates precedes and correlates with changes in synaptic plasticity in different regions of the hippocampus (Murphy *et al.* 2000).

The consequences of cellular dysfunction, most likely arising as a result of polyQ aggregate formation, have also been uncovered outside the CNS. The R6/2 mice develop insulin-responsive diabetes, described at 12.5 weeks (Hulbert *et al.* 1999). Nuclear inclusions can be detected in the islets by six weeks, and at 14 weeks there was no evidence of cell death (Sathasivam *et al.* 1999). However, immunohistochemical staining showed dramatic reductions in glucagon in the alpha-cells and insulin in the beta-cells at 12.5 weeks. Direct tissue assays showed that glucagon and insulin content were reduced to only 10% and 15% of controls (C57BL/6), respectively (Hulbert *et al.* 1999).

Discussion

The tight correlation between the pathogenic repeat threshold and the aggregation potential of the mutant exon 1 protein strongly suggests that the aggregation process, of which intracellular inclusions are the end stage, forms the molecular basis of HD. However, it is clear from the analysis of the R6/2 mice that polyQ aggregates do not kill cells easily. There is no evidence of neurodegeneration at 12 weeks of age, by which time the phenotype is pronounced and the R6/2 brains are loaded with polyglutamine. The process of neurodegeneration does not begin until 14 weeks of age and then only occurs in very restricted neuronal populations. Cell death may not necessarily occur because of the cell-autonomous intracellular accumulation of polyQ aggregation but, alternatively, as a result of aberrant intercellular interactions between dysfunctional neurons. In keeping with this second hypothesis, there is some evidence for an excitotoxic basis to the cell death that occurs in the R6/2 striatum. The decrease that has been observed in the cortical presynaptic mGluR2 and mGluR3 receptors would result in an increase in glutamate release (Cha *et al.* 1998). Biochemical analysis of R6/2 mouse brain at 12 weeks demonstrated a significant reduction in aconitase and complex IV activities in the striatum and complex IV activity in the cerebral cortex (Tabrizi *et al.*, 2000). Increased immunostaining for inducible nitric oxide synthase and nitrotyrosine was seen in the transgenic mouse model but not in the control brains. The decrease in aconitase activity and increased nitrotyrosine residues in the transgenic mouse brain may be caused by the excitotoxic activation of nitric oxide and superoxide generation in addition to iNOS (inducible nitric oxide synthase) induction which precedes cell death (Tabrizi *et al.*, 2000).

The prevention of polyglutamine aggregation is an important target for the development of therapeutic interventions in HD. The rate of aggregate formation is likely to be dependent on the concentration of the aggregate precursor. This is supported by the fact that the onset of the phenotype is accelerated in the R6 lines when they are bred to homozygosity or when the lines are interbred. Therefore, it would be expected that this process could be delayed by (i) down-regulating the expression of the mutant huntingtin protein; (ii) preventing or slowing down the processing of full-length huntingtin thought to be a necessary step in the pathogenesis of the human disease; (iii) stabilizing the soluble form of the protein; or (iv) preventing the aggregation process.

Recently, it was shown that the phenotype in line R6/2 could be delayed when R6/2 mice were crossed to mice that express a dominant-negative caspase 1 transgene (NSE M17Z) in which the NSE (neuron-specific enolase) promoter directs the expression of mutant caspase 1 to the neurons and glia of the CNS (Ona *et al.* 1999). In the double transgenics (R6/2-NSE M17Z), inclusion formation, motor dysfunction, weight loss and neurotransmitter changes were all delayed. For example, the appearance of nuclear inclusions was postponed from three weeks until after nine weeks in the CA1 region of the hippocampus (Ona *et al.*, 1999). Although the mechanism by which the mutant caspase 1 is exerting this effect is not clear, it is extremely exciting that a means of delaying the symptoms in the R6 lines has already been demonstrated.

A major effort will be invested in the identification of small molecules that can prevent or slow down the kinetics of polyQ aggregate formation. These will be selected initially for their ability to prevent aggregation *in vitro* (E. E. Wanker, Berlin) and then tested in the R6 lines as possible therapeutic agents. The identification of inclusions outside the CNS has been important as it will allow drugs to prevent aggregation *in vivo* without the requirement in the first instance that the molecule in question can cross the blood–brain barrier.

We wish to thank Stefan Buk, Jang Ho Cha, Steve Dunnett, Hans Lehrach, Xiao-Jiang Li, Jenny Morton, Kerry Murphy, Gavin Reynolds, Tony Schapira, Eberhard Scherzinger, Sarah Tabrizi, Erich Wanker, Nancy Wexler and Anne Young for discussions and collaborations. This work was supported by grants from the Wellcome Trust, European Union (BMH4 CT96 0244), Hereditary Disease Foundation (in the form of an award to G.P.B. from Harry Lieberman), the Huntington's Disease Society of America and the Special Trustees of Guy's Hospital.

References

Ambrose, C. M. (and 20 others) 1993 Structure and expression of the Huntington's disease gene: evidence against simple inactivation due to an expanded CAG repeat. *Somat. Cell Mol. Genet.* **20**, 27–38.

Bates, G. P., Mangiarini, L. & Davies, S. W. 1998 Transgenic mice in the study of polyglutamine repeat expansion diseases. *Brain Pathol.* **8**, 699–714.

Baxendale, S. (and 10 others) 1995 Comparative sequence analysis of the human and puffer fish Huntington's disease gene. *Nature Genet.* **10**, 67–75.

Becher, M. W., Kotzuk, J. A., Sharp, A. H., Davies, S. W., Bates, G. P., Price, D. L. & Ross, C. A. 1998 Intranuclear neuronal inclusions in Huntington's disease and dentatorubral pallidoluysian atrophy: correlation between the density of inclusions and IT15 CAG repeat length. *Neurobiol. Dis.* **4**, 387–395.

Carter, R. J., Lione, L. A., Humby, T., Mangiarini, L., Mahal, A., Bates, G. P., Morten, A. J. & Dunnett, S. B. 1999 Characterisation of progressive motor deficits in mice transgenic for the human Huntington's disease mutation. *J. Neurosci.* **19**, 3248–3257.

Cha, J.-H. J., Kosinski, C. M., Kerner, J. A., Alsdorf, S. A., Mangiarini, L., Davies, S. W., Penney, J. B., Bates, G. P.& Young, A. B. 1998 Altered brain neurotransmitter receptors in transgenic mice expressing a portion of an abnormal human Huntington's disease gene. *Proc. Natl Acad. Sci. USA* **95**, 6480–6485.

Cha, J.-H. J., Frey, A. S., Alsdorf, S. A., Kerner, J. A., Kosinski, C. M., Mangiarini, L., Penney, J. B., Davies, S. W., Bates, G. P. & Young, A. B. 1999 Altered neurotransmitter receptor expression in transgenic mouse models of Huntington's disease. *Philos. Trans. R. Soc. Lond. B* **354**, 981–989.

Davies, S. W., Turmaine, M., Cozens, B. A., DiFiglia, M., Sharp, A. H., Ross, C. A., Scherzinger, E., Wanker, E. E., Mangiarini, L. & Bates, G. P. 1997 Formation of neuronal intranuclear inclusions (NII) underlies the neurological dysfunction in mice transgenic for the HD mutation. *Cell* **90**, 537–548.

Davies S. W., Turmaine, M., Cozens, B. A., Raza, A. S., Mahal, A., Mangiarini, L. and Bates, G. P. 1999 From neuronal inclusions to neurodegeneration: neuropathological investigation of a transgenic mouse model of Huntington's disease *Philos. Trans. R. Soc. Lond. B. Biol. Sci.* **354**, 971–979

de Rooij, K. E., Dorsman, J. C., Smoor, M. A., den. Dunnen J. T. & van Ommen, G.-J. 1996 Subcellular localisation of the Huntington's disease gene product in cell lines by immunofluorescence and biochemical subcellular fractionation. *Hum. Mol. Genet.* **5**, 1093–1099.

DiFiglia, M. (and 11 others) 1995 Huntingtin is a cytoplasmic protein associated with vesicles in human and rat brain neurons. *Neuron* **14**, 1075–1081.

DiFiglia, M., Sapp, E., Chase, K. O., Davies, S. W., Bates, G. P., Vonsattel, J.-P., and Aronin, N. 1997 Aggregation of huntingtin in neuronal intranuclear inclusions and dystrophic neurites in brain. *Science* **277**, 1990–1993.

Farrer, L. A., Cupples, L. A., Kiely, D. K., Conneally, P. M. & Myers, R. H. 1992 Inverse relationship between age at onset of Huntington's disease and paternal age suggests involvement of genetic imprinting. *Am. J. Hum. Genet.* **50**, 528–535.

Gutekunst, C.-A., Levey, A. I., Heilman, C. J., Whaley, W. L., Yi, H., Nash, N. R., Rees, H. D., Madden, J. J. & Hersch, S. M. 1995 Identification and localisation of huntingtin in brain and human lymphoblastoid cell lines with anti-fusion protein antibodies. *Proc. Natl Acad. Sci. USA* **92**, 8710–8714.

Gutekunst, C.-A., Li., S.-H., Yi, H., Mulroy, J. S., Kuemmerle, S., Jones, R., Rye, D., Ferrante, R. J., Hersch, S. M. & Li, X.-J. 1999 Nuclear and neuropil aggregates in Huntington's disease: relationship to neuropathology. *J. Neurosci.* **19**, 2522–2534.

Harper, P. S. 1996 *Huntington's disease*, vol. 31. London: W. B. Saunders.

Hulbert, M. S., Zhou, W., Wasmeier, C., Kaddis, F. G., Hutton, J. C. & Freed, C. R. 1999 Mice transgenic for an expanded CAG repeat in the Huntington's disease gene develop diabetes. *Diabetes* **48**, 649–651.

Jackson, G. R., Salecker, I., Dong, X., Yao, X., Arnheim, N., Faber, P. W., MacDonald, M. E. & Zipursky, S. L. 1998 Polyglutamine-expanded human huntingtin transgenes induce degeneration of *Drosophila* photoreceptor neurons. *Neuron* **21**, 633–642.

Kerry, P. S. J., Carter, R. J., Lione, L. A., Mangiarini, L., Mahal, A., Dunnett, S. B., Bates, G. P. and Morton, A. J. 2000 Abnormal Synaptic Plasticity and Impaired Spatial Cognition in Mice Transgenic for Exon 1 of the Human Huntington's Disease Mutation *Journal of Neuroscience* **in press**.

Lansbury, P. T. 1997 Structural neurology: are seeds at the root of neuronal degeneration. *Neuron* **19**, 1151–1154.

Li, H., Li, S.-H., Cheng, A. L., Mangiarini, L., Bates, G. P. & Li, X.-J. 1999 Ultrastructural localisation and progressive formation of neuropil aggregates in Huntington disease transgenic mice. *Hum. Mol. Genet.* **8**, 1227–1236.

Lunkes, A. & Mandel, J.-L. 1998 A cellular model that recapitulates major pathogenic steps of Huntington's disease. *Hum. Mol. Genet.* **7**, 1355–1361.

Mangiarini, L. (and 10 others) 1996 Exon 1 of the Huntington's disease gene containing a highly expanded CAG repeat is sufficient to cause a progressive neurological phenotype in transgenic mice. *Cell* **87**, 493–506.

Mangiarini, L., Sathasivam, K., Mahal, A., Mott, R., Seller, M. & Bates, G. P. 1997 Instability of highly expanded CAG repeats in transgenic mice is related to expression of the transgene. *Nature Genet.* **15**, 197–200.

Martindale, D. (and 12 others) 1998 Length of huntingtin and its polyglutamine tract influences localisation and frequency of intracellular aggregates. *Nature Genet.* **18**, 150–154.

Nance, M. A., Mathias-Hagen, V., Breningstall, G., Wick, M. J. & McGlennen, R. C. 1999 Analysis of a very large trinucleotide repeat in a patient with juvenile Huntington's disease. *Neurology* **52**, 392–394.

Ona, V. O., Li, M., Vonsattel, J. P., Andrews, L. J., Khan, S. Q., Chung, W. M., Frey, A. S., Menon, A. S., Li, X.J., Stieg, P.E., Yuan, J., Penney, J. B., Young, A. B., Cha, J. H., and Friedlander, R. M. 1999 Inhibition of caspase-1 slows disease progression in a mouse model of Huntington's disease. *Nature* **399**, 263–267.

Perutz, M. F., Johnson, T., Suzuki, M. & Finch, J. T. 1994 Glutamine repeats as polar zippers: their possible role in inherited neurodegenerative diseases. *Proc. Natl Acad. Sci. USA* **91**, 5355–5358.

Robitaille, Y., Lopes-Cendes, I., Becher, M., Rouleau, G. & Clark, A. W. 1997 The neuropathology of CAG repeat diseases: review and update of genetic and molecular features. *Brain Pathol.* **7**, 877–881.

Rubinsztein, D. C. (and 36 others) 1996 Phenotypic characterisation of individuals with 30–40 CAG repeats in the Huntington's disease (HD) gene reveals HD cases with 36 repeats and apparently normal elderly individuals with 36–39 repeats. *Am. J. Hum. Genet.* **59**, 16–22.

Sanberg, P. R., Fibiger, H. C. & Mark, R. F. 1981 Body weight and dietary factors on Huntington's disease patients compared with matched controls. *Med. J. Aust.* **1**, 407–409.

Sathasivam, K., Hobbs, C., Turmaine, M., Mangiarini, L., Mahal, A., Bertaux, F., Wanker, E. E., Doherty, P., Davies, S. W. & Bates, G. P. 1999 Formation of polyglutamine inclusions in non-CNS tissue. *Hum. Mol. Genet.* **8**, 813–822.

Scherzinger, E., Lurz, R., Turmaine, M., Mangiarini, L., Hollenbach, B., Hasenbank, R., Bates, G. P., Davies, S. W., Lehrach, H. & Wanker, E. E. 1997 Huntingtin encoded polyglutamine expansions form amyloid-like protein aggregates *in vitro* and *in vivo*. *Cell* **90**, 549–558.

Scherzinger, E., Sittler, A., Heiser, V., Schweiger, K., Hasenbank, R., Bates, G. P., Lehrach, H. & Wanker, E. E. 1999 Self-assembly of polyglutamine-containing huntingtin fragments into amyloid-like fibrils: implications for Huntington's disease pathology. *Proc. Natl Acad. Sci. USA* **96**, 4604–4609.

Sharp, A. H. (and 15 others) (1995). Widespread expression of Huntington's disease gene (IT15) protein product. *Neuron* **14**, 1065–1074.

Stine, O. C., Pleasant, N., Franz, M. L., Abbott, M. H., Folstein, S. E. & Ross, C. A. 1993 Correlation between the onset age of Huntington's disease and length of the trinucleotide repeat in IT-15. *Hum. Mol. Genet.* **2**, 1547–1549.

Tabrizi, S. J., Workman, J., Hart, P. E., Mangiarini, L., Mahal, A., Bates, G., Cooper, J. M. and Schapira, A. H. 2000 Mitochondrial dysfuntion and free radical damage in the Huntington R6/2 transgenic mouse *Ann. Neurol.* **47**, 80–86.

Trottier, Y., Devys, D., Imbert, G., Sandou, F., An, I., Lutz, Y., Weber, C., Agid, Y., Hirsch, E. C. & Mandel, J.-L. 1995 Cellular localisation of the Huntington's disease protein and discrimination of the normal and mutated forms. *Nature Genet.* **10**, 104–110.

Turmaine, M., Raza, A., Mahal, A., Mangiarini, L., Bates, G. P. and Davies, S. W. 2000 Nonapoptotic neurodegeneration in a transgenic mouse model of Huntington's disease *Proc. Natl. Acad. Sci. USA* **97**, 8093–8097

Vonsattel, J.-P., Myers, R. H., Stevens, T. J., Ferrante, R. J., Bird, E. D. & Richardson, E. P. 1985 Neuropathological classification of Huntington's disease. *J. Neuropathol. Exp. Neurol.* **44**, 559–577.

Vonsattel, J.-P. & DiFiglia, M. 1998 Huntington's disease. *J. Neuropathol. Exp. Neurol.* **57**, 369–384.

Young, A. B. 1998 Huntington's disease and other trinucleotide repeat disorders. In *Molecular neurology* (ed. J. B. Martin) pp. 35–54. New York: Scientific American.

3

From neuronal inclusions to neurodegeneration: neuropathological investigation of a transgenic mouse model of Huntington's disease

Stephen W. Davies, Mark Turmaine, Barbara A. Cozens, Aysha S. Raza, Amarbirpal Mahal, Laura Mangiarini and Gillian P. Bates

Introduction

Huntington's disease (HD) is one of eight progressive inherited neurodegenerative disorders caused by a common genetic mechanism (reviewed in Harper 1996). In addition to HD, the seven other diseases comprise: spinal and bulbar muscular atrophy (SBMA, also known as Kennedy's disease), dentatorubral–pallidoluysian atrophy (DRPLA), and five different spinocerebellar ataxias (SCA-1, SCA-2, SCA-3 (also known as Machado–Joseph disease), SCA-6 and SCA-7). In each disease the expansion of a CAG trinucleotide repeat within unrelated genes gives rise to an expanded sequence of glutamine residues (polyQ) within the corresponding protein. Disease generally occurs when the number of glutamine residues exceeds 35–40, suggesting that the disease is caused directly by the expanded glutamine tracts rather than by an abnormal function of the protein with the repeats. Perutz *et al.* (1994) proposed that expanded glutamine repeats might self-associate into aggregates with a β-pleated sheet structure. However, until 1997 such structures had not been reliably observed in any of these polyQ expansion diseases.

Ever since the identification of the HD gene in 1993 (Huntington's Disease Collaborative Research Group 1993), attempts have been made to model HD in transgenic mice (see Bates *et al.* 1998; Sathasivam *et al.*, 1999; Mangiarini *et al.*, Chapter 2). Several lines of mice with a neurological phenotype reminiscent of the human disease that they seek to model have been produced by expressing a construct comprising exon 1 of human huntingtin with $(CAG)_{115}$ to $(CAG)_{156}$ repeats under the control of the human huntingtin promoter (Mangiarini *et al.* 1996). Immunocytochemical staining of sections from the brains of these mice with antibodies specific for the N-terminus of huntingtin and with anti-ubiquitin antibodies revealed the presence of large numbers of neuronal intranuclear inclusions (NIIs) (Davies *et al.* 1997). Ultrastructurally these inclusions seem to have a diameter of *ca.* 2 μm, to have a pale granular and fibrous appearance and to be devoid of an encircling membrane. Experiments *in vitro* (see Hollenbach *et al.*, Chapter 11) have further shown that the transgene-encoded huntingtin protein assembles into fibrils that show birefringence when stained with Congo red, and is indicative of a β-pleated sheet structure (Scherzinger *et al.* 1997).

After their discovery in transgenic mice, NIIs have now been identified in post-mortem Huntington's disease brain (DiFiglia *et al.* 1997; Becher *et al.* 1998; Gourfinkel-An *et al.* 1998). They were found to be immunoreactive with antibodies directed against the N-terminus of huntingtin and with anti-ubiquitin antibodies, but not with antibodies directed against the middle region or the C-terminus of huntingtin. This suggested that it is a truncated N-terminal portion of huntingtin that accumulates in the inclusions. This was supported by the finding that an N-terminal fragment of *ca.* 40 kDa is enriched in nuclear fractions from HD brain (DiFiglia *et al.* 1997). A nuclear localization of huntingtin protein is unlike the exclusively cytoplasmic localization found in previous studies of mouse, monkey or human brain (Gutekunst *et al.* 1995; DiFiglia *et al.* 1995; Trottier *et al.* 1995; Bhide *et al.* 1996), although a single report has documented the localization of huntingtin within the nucleus of mouse embryonic fibroblasts and mouse neuroblastoma cells (de Rooij *et al.* 1996). A gain of function, conferred on the N-terminal fragment of huntingtin by the pathological expansion of the polyQ sequence, therefore seems to be entry into (and aggregation within) the neuronal nucleus. The ultrastructural appearance of NIIs in HD (DiFiglia *et al.* 1997) was found to be similar to those found in the HD transgenic mouse; the inclusions were highly reminiscent of structures found in an earlier ultrastructural study of biopsy samples of HD brain (Roizin *et al.* 1979).

The discovery of NIIs in Huntington's disease prompted studies investigating their presence in other diseases due to the expansion of glutamine repeats. Within the past few years, inclusions of aggregated protein have been found in the neuronal nucleus in several other polyQ diseases.

In SCA-3, the 42 kDa cytoplasmic protein ataxin-3 is found in ubiquitinated nuclear inclusions within those nerve cells that are known to degenerate in the disease (Paulson *et al.* 1997*a*,*b*; Schmidt *et al.* 1998). Ataxin-1 is an 87 kDa nuclear protein that contains expanded glutamine repeats in the disease SCA-1. In SCA-1 transgenic mice, the wild-type human protein localizes to several nuclear structures 0.5 μm in diameter, whereas the protein with an expanded polyQ sequence is found in a single ubiquitinated nuclear inclusion 2 μm in diameter (Skinner *et al.* 1997). In post-mortem brain of patients with SCA-1, ubiquitinated 2 μm inclusions were present in neuronal nuclei in the regions of the brain affected by the disease. Similarly in the spinocerebellar ataxia SCA-7, caused by the expansion of a polyQ sequence in a 95 kDa protein called ataxin-7, numerous intranuclear inclusions 3 μm in diameter were observed in neurons specifically affected by the disease (Holmberg *et al.* 1998). NIIs were immunoreactive with an antibody specific for long polyQ stretches, and were also stained with an anti-ubiquitin antibody.

DRPLA is caused by an expanded polyQ sequence in a 124 kDa protein termed atrophin-1. In patients with DRPLA, neurons in the dentate nucleus of the cerebellum have been found to contain intranuclear inclusions 2 μm in diameter and are of a similar ultrastructural appearance to those found in HD (Becher *et al.* 1998; Hayashi *et al.* 1998; Igarishi *et al.* 1998). These inclusions contain both atrophin-1 and ubiquitin.

SBMA is caused by the expansion of a polyQ sequence in the 104 kDa androgen receptor. In patients with SBMA, spinal and bulbar motor neurons contain ubiquitinated intranuclear inclusions 1–5 μm in diameter (Li *et al.* 1998*a,b*). They are immunoreactive with antibodies against the N-terminal region of the androgen receptor containing the polyQ sequence, but not with antibodies against the C-terminal region. In addition to these NIIs, immunoreactive inclusions can be found in several peripheral tissues (Li *et al.* 1998*b*). Both the cleavage and location within peripheral tissues are similar to results found in the HD brain (DiFiglia *et al.* 1997) and the transgenic mouse model of HD respectively (see Sathasivam *et al.*, 1999; Mangiarini *et al.*, Chapter 2).

These findings indicate that the formation of NIIs is the pathogenic mechanism underlying trinucleotide repeat disorders with expanded glutamine repeats (Davies *et al.* 1998). They also suggest that glutamine tracts over a certain length might be inherently toxic to neurons, irrespective of the proteins in which they are located. This has been tested in transgenic mice with a 146 glutamine repeat stretch inserted into the hypoxanthine phosphoribosyltransferase protein (Ordway *et al.* 1997). These mice developed a neurological phenotype that was preceded by the appearance of numerous ubiquitinated intranuclear inclusions in a number of regions of the brain. Expression of either C-terminal fragments of ataxin-3 or N-terminal fragments of huntingtin containing expanded polyQ sequences in neurons in *Drosophila* have also been shown to lead to the formation of NIIs (Jackson *et al.* 1998; Warrick *et al.* 1998).

We have reviewed briefly the wealth of recent data that have documented the formation of NIIs in both human disease and transgenic animal models. We shall now describe the temporal sequence of neuropathological changes that occur, after the appearance of NIIs, within neurons of the brain of HD transgenic mice (see Sathasivam *et al.*, 1999; Mangiarini *et al.*, Chapter 2) for a full description of the various lines of R6 transgenic mice). We relate these changes to those found in the human HD post-mortem brain.

NIIs and nuclear changes

NIIs can first be detected in sections of brain from R6 transgenic mice with antibodies against the N-terminal region of huntingtin. Initially the nucleus contains a diffuse reaction product consisting of several discrete aggregates of increased staining. Additionally there is increased huntingtin immunoreactivity associated with nuclear pores. Reaction product from immunoreactivity can be found on either face of the pore as well as traversing the lumen. This diffuse nuclear staining condenses, over time, to form a single NII with a loss of immunoreactivity from the surrounding nucleus. These single NIIs can then be detected with antibodies against ubiquitin (Davies *et al.* 1997); some time later they can be detected by conventional transmission electron microscopic studies. The initial stages of the formation of NIIs cannot be recognized as a discrete structure by electron microscopy. The initial protein components of the NIIs are subsequently added to, as inclusions are later found to be immunoreactive for the 20S proteasome and the molecular chaperone HDJ-2/HSDJ (B. A. Cozens and S. W. Davies, unpublished data; Cummings *et al.* 1998). Sequestration of the proteasome has been found within cytoplasmic inclusions in several neurodegenerative diseases (Ii *et al.* 1997). The high concentrations of the proteasome found in the neuronal nucleus (Mengual *et al.* 1996) suggest that proteasome components might become trapped in the intranuclear inclusions, thereby contributing to the demise of the neuron. After the formation of an NII, the nucleolus enlarges progressively (Davies *et al.* 1997). The appearance of an NII within a neuronal nucleus is followed by the invagination of the nuclear membrane (figure 3.1); this can often be quite marked, with the formation of labyrinthine membrane complexes in cerebellar Purkinje cells (S. W. Davies, unpublished data). A similar feature of nuclear membrane indentation in cerebellar Purkinje cells, after the formation of NIIs, has been observed in a transgenic mouse model of SCA-1 (Skinner *et al.* 1997).

Invagination of the nuclear membrane seems to be accompanied by an increase in the density of nuclear pores in both conventional transmission electron microscopy and in freeze–fracture preparations (figure 3.1). The increased number of nuclear pores is not associated with any peripheral clumping of the chromatin adjacent to the nuclear membrane, because although chromatin forms dispersed aggregates throughout the nucleus, it does not marginate to the periphery. These prominent ultrastructural changes within the neuronal nucleus (formation of NIIs, nuclear membrane invagination, condensation of chromatin and increased nuclear pore density) have all been reported to occur in post-mortem HD brain (see Davies *et al.* 1997).

Dystrophic neurite inclusions

In addition to the nuclear inclusions, we find inclusions of seemingly identical granular and fibrous structures within dystrophic neuronal processes (Bates *et al.* 1998). These dystrophic neurite inclusions (DNIs) occur within neurites, myelinated axons, vesicle-containing preterminal processes or nerve terminals. Immunocytochemical studies show that these inclusions are, similarly

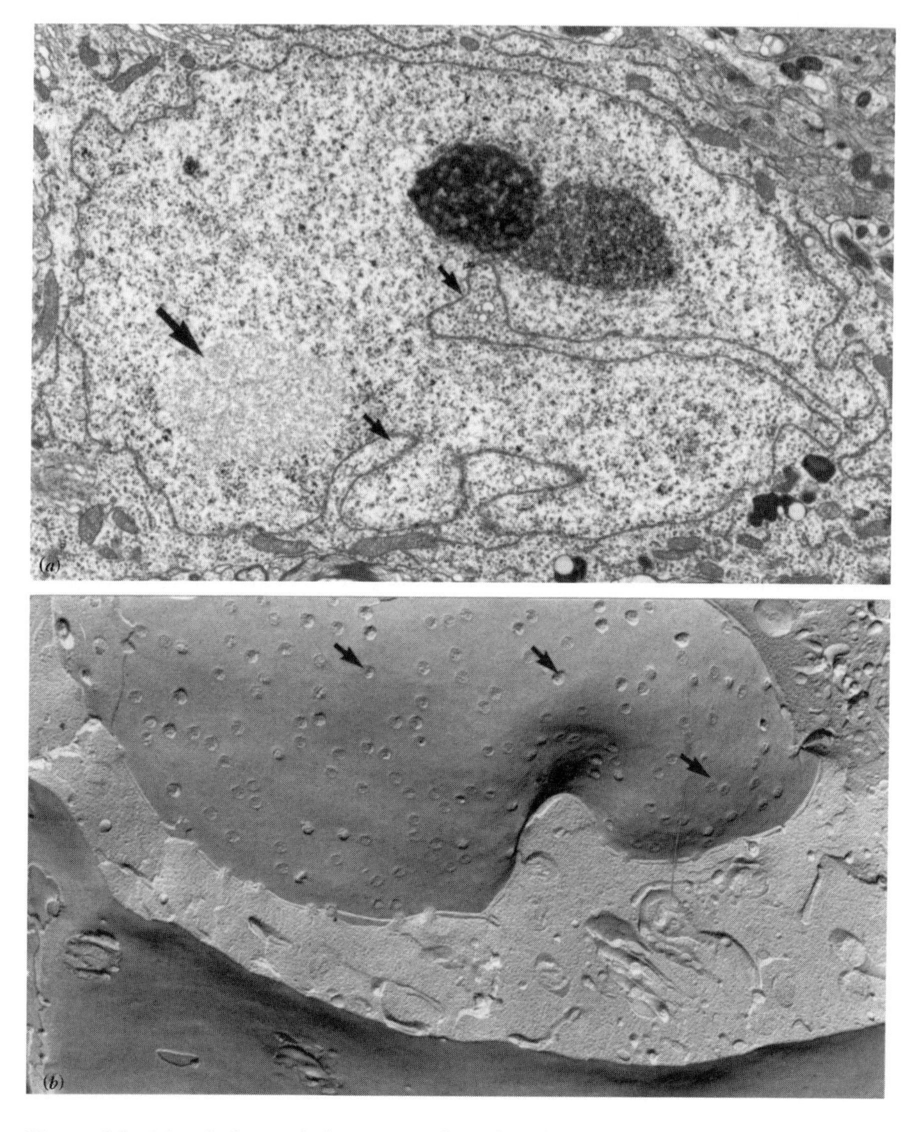

Figure 3.1 Morphology of the neuronal nucleus in the HD transgenic mouse. (*a*) Transmission electron micrograph of a striatal neuron exhibiting prominent indentation of the nuclear membrane (small arrows) and containing a pale-staining granular and fibrous NII, 3 μm in diameter (large arrow in bottom left-hand corner). A more darkly stained nucleolus is also present (top right-hand corner). (*b*) Freeze–fracture of a nuclear membrane of a striatal neuron from an R6/2 transgenic mouse. Note the density of nuclear pores (small arrows). Magnification ×10 000.

to nuclear inclusions, composed of huntingtin and ubiquitin (figure 3.2). The ultrastructural appearance of DNIs in the transgenic mouse is similar to the DNIs originally described by DiFiglia *et al.* (1997) in HD post-mortem tissue. DNIs are few in number and appear only some time after NIIs in the transgenic mouse brain. This is the first description of the presence of huntingtin immunoreactivity in inclusions in dystrophic neurites in a transgenic mouse model of HD. DNIs have not been reported in SBMA, DRPLA or any of the spinocerebellar ataxias.

Glial intranuclear inclusions

In addition to the more prevalent NIIs and DNIs, we additionally observe nuclear inclusions within glial cells. Glial intranuclear inclusions (GIIs) can be found within astrocytes, oligodendrocytes and microglia of the R6/2 and R6/1 lines of transgenic mice (figure 3.3); however, these structures are extremely rare. We have identified only a few examples of these structures in neuroglial cells. GIIs have not to our knowledge been described in post-mortem human HD, DRPLA, SBMA or SCA brain, although it is of interest that inclusions composed of α-synuclein have been reported in nuclei of oligodendrocytes in post-mortem brain in multiple system atrophy (Papp & Lantos 1994; Spillantini *et al.* 1998).

Accumulation of lipofuscin

After the formation of NIIs and the other ultrastructural changes within the nucleus and cytoplasm described above, we also see the appearance of frequent accumulations of lipofuscin within the cytoplasm of neurons (figure 3.4*a*; Davies *et al.* 1997). This can be seen even in juvenile mice from 12 weeks old (R6/2 line). Lipofuscin accumulation within neurons is usually associated with normal ageing but has also been found in post-mortem HD brain (Vonsattel & DiFiglia 1998).

Neuronal shrinkage

The brains of R6/2 transgenic mice are 19% smaller by weight than those of their littermates (Davies *et al.* 1997). This decrease in size is progressive; shrinkage is not apparent until six weeks of age, and occurs in all regions of the brain. Changes within the brain precede any of the changes that occur in body weight (Davies *et al.* 1997), although this might not remain true when individual tissues are examined (see Sathasivam *et al.*, 1999; Mangiarini *et al.*, Chapter 2). We have begun a morphometric analysis of the transgenic mouse brain and find that a marked atrophy occurs in neurons of both

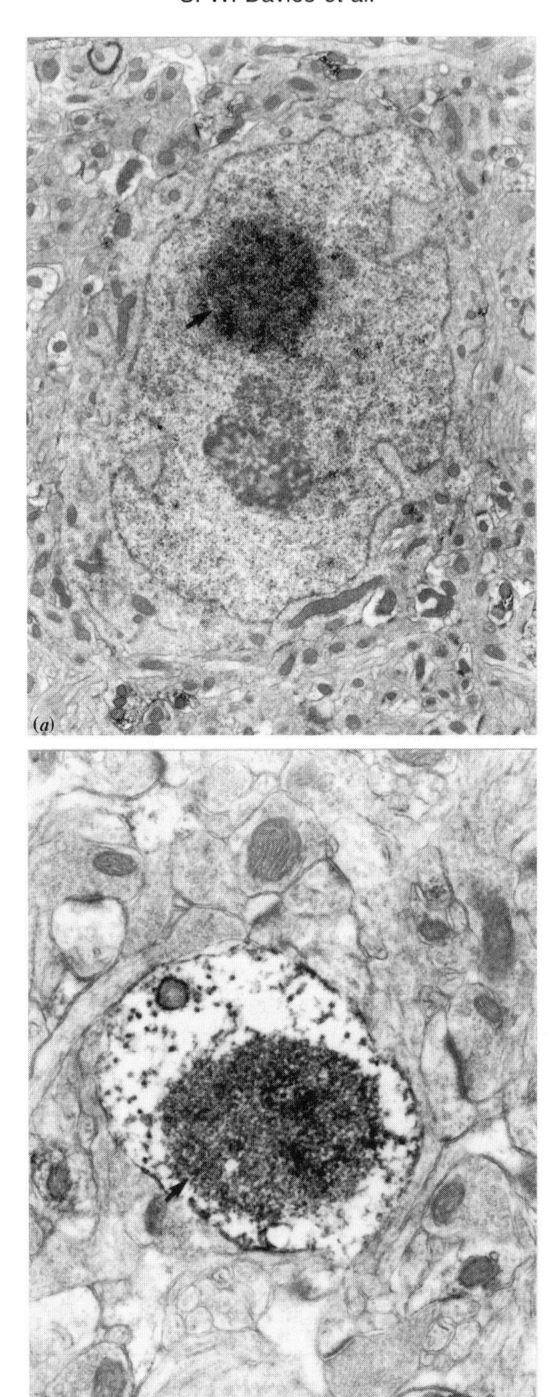
(a)
(b)

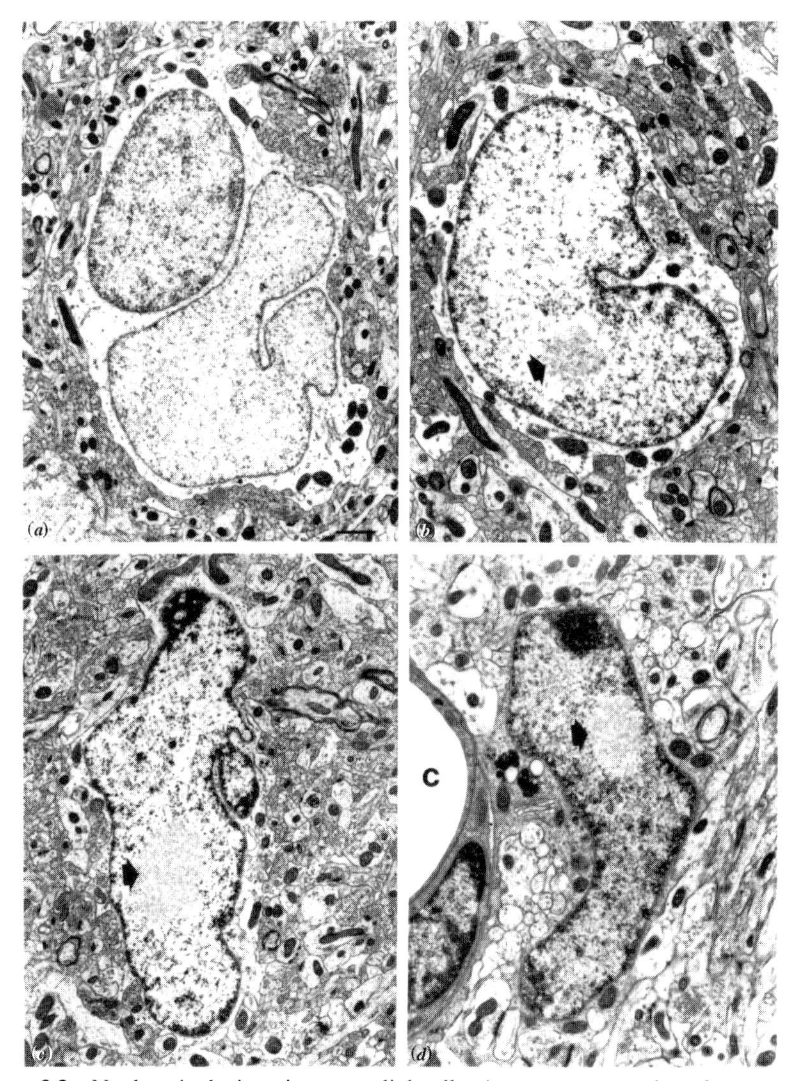

Figure 3.3 Nuclear inclusions in neuroglial cells. An astrocyte nucleus from control mouse brain (*a*) is compared with astrocyte nuclei (*b*,*c*) and the nucleus of a microglial cell (*d*). The glial intranuclear inclusions are arrowed in (*b*)–(*d*). In (*d*) the profile labelled c is a small capillary. Scale bar, 500 nm.

Figure 3.2 Comparison of nuclear and neurite, huntingtin-immunoreactive, inclusions. A dense granular reaction product defines the area of huntingtin immunoreactivity arrowed within the nucleus (*a*) and dystrophic neurite (*b*). Magnification (*a*) ×10 000 and (*b*) ×50 000

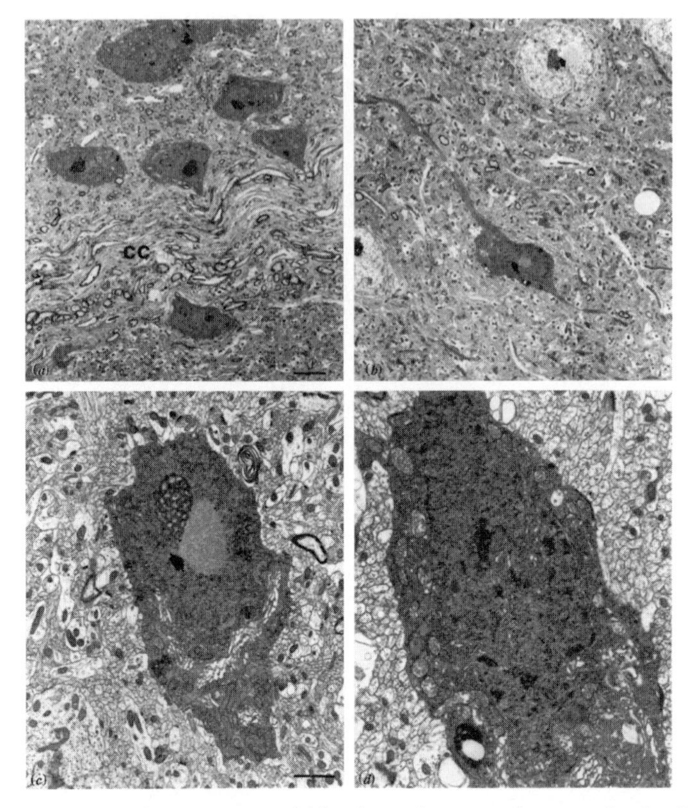

Figure 3.4 Degenerating neurons within the striatum and cortex in the R6/2 HD transgenic mouse. (*a*) Degenerating neurons located on either side of the corpus callosum (cc); (*b–d*) degenerating cells are shown at higher magnification with the NIIs arrowed. Scale bars, (*a*) 5 μm and (*c*) 1.5 μm.

neocortex and striatum beginning at *ca*. six weeks of age (A. S. Raza and S. W. Davies, unpublished data). This is progressive until 13 weeks of age. Vonsattel & DiFiglia (1998) have commented that neurons in the HD post-mortem brain 'contain more lipofuscin and may be smaller than usually expected'; similar observations have been made in the other polyQ diseases.

Neurodegeneration

Within the anterior cingulate cortex, dorsal striatum and Purkinje cells of the cerebellar vermis we can find degenerating neurons (Turmaine *et al.* 1999). These cells are of a distinctive darkened appearance and can best be studied either by light microscopy in semi-thin sections stained with toluidine blue or by electron microscopy in thin sections stained with osmium and lead acetate. Neurons within these areas of the brain are darkly stained in either preparation and exhibit both nuclear and cytoplasmic condensation and clumping of chromatin. Subcellular organelles, including mitochondria, seem intact although there is some swelling of the lumen of the Golgi network. The plasma membrane is ruffled, giving a scalloped appearance to the darkened cell. All degenerating cells contain a distinctive NII and exhibit extensive invagination of the nuclear membrane; however, there is no fragmentation of either the cytoplasm or the nucleus. We cannot find any characteristic apoptotic bodies. A detailed analysis of the number of neurons degenerating within the anterior cingulate cortex suggests that degenerating cells persist in this darkened atrophic form for several weeks.

We have repeatedly tried to identify DNA fragmentation within degenerating neurons by labelling them with the terminal deoxynucleotidyl transferase (TdT)-mediated deoxyuridine triphosphate (dUTP)-biotin nick end labelling (TUNEL) method for the determination of DNA strand breaks *in situ*, but we were unable to find any labelling of these neurons (Turmaine *et al.* 1999).

A uniform ultrastructural appearance of degenerating neurons within defined areas in the brains of HD transgenic mice prompted us to look within these same areas in post-mortem HD brain. We again found dark degenerating neurons, of identical ultrastructural appearance to those found in the transgenic mouse brain. These cells, containing NIIs, were never found to be fragmented into apoptotic bodies or to show blebbing of the nucleus or cytoplasm.

The glial response

Neurons undergoing dark cell degeneration are eventually surrounded and penetrated by the processes of adjacent astrocytes. These neuroglial cells are similar in appearance to immature astrocytes: they lack prominent glial filaments, have a large pale irregular-shaped nucleus and have numerous small mitochondria. The ultrastructural feature of a lack of glial filaments probably explains the absence of glial fibrillary acidic protein (GFAP) immunoreactivity in immunocytochemical investigations seeking reactive astrocytes in the transgenic mouse brain. We can find no inflammatory response and no instance of engulfment of fragmented degenerating neurons by macrophages. Rapid engulfment of fragmented cells is another cardinal feature of cell death by apoptosis (Kerr *et al.* 1972; Wylie *et al.* 1980).

Discussion

The formation of an intranuclear inclusion within a neuron is followed by a defined series of morphological changes, initially within the nucleus and later within the cytoplasm. These can eventually lead to the death of specific populations of cells. Neurons undergoing degeneration do so by a characteristic process that is dissimilar to the well-documented processes of either apoptosis or necrosis. All of these changes that we have so far identified in the brains of transgenic mice, expressing exon 1 of the human HD gene containing an expanded glutamine repeat, have similarly been found in the human HD brain.

The initial stages of protein aggregation, well defined *in vitro* (Scherzinger *et al.* 1997), have not been fully characterized *in vivo*. The development of an NII, recognizable in the electron microscope, clearly requires the recruitment of many additional proteins. We currently know the identity of a few of these: the subunits of the 20S proteasome and associated 19S and 11S activator complexes and the molecular chaperones HDJ-2/HSDJ have been identified in nuclear inclusions in transgenic mouse models and in post-mortem brain of both SCA-1 and HD (Cummings *et al.* 1998; B. A. Cozens and S. W. Davies, unpublished data).

Once fully formed, NIIs seem to provoke a wide spectrum of pathological changes within the cell. These include the complex series of nuclear and cytoplasmic changes that we have described. Within the brain of the HD transgenic mouse, specific populations of neurons eventually die; importantly, these are the same populations of cells that degenerate in juvenile-onset HD (Vonsattel *et al.* 1985; Robitaille *et al.* 1997; Ross *et al.* 1997; Vonsattel & DiFiglia 1998; Young 1998). Thus significant pathological change can be found not only in the striatum (Vonsattel *et al.* 1985) but additionally in the cerebellum (Ross *et al.* 1997; Vonsattel & DiFiglia 1998; Young 1998) and in the cortex of the frontal lobes (de la Monte *et al.* 1988; Sotrel *et al.* 1991; Jackson *et al.* 1995). The mechanism of cell death is unusual: degenerating neurons lack the classical features that characterize either necrosis or apoptosis (Kerr *et al.* 1972; Wylie *et al.* 1980; Clarke 1990). The progressive atrophy of the cell with condensation of the nucleus and cytoplasm is unlike necrosis, but reminiscent of apoptosis. However, it is significant that despite our observation that most of the pathological changes occur in the nucleus, the defining nuclear changes of apoptosis are not found (figure 3.4).

The nuclear events of apoptosis begin with the collapse of the chromatin against the nuclear periphery and into one or a few large clumps within the nucleus. The chromatin becomes progressively more condensed, whereas the nuclear envelope remains morphologically intact. In many cases the entire nucleus condenses into a single dense ball, whereas in others the chromatin buds into smaller balls resembling a bunch of grapes, with each 'grape' surrounded by nuclear membrane (Earnshaw 1995). Changes in the chromatin are accompanied by two changes in the nuclear envelope. The nuclear pores redistribute by sliding away from the surface of the condensed chromatin domains and accumulating in the regions between them, and the nuclear lamina disassembles. None of these ultrastructural changes can be found in the nuclei of degenerating neurons in the HD transgenic mouse.

A defining feature of the changes that occur in the chromatin in apoptosis is the formation of oligosomal fragments that can be labelled by the TUNEL method. However, we cannot find any evidence of fragmentation of the DNA in this manner. This is in contrast to the reports that have documented TUNEL-positive staining of both glia and neurons in post-mortem HD brain (Brannon Thomas *et al.* 1995; Dragunow *et al.* 1995; Portera-Cailliau *et al.* 1995). However, TUNEL-positive profiles in post-mortem tissue are notoriously difficult to interpret (Lucassen *et al.* 1997). Finally, apoptotic neurons are rapidly engulfed (within days) and phagocytosed by macrophages, whereas the degenerating neurons that we have described persist for long periods (several weeks) and are eventually ensheathed by astrocytic, not microglial, processes. We conclude that this slow pathological process, leading to neuronal death, is different from either of the mechanisms of cell death described previously.

The transgenic mice that we have reported (Mangiarini *et al.* 1996; Davies *et al.* 1997) have helped to elucidate the pathological features of HD. The results presented suggest that these mice demonstrate all of the key pathological features of the disease. They provide us with a unique opportunity to test novel therapeutic agents and to elucidate further the pathway from protein aggregation to neurodegeneration (Goedert *et al.* 1998). This novel progressive non-apoptotic process still remains to be fully characterized.

This work was supported by grants from the Wellcome Trust, the Huntington's Disease Society of America and the Hereditary Disease Foundation.

References

Bates, G. P., Mangiarini, L. & Davies, S. W. 1998 Transgenic mice in the study of polyglutamine repeat expansion diseases. *Brain Pathol.* **8**, 699–714.

Becher, M. W., Kotzuk, J. A., Sharp, A. H., Davies, S. W., Bates, G. P., Price, D. L. & Ross, C. A. 1998 Intranuclear neuronal inclusions in Huntington's disease and dentatorubral and pallidoluysian atrophy: correlation between the density of inclusions and IT15 CAG triplet repeat length. *Neurobiol. Dis.* **4**, 387–398.

Bhide, P. G., Day, M., Sapp, E., Schwarcz, C., Sheth, A., Kim, J., Young, A. B., Penney, J., Golden, J., Aronin, N. & DiFiglia, M. 1996 Expression of normal and mutant huntingtin in the developing brain. *J. Neurosci.* **16**, 5523–5535.

Brannon Thomas, L., Gates, D. J., Richfield, E. K., O'Bien, T. F., Schweitzer, J. B. & Steindler, D. A. 1995 DNA end labelling (TUNEL) in Huntington's disease and other neuropathological conditions. *Exp. Neurol.* **133**, 265–272.

Clarke, P. G. H. 1990 Developmental cell death: morphological diversity and multiple mechanisms. *Anat. Embryol.* **181**, 195–213.

Cummings, C. J., Mancini, M. A., Antalffy, B., De Franco, D. B., Orr, H. T. & Zogbi, H. Y. 1998 Chaperone suppression of aggregation and altered subcellular proteasome localisation imply protein misfolding in SCA1. *Nature Genet.* **19**, 148–154.

Davies, S. W., Turmaine, M., Cozens, B. A., DiFiglia, M., Sharp, A. H., Ross, C. A., Scherzinger, E., Wanker, E. E., Mangiarini, L. & Bates, G. P. 1997 Formation of neuronal intranuclear inclusions underlies the neurological dysfunction in mice transgenic for the HD mutation. *Cell* **90**, 537–548.

Davies, S. W., Beardsall, K., Turmaine, M., DiFiglia, N., Aronin, N. & Bates, G. P. 1998 Neuronal intranuclear inclusions (NII), the common neuropathology of triplet repeat disorders with polyglutamine repeat expansions. *Lancet* **351**, 131–133.

de la Monte, S. M. de la, Vonsattel, J.-P. G. & Richardson, E. P. 1988 Morphometric demonstration of atrophic changes in the cerebral cortex, white matter and neostriatum in Huntington's disease. *J. Neuropathol. Exp. Neurol.* **47**, 516–525.

de Rooij, K. E., Dorsman, J. C., Smoor, M. A., den Dunnen, J. T. & Van Ommen, G.-J. B. 1996 Subcellular localisation of the Huntington's disease gene product in cell lines by immunofluorescence and biochemical subcellular fractionation. *Hum. Mol. Genet.* **5**, 1093–1099.

DiFiglia, M. (and 11 others) 1995 Huntingtin is a cytoplasmic protein associated with vesicles in human and rat brain neurons. *Neuron* **14**, 1075–1081.

DiFiglia, M., Sapp, E., Chase, K. O., Davies, S. W., Bates, G. P., Vonsattel, J. P. & Aronin, N. 1997 Aggregation of huntingtin in neuronal intranuclear inclusions and dystrophic neurites in brain. *Science* **277**, 1990–1993.

Dragunow, M., Faull, R. L. M., Lawlor, P., Beihartz, E. J., Singleton, K., Walker, E. B. & Mee, E. 1995 *In situ* evidence for DNA fragmentation in Huntington's disease striatum and Alzheimer's disease temporal lobes. *Clin. Neurosci. Neuropathol.* **6**, 1053–1057.

Earnshaw, W. C. 1995 Nuclear changes in apoptosis. *Curr. Opin. Cell Biol.* **7**, 337–343.

Goedert, M., Spillantini, M. G. & Davies, S. W. 1998 Filamentous nerve cell inclusions in neurodegenerative diseases. *Curr. Opin. Neurobiol.* **8**, 619–632.

Gourfinkel-An, I., Cancel, G., Duyckaerts, C., Faucheaux, B., Hauw, J.-J., Trottier, Y., Brice, A., Agid, Y. & Hirsch, E. C. 1998 Neuronal distribution of intranuclear inclusions in Huntington's disease with adult onset. *NeuroReport* **9**, 1823–1826.

Gutekunst, C.-A., Levey, A. I., Heilman, C. J., Whalley, W. L., Yi, H., Nash, N. R., Rees, H. D., Madden, J. J. & Hersch, S. M. 1995 Identification and localisation of huntingtin in brain and human lymphoblastoid cell lines with anti-fusion protein antibodies. *Proc. Natl Acad. Sci. USA* **92**, 8710–8714.

Harper, P. S. 1996 *Huntington's disease*, 2nd edn (*Major problems in neurology*, vol. 31). London: W. B. Saunders.

Hayashi, Y. (and 10 others) 1998 Hereditary dentatorubral–pallidoluysian atrophy: detection of widespread ubiquitinated neuronal and glial intranuclear inclusions in the brain. *Acta Neuropathol.* **96**, 547–552.

Holmberg, M. (and 10 others) 1998 Spinocerebellar ataxia 7 (SCA7): a neurodegenerative disorder with neuronal intranuclear inclusions. *Hum. Mol. Genet.* **7**, 913–918.

Huntington's Disease Collaborative Research Group 1993 A novel gene containing a trinucleotide repeat that is unstable on Huntington's disease chromosomes. *Cell* **72**, 971–983.

Ii, K., Itoh, H., Tanaka, K. & Hirano, A. 1997 Immunocytochemical colocalisation of the proteasome in ubiquitinated structures in neurodegenerative diseases and the elderly. *J. Neuropathol. Exp. Neurol.* **56**, 125–131.

Jackson, G. R., Salecker, I., Dong, X., Yao, X., Arnheim, N., Faber, P. W., MacDonald, M. E. & Zipursky, S. L. 1998 Polyglutamine-expanded human huntingtin transgenes induce degeneration of *Drosophila* photoreceptor neurons. *Cell* **21**, 633–642.

Jackson, M., Gentleman, S., Lennox, G., Ward, L., Gray, T., Randall, K., Morell, K. & Lowe, J. 1995 The cortical neuritic pathology of Huntington's disease. *Neuropathol. Appl. Neurobiol.* **21**, 18–26.

Kerr, J. F. R., Wyllie, A. H. & Currie, A. R. 1972 Apoptosis: a basic biological phenomenon with wide ranging implications in tissue kinetics. *Br. J. Cancer* **26**, 239–257.

Li, M., Miwa, S., Kobayshi, Y., Merry, D. E., Yamamoto, M., Tanaka, F., Doyu, M., Hashizume, Y., Fischbeck, K. H. & Sobue, G. 1998*a* Nuclear inclusions of the

androgen receptor protein in spinal and bulbar muscular atrophy. *Ann. Neurol.* **44**, 249–254.
Li, M., Nakagomi, Y., Kobayshi, Y., Merry, D. E., Tanaka, F., Doyu, M., Mitsume, T., Hashizume, Y., Fischbeck, K. H. & Sobue, G. 1998*b* Nonneuronal nuclear inclusions of androgen receptor protein in spinal and bulbar muscular atrophy. *Am. J. Pathol.* **153**, 695–701.
Lieberman, A. P., Robitaille, Y., Trojanowski, J. Q., Dickson, D. W. & Fischbeck, K. H. 1998 Polyglutamine-containing aggregates in neuronal intranuclear inclusion disease. *Lancet* **351**, 884.
Lucassen, P. J., Chung, W. C. J., Kamphorst, W. & Swaab, D. F. 1997 DNA damage distribution in the human brain as shown by in situ end labelling; area-specific differences in ageing and Alzheimer disease in the absence of apoptotic morphology. *J. Neuropathol. Exp. Neurol.* **56**, 887–900.
Mangiarini, L. (and 10 others) 1996 Exon 1 of the HD gene with an expanded CAG repeat is sufficient to cause a progressive neurological phenotype in transgenic mice. *Cell* **87**, 493–506.
Mengual, E., Arizti, P., Rodrigo, J., Giménez-Amaya, J. M. & Castaño, J. G. 1996 Immunohistochemical distribution and electron microscopic subcellular localisation of the proteasome in the rat CNS. *J. Neurosci.* **16**, 6331–6341.
Ordway, J. M. (and 11 others) 1997 Ectopically expressed CAG repeats cause intranuclear inclusions and a progressive late onset neurological phenotype in the mouse. *Cell* **91**, 753–763.
Papp, M. I. & Lantos, P. L. 1994 The distribution of oligodendroglial inclusions in multiple system atrophy and its relevance to clinical symptomatology. *Brain* **117**, 235–243.
Paulson, H. L., Das, S. S., Crino, P. B., Perez, M. K., Patel, S. C., Gotsdiner, D., Fischbeck, K. H. & Pittman, R. N. 1997*a* Machado–Joseph disease gene product is a cytoplasmic protein widely expressed in brain. *Ann. Neurol.* **41**, 453–462.
Paulson, H. L., Perez, M. K., Trottier, Y., Trojanowski, J. Q., Subramony, S. H., Das, S. S., Vig, P., Mandel, J. L., Fischbeck, K. H. & Pittman, R. N. 1997*b* Intranuclear inclusions of expanded polyglutamine protein in spinocerebellar ataxia type 3. *Neuron* **19**, 333–344.
Perutz, M. F., Johnson, T., Suzuki, M. & Finch, J. T. 1994 Glutamine repeats as polar zippers: their possible role in inherited neurodegenerative diseases. *Proc. Natl Acad. Sci. USA* **91**, 5355–5358.
Portera-Cailliau, C., Hedreen, J. C., Price, D. L. & Koliatsos, V. E. 1995 Evidence for apoptotic cell death in Huntington disease and excitotoxic animal models. *J. Neurosci.* **15**, 3775–3787.
Robitaille, Y., Lopes-Cendes, I., Becher, M., Rouleau, G. & Clark, A. W. 1997 The neuropathology of CAG repeat diseases: review and update of genetic and molecular features. *Brain Pathol.* **7**, 877–881.
Roizin, L., Stellar, S. & Liu, J. C. 1979 Neuronal nuclear–cytoplasmic changes in Huntington's chorea: electron microscope investigations. *Adv. Neurol.* **23**, 95–122.
Ross, C. A., Becher, M. W., Coloner, V., Engelender, S., Wood, J. D. & Sharp, A. H. 1997 Huntington's disease and dentatorubral–pallidoluysian atrophy: proteins, pathogenesis and pathology. *Brain Pathol.* **7**, 1003–1017.
Sathasivan, K., Hobbs, C., Mangiarini, L., Mahal, A., Turmaine, M., Doherty, P., Davies, S. W. and Bates, G. P. 1999 Transgenic models of Huntington's disease. *Royal Society* **354**, 963–970
Scherzinger, E., Lurz, R., Turmaine, M., Mangiarini, L., Hollenbach, B., Hasenbank, R., Bates, G. P., Davies, S. W., Lehrach, H. & Wanker, E. E. 1997 Huntingtin encoded polyglutamine expansions form amyloid-like protein aggregates *in vitro* and *in vivo*. *Cell* **90**, 549–558.
Schmidt, T. (and 10 others) 1998 An isoform of ataxin-3 accumulates in the nucleus of neuronal cells in affected brain regions of SCA3 patients. *Brain Pathol.* **8**, 669–681.

Skinner, P. J., Koshy, B. T., Cummings, C. J., Klement, I. A., Helin, K., Servadio, A., Zoghbi, H. Y. & Orr, H. T. 1997 Ataxin-1 with an expanded glutamine tract alters nuclear matrix-associated structures. *Nature* **389**, 971–974.

Sotrel, A., Paskevitch, P. A., Kiely, D. K., Bird, E. D., Williams, R. S. & Meyers, R. H. 1991 Morphometric analysis of the prefrontal cortex in Huntington's disease. *Neurology* **41**, 1117–1123.

Spillantini, M. G., Crowther, R. A., Jakes, R., Cairn, N. J., Lantos, P. L. & Goedert, M. 1998 Filamentous α-synuclein inclusions link multiple system atrophy with Parkinson's disease and dementia with Lewy bodies. *Neurosci. Lett.* **251**, 205–208.

Trottier, Y., Devys, D., Imbert, G., Sadou, F., An, I., Lutz, Y., Weber, C., Agid, Y., Hirsch, E. C. & Mandel, J.-L. 1995 Cellular localisation of the Huntington's disease protein and discrimination of the normal and mutated forms. *Nature Genet.* **10**, 104–110.

Turmaine, M., Raza, A. S., Mahal. A, Mangiarini, L., Bates, G. P. & Davies S. W. 1999 Non apoptopic dark cell degeneration in a transgenic mouse model of Huntington's disease.*Nature.* (Submitted.)

Vonsattel, J.-P. G. & DiFiglia, M. 1998 Huntington disease. *J. Neuropathol. Exp. Neurol.* **57**, 369–384.

Vonsattel, J.-P. G., Meyers, R. H., Stevens, T. J., Ferrante, R. J., Bird, E. D. & Richardson, E. P. 1985 Neuropathological classification of Huntington's disease. *J. Neuropathol. Exp. Neurol.* **44**, 559–577.

Warrick, J. M., Paulson, H. L., Gray-Board, G. L., Bui, Q. T., Fischbeck, K. H., Pittman, R. N. & Bonini, N. M. 1998 Expanded polyglutamine protein forms nuclear inclusions and causes neural degeneration in *Drosophila. Cell* **93**, 939–949.

Wyllie, A. H., Kerr, J. F. R. & Currie, A. R. 1980 Cell death: the significance of apoptosis. *Int. Rev. Cytol.* **68**, 251–305.

Young, A. B. 1998 Huntington's disease and other trinucleotide repeat disorders. In *Molecular neurology* (ed. J. B. Martin), pp. 35–54. New York: Scientific American.

4

Behavioural changes and selective neuronal loss in full-length transgenic mouse models for Huntington's disease

P. Hemachandra Reddy, Vinod Charles, William O. Whetsell, Jr and Danilo A. Tagle

Introduction

Huntington's disease (HD) is an autosomal dominant, inherited human neurodegenerative disorder characterized by hyperkinetic involuntary movements including motor restlessness and chorea, slowed voluntary movements and psychological and intellectual impairment. Selective and progressive neuronal loss and gliosis in striatum, cerebral cortex, thalamus, subthalamus and hippocampus (Vonsattel *et al.* 1985; Folstein 1990; Hedreen *et al.* 1991; Spargo *et al.* 1993; Hedreen & Folstein 1995) are well recognized as neuropathological correlates for the clinical manifestation of HD. Neuropathological examinations of HD patients show significant neuronal loss and reactive astrocytosis in the neostriatum (Hedreen & Folstein 1995; Vonsattel *et al.* 1985). Neurodegeneration has also been reported in both segments of the pallidum and the medial and lateral compartments of the subthalamic nucleus (Folstein 1990), and more recently in the hippocampal area (Spargo *et al.* 1993). There is also neuronal loss in the deep layers (V and VI) of the cerebral cortex (Hedreen *et al.* 1991; Wagster *et al.* 1994).

The identification of the *HD* gene in 1993 (Huntington's Disease Collaborative Research Group 1993) demonstrated that the underlying mutation is in the expansion of CAG trinucleotide repeats in exon 1 of the gene (Andrew *et al.* 1993; Duyao *et al.* 1993; Huntington's Disease Collaborative Research Group 1993; Stine *et al.* 1993). In the normal human population, the CAG repeat length ranges from five to 35 copies, whereas repeat lengths of 36–75 CAGs can lead to adult-onset HD and 48–121 repeats give rise to juvenile HD (Andrew *et al.* 1993; Duyao *et al.* 1993; Stine *et al.* 1993; Rubinsztein *et al.* 1996; Brinkman *et al.* 1997). The CAG repeat tract is translated as polyglutamines in the protein product (Jou & Myers 1995; Persichetti *et al.* 1995; Sharp *et al.* 1995). *HD* mRNA and the protein product show a widespread distribution (Li *et al.* 1993; Strong *et al.* 1993; Sharp *et al.* 1995), and thus much remains to be understood about the selective and progressive neurodegeneration described in HD.

A number of attempts have been made to generate animal models for HD either to determine the normal function of huntingtin or to model the disease process. Excitotoxic models involved chemical lesioning either by direct striatal injections with excitatory amino acids (Mason & Fibiger 1979; Schwarcz *et al.* 1983; Beal *et al.* 1988) or systemic injections of mitochondrial metabolic inhibitors (Borlongan *et al.* 1995; Kodsi & Swerdlow 1997; Miller & Zaborszky 1997) in rodent and non-human primates. Use of these techniques has been shown to model either acute or chronic neurochemical, neuroanatomical and behavioural changes seen in HD, but do not account for the progressive nature of the disease nor for the genetic and molecular aspects of this inherited disorder. Recent studies (Duyao *et al.* 1995; Nasir *et al.* 1995; Zeitlin *et al.* 1995) involving inactivation of the homologous gene (*Hdh*) in mice have shown early embryonic lethality, suggesting an essential role for huntingtin in gastrulation (see table 4.1). These studies suggest a gain-in-function effect for the mutant huntingtin protein. More recent results (White *et al.* 1997) using targeted introduction of expanded CAG repeats into endogenous *Hdh* imply a role for huntingtin in neurogenesis. Moreover, White *et al.* (1997) showed that the presence of an expanded polyglutamine tract in the protein does not appear to impede normal functioning of huntingtin, and the mice failed to show any abnormal phenotype. Recently, a number of laboratories have generated transgenic mice for HD (table 4.2) but thus far, only a few models exist where the full-length huntingtin protein is expressed from a full-length cDNA (Reddy *et al.* 1998, 1999) or from a yeast artificial chromosome (YAC) transgene (Hodgson *et al.* 1999). Mice transgenic for the mutated full-length huntingtin protein exhibited both progressive behavioural and neuropathological changes analogous to that of HD (Reddy *et al.* 1998). A more detailed analysis and characterization of these mice expressing full-length human huntingtin is presented here.

Table 4.1 *Summary of knock-out mouse models for Huntington disease*

targeted region	*embryonic lethality*	*ES cell strain*	*genetic background*	*reference*
exon 5	+	129/sv	C57BL/6J	Nasir *et al.* 1995
exon 4	+	129/sv	CD1, C57BL/6J	Duyao *et al.* 1995
promoter	+	129/sv	MF1, C57BL/6J	Zeitlin *et al.* 1995

Table 4.2 *Summary of transgenic mouse models for Huntington's disease*

HD gene	*strain*	*promoter*	*no. of CAGs*	*expression*	*phenotype*	*neuro-degeneration*	*NIIs*	*reference*
exon 1	C57BL/6	HD	18	yes	no	no	no	Mangiarini *et al.* 1996
exon 1	C57BL/6	HD	115–156	yes	yes	no	yes	Mangiarini *et al.* 1996
full-length cDNA	FVB/N	CMV	44	no	no	no	not reported	Goldberg *et al.* 1996
full-length cDNA	FVB/N	CMV	16	yes	no	no	no	Reddy *et al.* 1998
full-length cDNA	FVB/N	CMV	48	yes	yes	yes	yes	Reddy *et al.* 1998
full-length cDNA	FVB/N	CMV	89	yes	yes	yes	yes	Reddy *et al.* 1998
1–1073 amino-acids	SJL/B6	NSE	46 and 100	yes	yes	yes	yes	Laforet *et al.* 1998
exons 1–3	C3H/HEJXC-57BL6/6JF1	PrP	18	yes	no	no	no	Schilling *et al.* 1999
exons 1–3	C3H/HEJXC-57BL6/6JF1	PrP	44	yes	yes	no	yes	Schilling *et al.* 1999
exons 1–3	C3H/HEJXC-57BL6/6JF1	PrP	82	yes	yes	no	yes	Schilling *et al.* 1999
full-length (genomic)	FVB/N	HD prom. (YAC)	18	yes	no	no	no	Hodgson *et al.* 1999
full-length (genomic)	FVB/N	HD prom. (YAC)	46	yes	no	no	yes	Hodgson *et al.* 1999
full-length (genomic)	FVB/N	HD prom. (YAC)	72	yes	yes	yes	yes	Hodgson *et al.* 1999

Material and methods

Construction of HD full-length clones

The full-length construct was made by ligating IT16L (bp 932–3018) with RT-PCR product C (from bp 2401–3270) at the *Bsm*I site. A three-way ligation of this product was performed with PCR products from bp 637–1429 and bp 187–858 using the *Ssp*I and *Xho*I sites, respectively. The resulting 3027 bp *Eag*I to *Ava*III restriction fragment was ligated to the cDNA clone IT15B (bp 3024–10366) to generate a full-length clone with 16 CAG repeats. The 9.9 kb *Eag*I to *Sph*I fragment was ligated into the expression vector pCDNA1.1. An *Rsr*II site at positions 549–555 bp was introduced by *in vitro* mutagenesis, and PCR products flanked by *Rsr*II and *Eag*I sites from genomic DNA of an adult-onset case with 48 repeats and a juvenile-onset case with 89 repeats were used to replace the 16 CAGs.

Transgenesis

The constructs were restricted with *Pvu*I and *Drd*I to release vector sequences prior to microinjection into sperm pronuclei of fertilized eggs derived from the FVB/N mouse strain. Genomic DNA from animals was isolated from tail biopsy and genotyped for the transgene by Southern blot analysis or PCR. The CAG repeat size was determined by PCR using HD1 and HD2 primers (Huntington's Disease Collaborative Research Group 1993). Genomic DNA from tail biopsy of 58 animals (HD16, n = 8; HD48, n = 26; and HD89, n = 24) was examined.

Phenotypic analysis

A total of 198 mice consisting of wild-type (n = 23); HD16 (n = 28); HD48 (n = 58); and HD89 (n = 89) were monitored for behavioural changes. Of the animals included in this part of the study, four were homozygotes for HD16, 24 were homozygotes for HD48 and 40 were homozygotes for HD89. The care and use of animals in this study was in accordance with institutional guidelines.

Histology and immunohistochemistry

Neuropathologic examinations were carried out on wild-type mice (n = 10), HD16 transgenics (n = 12), HD48 transgenics (n = 20) and HD89 transgenics (n = 9). Of these animals, two were homozygotes from HD16, three were homozygotes from HD48 and two were homozygotes from HD89 lines. Wild-type and transgenic animals were sacrified at various ages and brains were removed and fixed overnight in 4% paraformaldehyde in phosphate-buffered saline (PBS). Fixed brains were cut in the coronal plane at *ca.* 0.2 mm intervals (from anterior to posterior extent), and resulting tissue blocks were embedded in paraffin. From these, 4–10 μm sections were cut and stained with either haematoxylin and eosin (H&E) or Nissl (cresyl violet). For immunohistochemistry, sections were deparaffinized in xylene and incubated with either anti-huntingtin antibody (1:800, P. H. Reddy, M. Williams and D. A. Tagle, unpublished data), anti-ubiquitin (1:50, Dako) or anti-GFAP (1:5000, Dako) then exposed (10 min) to streptavidin–biotin–horseradish peroxidase (HRP) complex (Dako) and lightly counterstained with haematoxylin. Anti-Neu-N monoclonal antibody (Chemicon) was used at 1:3000. Sections were stained according to the ABC method. Briefly, sections were washed and placed in the appropriate biotinylated secondary antibody solution for 1 h. Following washes, the sections were incubated for 1.5 h in ABC (avidin–biotin complex). Finally, sections were placed in a chromagen solution containing 3,3'-diaminobenzidine (DAB)–hydrogen peroxide with and without nickel 2 sulphate. Light microscopic evaluation and photography utilized bright-field optics. Cell counts were performed on Nissl-stained sections using a calibrated grid (Olympus). Each area counted was 0.25 mm^2 at a magnification of 200×. For these counts, sections of anatomically distinct striatum (caudatoputamenal complex) at approximately the mid-striatal level (coronal plane) were used. For each brain, ten randomly selected fields were counted to determine the number and the ratio of small or medium neurons and large neurons within the grid square; neurons were distinguishable from astrocytes and oligodendrocytes based on nuclear morphology. Averages of the ten field counts from the mid-striatal section from each animal were used for comparison of neuronal populations in the mid-striatum in wild-type and transgenic animals.

Results

Generation of HD mouse transgenics

The expression constructs consisted of 15 kb of DNA that included 10 179 bp of HD cDNA sequence (Huntington's Disease Collaborative Research Group 1993), modified to include either 16, 48 or 89 CAG repeats (Reddy *et al.* 1998). The cytomegalovirus (CMV) promoter was used in conjunction with an SV40 enhancer to drive high level and widespread expression of the transgene. Transgenic animals were generated by microinjection into the sperm pronuclei of fertilized FVB/N mouse eggs. Southern blot analysis (figure 4.1) and PCR (data not shown) were used to genotype the animals. A total of five founder lines (16A–16E) were obtained for the 16 repeat transgenes and three transgenic lines each for the 48 (48B–48D) and 89 (89A–89C) constructs. The copy number and integration site of the transgene in each line were determined by Southern blot analysis (figure 4.1). The number of transgenes inserted in the founder lines ranged from two to 22 copies (Reddy *et al.* 1998).

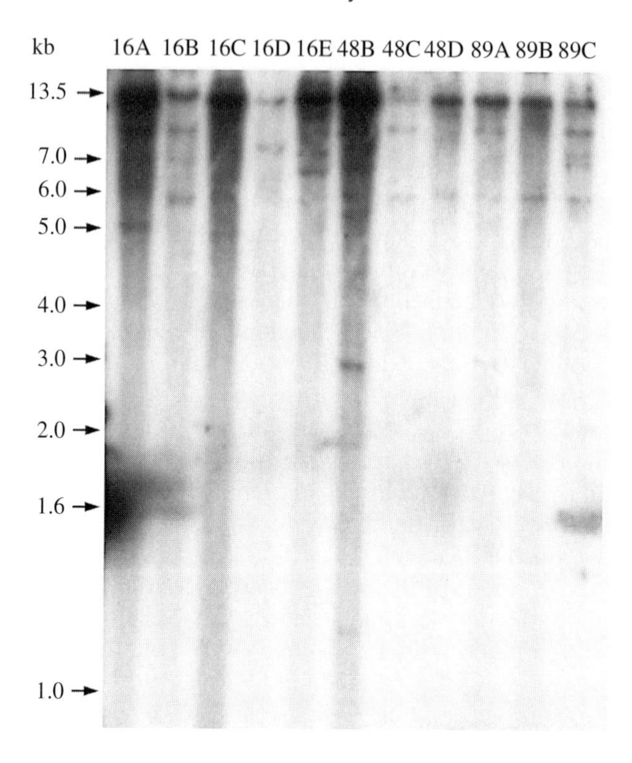

Figure 4.1 Southern blot analysis of founder lines. Ten micrograms of genomic tail DNA from each founder line was restricted with *Bgl*II and electrophoresed on a 0.8% agarose gel for Southern blot analysis with a probe derived from human IT15 cDNA (Huntington's Disease Collaborative Research Group 1993). Densitometric analysis (data not shown) indicated that the HD48B line had the highest copy number and the HD48C line had the least number of transgenes integrated. Each founder line has a different genomic integration site. The core 13.5 kb band represents *Bgl*II-restricted transgene concatamers, and the additional bands are the flanking sequences from mouse that are unique for each line. More than two junction fragments were found for 16A, 16B, 48B and 89C.

Widespread expression of the transgene

Western blot analysis of mouse tissue homogenates from each line was used to determine the level and site of expression of the transgene. A polyclonal and a monoclonal antibody (HD48Ab and mAb HD48; P. H. Reddy, M. Williams and D. A. Tagle, unpublished data) directed against the N-terminal end of huntingtin was used to detect the expression of the transgene. The transgene was expressed in several different tissues tested including the brain, heart, spleen, kidney, lung, liver and gonads (Reddy *et al.* 1998). The widespread expression of the transgene is similar to that observed for endogenous huntingtin (Wood *et al.* 1996). A C-terminal antibody, HF1 (White *et al.* 1997), that recognizes both human and mouse huntingtin, was used to compare the level of transgene expression relative to mouse endogenous huntingtin. Western analyses indicated that the HD48B line had the highest expression level in the brain which was fivefold higher relative to endogenous levels, whereas the expression of the transgene in the brain of HD48C mice was the lowest, and was roughly equivalent to endogenous levels (Reddy *et al.* 1998). In the brain, the transgene was expressed in the major regions examined including cortex, striatum, hippocampus and cerebellum. Brain expression was highest in the striatum and in the motor cortex (Reddy *et al.* 1998).

Behavioural phenotype

Transgenic animals and wild-type littermates were monitored from birth until death in order to identify age of onset and progression of any abnormal phenotype. The earliest detectable abnormality was seen as early as eight weeks of age in HD48(B,D) and HD89(A–C) transgenic mice. Beginning at this time, transgenic animals expressing the repeat expansion exhibited a feet-clasping and/or a trunk-curling posture when suspended by their tails (figure 4.2*b*). This behaviour persisted throughout each animal's life. However, compared with mice from other lines with expanded repeats, HD48C animals lagged in the onset of feet clasping which was observed instead at 25 weeks of age. In contrast, observations of wild-type and HD16 transgenic animals ranging in age from eight to 36 weeks demonstrated normal limb posture when suspended by their tails (figure 4.2*a*).

At about 20 weeks of age, stereotypic behaviour patterns were observed in at least 37% of HD48(B–D) and HD89(A–C) transgenic mice. Stereotypy consisted of generalized hyperactive behaviour in the form of unidirectional rotations, backflipping and excessive grooming. Rotational behaviour was frequently observed in these mice during this hyperactive phase. The rotations consisted of a rapid circling at *ca.* 1 rev s^{-1} within a diameter of 12–20 cm (figure 4.2*c*), irrespective of the cage size in which the animals were housed. The duration of the hyperactive phase was highly variable and can last as long as up to 36–44 weeks of age. Animals that exhibited stereotypy and those that did not show overt hyperactive behaviour were found to have heightened exploratory activity when tested in an open field monitor (data not shown). None of the wild-type and HD16 transgenic animals in the control group exhibited hyperactivity at the times tested.

By 24 weeks of age, HD48(B,D) and HD89(A–C) animals started to become less active and less alert than the control group. Urine retention or incontinence was also frequently observed in these mice. The hypokinetic phase lasted typically for four to six weeks and progressed to locomotor deterioration, whereby the animals showed akinesia. Akinetic animals exhibited paucity of volitional movement and lack of responsiveness to sensory stimuli. Death followed usually within ten days after the animals became akinetic. The progression and staging of the abnormal phenotype observed in these transgenic mice relative to its time-course is shown in figure 4.2*d*.

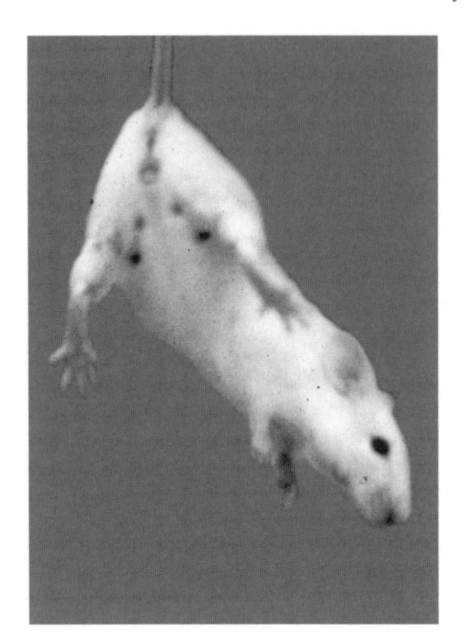

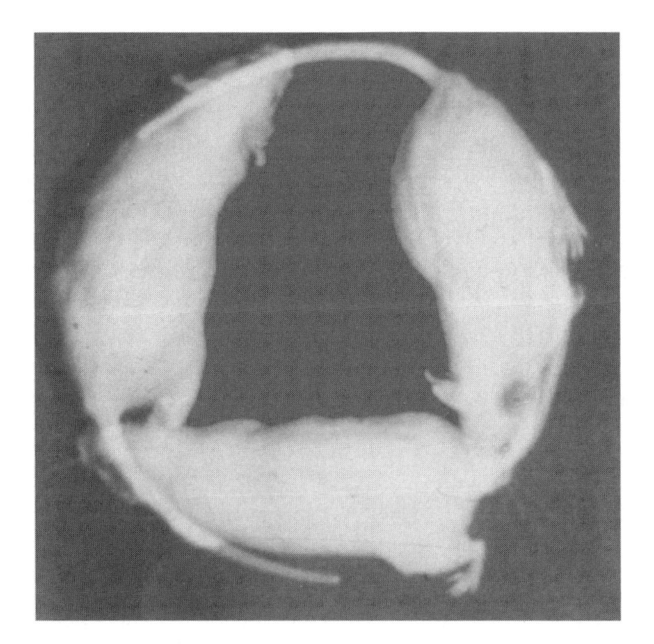

Figure 4.2 Behavioural phenotype of HD transgenics. (*a*) The normal response of a wild-type mouse during tail suspension with limbs extended out and (*b*) an HD48 homozygote mouse in a feet-clasping posture. Clasping of the fore- and hindlimbs was observed within 10 s to 1 min of suspension, and only in transgenics for the expanded

No significant differences were found in the age of onset between animals transgenic for the 48 CAG repeat expansion and those with 89 repeats

Dosage effect and repeat size

In order to study the instability of the repeats in these animals, tail DNA from 58 animals from the different transgenic lines was examined by PCR using primers flanking the CAG repeat sequences (Huntington's Disease Collaborative Research Group 1993). Intergenerational repeat instability was not detected in DNA of these mice from three generations (data not shown).

Homozygote animals were generated from the high expressor lines, HD48B and HD89A, in order to study the effects of gene dosage in these mice. In general, HD48(B,D) and HD89(A–C) animals homozygous for the transgene were observed to have onset of symptoms as much as eight weeks earlier than the heterozygotes (Figure 4.2*d*). While this difference was observed between homozygotes and heterozygotes, the same pattern and progression of behavioural changes ensued and homozygotes did not appear to have any more severe phenotype than the heterozygotes.

Neuronal loss

To determine if the progressive neurological phenotype observed in these mice correlated with neuropathological changes in the brain, wild-type, HD16, HD48 and HD89 animals at various stages were examined for neuropathological changes on brain sections stained with H&E (Reddy *et al.* 1998). Degenerating neurons were observed in brain sections from HD48(B,D) and HD89(A–C) mice. However, none of these changes were evident in wild-type or HD16 mice during the course of this study. Neurodegeneration was most evident in transgenic animals that were in the hypokinetic and akinetic phases, with no visible changes in animals at earlier stages (figure 4.3 (plate 3)). Sections from HD48B and HD89(A–C) mice exhibited scattered dark, shrunken neurons with pyknotic, densely staining nuclei and eosinophilic cytoplasm in the striatum (figure 4.3*b* (plate 3)) and

repeats. (*c*) Circling behaviour is depicted in an HD89 mouse using time-lapse photography during a 1 s interval. On occasion, animals run in very tight circles as shown. Animals have been observed to circle either clockwise or counterclockwise and remain fixated in that direction throughout the hyperactive phase. (*d*) The time-course of phenotypic changes in HD48(B,D) and HD89(A–C) lines was determined (HD48 heterozygotes, $n = 8$; HD48 homozygotes, $n = 6$; HD89 heterozygotes, $n = 19$; HD89 homozygotes, $n = 13$). Feet-clasping was observed starting at two months of age in both homozygote and heterozygote animals. The onset of hyperactivity generally started at four months of age in heterozygotes, but started two months earlier in homozygotes. Hypoactivity and then akinesia follows the hyperactive period. The mode is shown as diamonds for both the hyperactive and hypoactive stages.

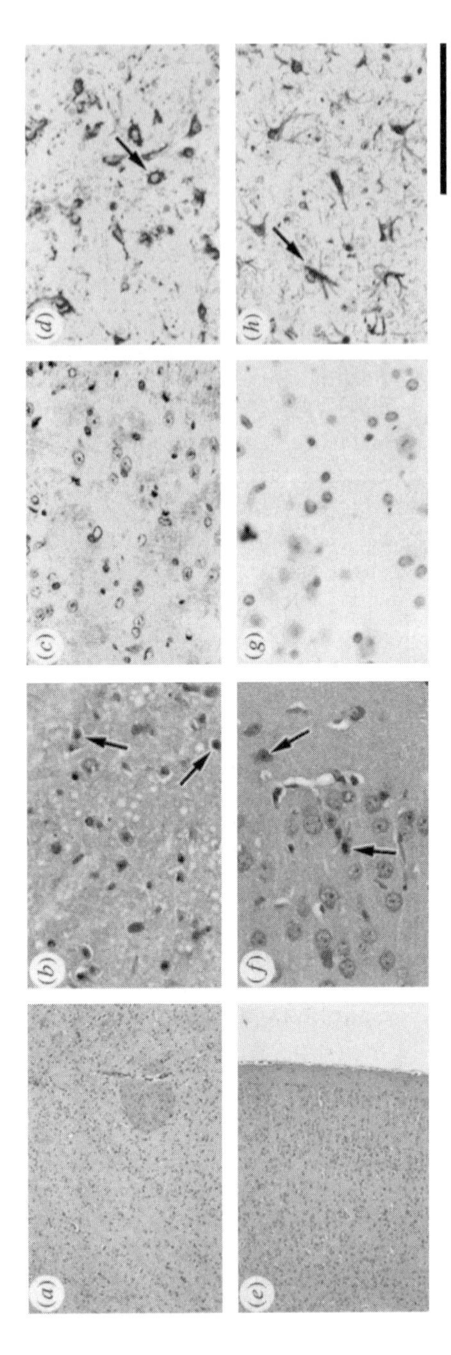

Figure 4.3 Neuronal loss and gliosis in HD transgenic striatum and cortex. Light micrographs at 630 times the original magnification are shown from the striatum and cerebral cortex of wild-type (*a,c,e,g*) and transgenic HD48 animals (*b,d,f,h*). Neurodegeneration in the striatum (*b*) and cortex (*f*) was observed after staining with haematoxylin and eosin. Astrocytosis in the striatum (*d*) and cortex (*h*) was evident after GFAP immunostaining. Scale bar, 100 μm. (See also colour plate section.)

in the cerebral cortex (figure 4.3*f* (plate 3)), whereas wild-type striatum and cortex appeared normal (figures 4.3*a*,*e* (plate 3)). Neurodegeneration was also found in the CA1 and CA3 regions of the hippocampus as well as in the thalamus (Reddy *et al.* 1998). Despite expression of the transgene in the cerebellum (Reddy *et al.* 1998), there was no visible degeneration of Purkinje or granule cells. Sections from HD16 animals were indistinguishable from those from wild-type mice.

Reactive astrocytosis is characteristically observed in HD concomitant with neuronal loss (Hedreen & Folstein 1995). In order to identify regions of gliosis, brain sections were immunostained with an antibody for glial fibrillary acidic protein (GFAP). Immunohistochemical analysis showed prominent GFAP staining of reactive astrocytes in the striatum and cerebral cortex (figures 4.3*d*,*h*, respectively (plate 3)) but not in wild-type sections (figures 4.3*c*,*g* (plate 3)).

To quantify the amount of neuronal loss, comparative cell counts were performed on Nissl-stained sections. Nissl (cresyl violet) staining of the striatum showed a normal density of neurons in brain sections from wild-type mice (Figure 4.4*a* (plate 4)) compared with diffuse staining with small darkly staining neurons in HD48B and HD89(A–C) mice (figure 4.4*b* (plate 4)). The ratio of small and medium neurons relative to large interneurons was determined on ten randomly selected fields from a mid-striatal section of each animal analysed. An overall reduction by approximately 20% was seen in the number of small/medium neurons that were found in the striatum of HD48B and HD89(A–C) transgenic animals compared with wild-type (figure 4.4*c* (plate 4)) while the number of large neurons remained relatively unchanged (data not shown). Generalized loss of as much as 20–40% of neurons in the frontal cortex was also observed in HD48B and HD89(A–C) mice compared with wild-type and HD16 animals (data not shown).

Appreciable differences in the number and morphology of neurons in the striatum and cerebral cortex of HD48B and HD89(A–C) mice were also observed after immunostaining with anti-neuronal nuclei (Neu-N) antibody. The Neu-N antibody is specific for neurons and does not label glial cells. Extensive neuronal atrophy and loss was seen in the striatum (figure 4.5*b* (plate 5)) when compared to wild-type (figure 5*a* (plate 5)). In sections from wild-type mice, Neu-N-ir neurons show the normal laminar organization within the cerebral cortex (figure 4.5*c* (plate 5)). However, several HD48 and HD89 mice showed appreciable neuronal loss, especially in layer V pyramidal cells, as well as in layers II, III and VI of the cerebral cortex (figure 4.5*d* (plate 5)). Pyramidal cells in layer V also appear to lose their normal dendritic morphology (figure 4.5*d* (plate 5)). Neurons undergoing degeneration appeared to be undergoing apoptosis upon staining by the TUNEL (terminal transferase (TdT)-mediated deoxyuridine triphosphate (dUTP)-biotin nick end labelling) method (Reddy *et al.* 1998).

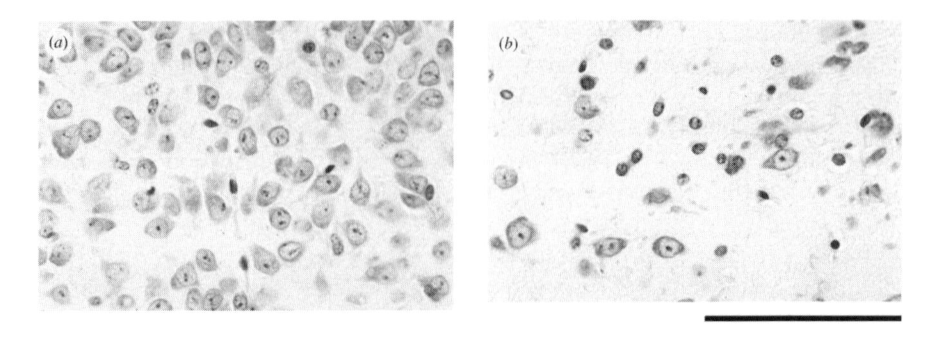

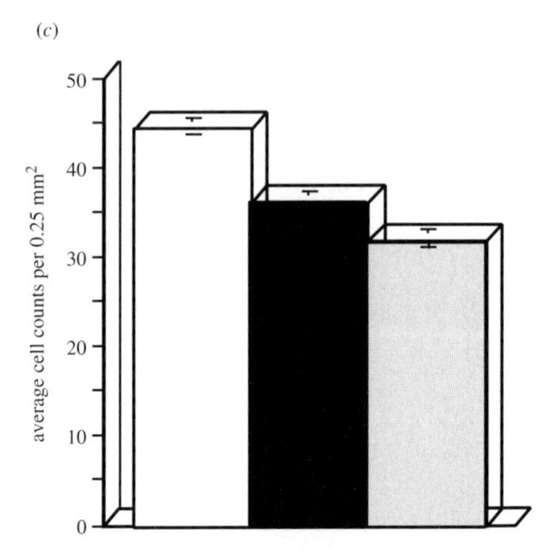

Figure 4.4 Neuronal loss in HD transgenic mice. Panels (*a*) and (*b*) are cresyl violet (Nissl)-stained sections from a wild-type and an HD89 mouse showing decreased striatal staining in the transgenic animal. (*c*) Comparative cell counts indicate significantly reduced small/medium neurons in HD48 (black column) and HD89 (shaded column) transgenic mice compared with wild-type (white column) ($p < 0.0001$ and $p < 0.0001$, respectively). A significant difference in cell counts between HD48 and HD89 animals was seen ($p < 0.0005$), despite no obvious differences in onset and disease progression. (See also colour plate section.)

Neuronal intranuclear inclusions

Ubiquitinated neuronal intranuclear inclusions (NIIs) have been found recently in the affected brain regions of patients with HD (DiFiglia *et al.* 1997) as well as in neurons of mice expressing polyglutamine expanded proteins (Davies *et al.* 1997; Ordway *et al.* 1997). It has been suggested that the presence of NIIs may be crucial to the pathogenesis of HD. Interestingly, NIIs have also been described in other CAG repeat disorders: spinocerebellar ataxia types 1, 3 and 7 (Paulson *et al.* 1997; Skinner *et al.* 1997; Holmberg *et al.* 1998). While the formation of NIIs poses an intriguing

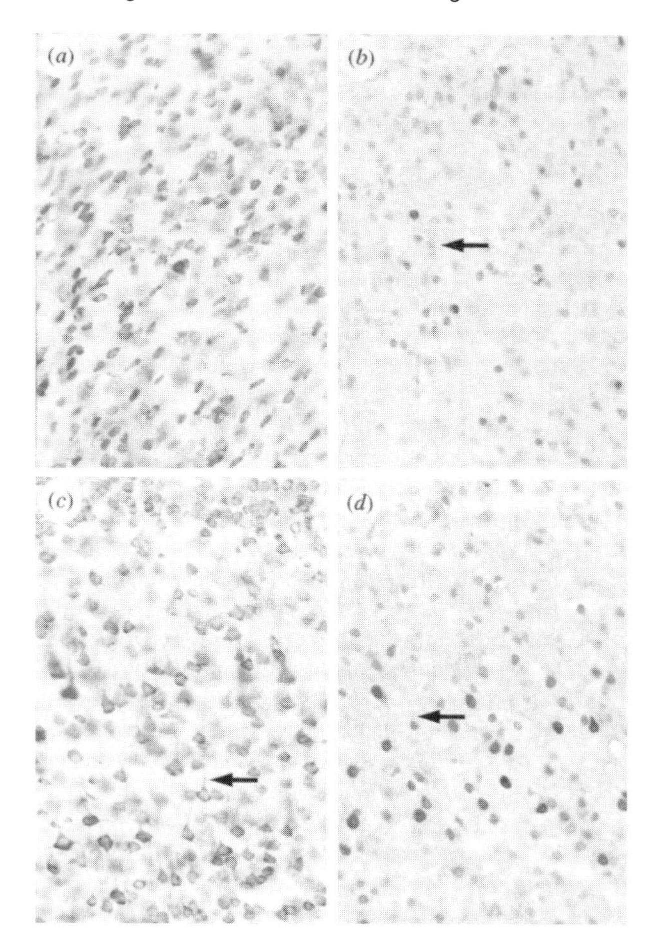

Figure 4.5 Decreased immunoreactivity for Neu-N antibody in HD transgenic mice. Neu-N labelling in sections from a wild-type (*a,c*) and a HD89 transgenic mouse (*b,d*). (*a*) and (*b*) are photomicrographs from striatum, and (*b*) and (*d*) are sections from the cortex. A decrease in immunoreactivity was evident in striatal projection neurons of transgenic mice compared with wild-type. The laminar organization of the cortex is well preserved in wild-type animals; however, this layering is lost, especially in layer V pyramidal cells of HD transgenic animals. The pyramidal cells also appear to show appreciable loss of dendritic morphology. Scale bar, 100 μm. (See also colour plate section.)

possibility for pathogenesis, questions remain as to whether NIIs are coincidental to the disease process or may even act as a protective response of the cells (Ordway *et al.* 1997) against toxicity by the sequestration of the polyglutamine products.

Polyglutamine aggregates in the form of NIIs, perinuclear aggregation and diffuse nuclear staining were observed in neurons of the striatum (figure

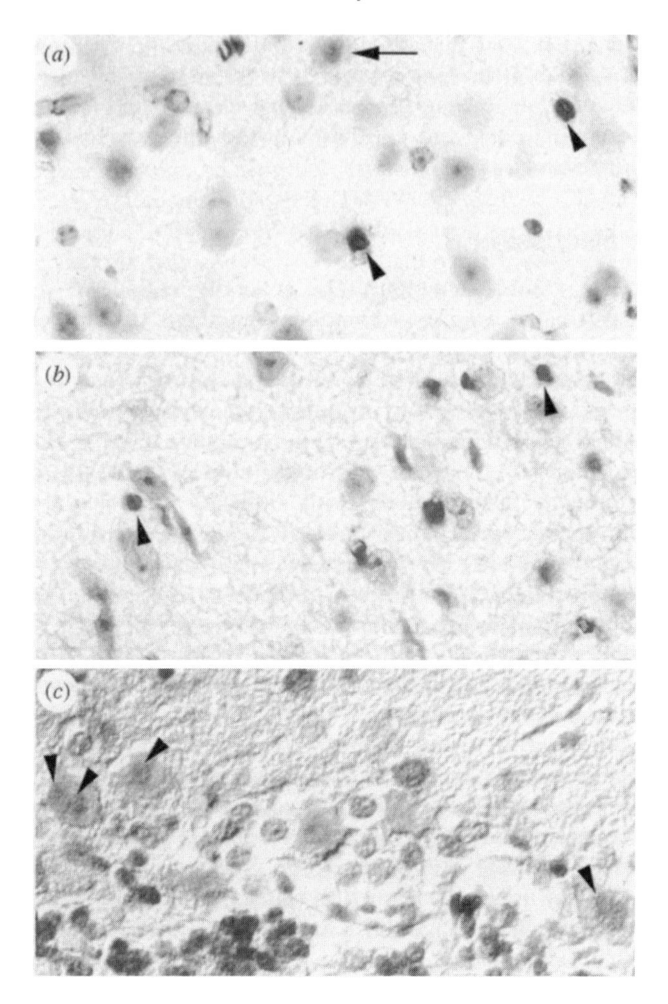

Figure 4.6 Immunoreactivity with anti-huntingtin and anti-ubiquitin. Sections from the striatum, cortex and cerebellum of an HD89 transgenic were immunolabelled with either anti-huntingtin or anti-ubiquitin and shown in (*a*), (*b*) and (*c*), respectively. Sections were developed by the immunoperoxidase method to detect polyglutamine aggregates, and lightly counterstained with haematoxylin to highlight the nucleus. In addition to NIIs, polyglutamine aggregates were also seen as diffuse nuclear or perinuclear staining. In addition, a number of neuropil aggregates were also identified. Immunoreactivity was also detected in neurons found in the cerebellum shown in (*c*), indicating that NIIs can also form in cells that are not known targets in HD. Scale bar, 100 μm. (See also colour plate section.)

4.6*a* (plate 6)) and cerebral cortex (figure 4.6*b* (plate 6)) of HD48B and HD89(A–C) mice (Reddy *et al.* 1998) using anti-huntingtin mAb48 monoclonal antibody. Hippocampal and thalamic neurons were also observed to exhibit polyglutamine aggregates (data not shown). An estimated 20% of neurons from the striatum of these mice possessed polyglutamine aggregates of which approximately 1% represents NIIs. This number is close to the 2–3% of neurons with NIIs in the striatum of juvenile HD patients (DiFiglia *et al.* 1997).

Purkinje cells of the cerebellum were also observed to have polyglutamine aggregates (figure 4.6*c* (plate 6)). Since these neurons are typically spared in HD, this observation would suggest that the presence of aggregates might not necessarily lead to neuronal loss. In addition, the same proportions of immunoreactive neurons were observed in animals that were at 12 weeks of age prior to any visible loss of neurons and in those at the hypokinetic and akinetic stages.

Discussion

HD clinical symptoms are generally considered as consisting of a triad of emotional, cognitive and motor disturbances. In adult-onset HD, early manifestations of the disease include subtle changes in coordination and perhaps in some minor involuntary movements, accompanied by mild depression and irritability. During the mid-stage of the disease process, chorea is usually a prominent feature along with difficulty in voluntary movements and cognitive deficits. Patients late in the disease may have severe chorea but are more often rigid and bradykinetic with more pronounced dementia. The full-length HD cDNA transgenic mouse models described here exhibited progressive motor abnormalities and neuropathological changes that are analogous to known features of the disease.

Although it is not clear how the overt chorea that is such a classic feature of HD will manifest in a quadruped, the stereotypic behaviour observed in these mice resembles the hyperkinesia seen in rodents after striatal damage due to direct or systemic injections of excitatory amino acids, such as kainic acid and quinolinic acid (Mason & Fibiger 1979; Sanberg *et al.* 1989) or mitochondrial metabolic inhibitors, such as 3-nitropropionic acid (Ludolph *et al.* 1991; Borlongan *et al.* 1995). Thus the stereotypic behaviour in these mice can be attributed in part to striatal lesions. It is possible that the stereotypic behaviour in rodent models and the chorea seen in primates are phenotypic correlates of the hyperkinetic dysfunction typical of basal ganglia disorders. Though the feet-clasping posture has been observed in other mouse models for other neurological dysfunctions, it is possible that this behavioural abnormality is analogous to the dystonic or abnormal limb posturing that has been described in HD patients (Folstein 1990). Locomotor deterioration is seen in HD48(B,D) and HD89(A–C) mice going from hyperkinesia to hypokinesia that eventually leads to akinesia. HD patients also progress from a choreic dyskinesia to a more disabling akinetic and Parkinson-like syndrome (Folstein 1990).

Unexpectedly, no apparent differences in onset or severity of abnormal phenotype were seen between mice with 48 and 89 CAG repeats (Reddy *et al.* 1998). Within each line and within each sibship, variability in the range of onset has also been observed. In humans, repeat lengths of 36–75 CAGs can lead to adult-onset HD while 48–121 repeats give rise to juvenile HD (Andrew *et al.* 1993; Duyao *et al.* 1993; Stine *et al.* 1993), implying that there are factors other than repeat length that can influence its penetrance in HD (Rubinsztein *et al.* 1996; McNeil *et al.* 1997). It is possible that mice respond to a different physiological threshold than humans toward polyglutamine expansions, where 48 repeats or greater can produce similar effects in mice. Another possibility is that huntingtin expression levels can influence HD pathogenesis. HD48(B,D) transgenic animals have relatively higher levels of expression than any animals from the HD89 lines, and this may exaggerate the effects of the 48 polyglutamines (Reddy *et al.* 1998). In comparison, HD48C line with transgene protein expression close to endogenous level had a much later onset and no evidence of neurodegeneration (Reddy *et al.* 1998). At the cellular level, it appears that vulnerable striatal neurons within the striosomes also show significantly higher expression of the huntingtin protein (Ferrante *et al.* 1997; Kosinski *et al.* 1997). The results presented here are compatible with this observation, and the variability in expression of the transgene may account for some of the observed differences.

Overexpression of mutated huntingtin theoretically can result in a much more rapid accumulation of the toxic effects of the polyglutamine expansion, consequently leading to an earlier age of onset or a faster course of the disease. No significant differences in clinical features have been observed between HD heterozygote and homozygote individuals (Wexler *et al.* 1987). This is unlike patients with dentatorubral–pallidoluysian atrophy (DRPLA) and Machado–Joseph disease (MJD) where homozygosity for the expanded alleles appears to contribute to the differences in onset and course of the disease compared with heterozygotes (Kawakami *et al.* 1995; Sato *et al.* 1995; Kurohara *et al.* 1997). In mice, expression levels of the mutated protein may also explain the apparent lack of phenotype thus far in knock-in HD models (table 4.3). Three groups have generated knock-in mouse models using targeted introduction of expanded CAG repeats into the endogenous *Hdh* gene. White *et al.* (1997) and Wheeler *et al.* (1999) have introduced either 50, 92 or 111 CAG repeats into the mouse *Hdh* gene and observed no phenotypic and histological changes in heterozygote mice. In an another knock-in study by Shelbourne *et al.* (1999), mice expressing full-length mutant protein (with 72–80 polyQs) have a normal lifespan and displayed abnormal social behaviour without any neuronal loss. Recently, Ishiguro and colleagues (Ishiguro *et al.* 1999) have introduced exon 1 of the *HD* gene with 77 CAGs into the mouse endogenous gene and did not observe any behavioural phenotype or pathological changes in the brain. These results suggest that elevated expression of the mutant huntingtin may also be critical in the onset of a progressive phenotype and pathological changes in mice. In these models, it

Table 4.3 *Summary of knock-in mouse models for Huntington disease*

Gene	*no. of CAGs*	*pheno-type*	*expression*	*neuro-degeneration*	*inclusions*	*reference*
Hprt	146	yes	yes	no	yes	Ordway *et al.* 1997
Hdh	72–80	no	yes	no	not reported	Shelbourne *et al.* 1999
Hdh	92 and 111	no	yes	no	yes	Wheeler *et al.* 1999
Hdh	77	no	yes	no	not reported	Ishiguro *et al.* 1999

may be necessary to allow additional time for the build-up of the toxic effects of mutated huntingtin.

HD neuropathological studies (Hedreen & Folstein 1995) indicate gradual neuronal loss accompanied by gliosis. TUNEL labelling of neurons in the striatum of HD patients has also been used to indicate apoptotic neurodegeneration (Dragunow *et al.* 1995; Portera-Cailliau *et al.* 1995; Thomas *et al.* 1995) as a possible means of cell death. Mice expressing full-length mutant huntingtin showed focal neuronal loss, specifically in the striatum and cortex but also including regions of the hippocampus and thalamus. Glial infiltrations in the same regions have been shown to occur with the neuronal loss. Neurodegeneration was best evident in HD transgenic mice with 48 and 89 repeats that were in the hypokinetic and akinetic stages. Furthermore, these dying neurons were also shown to be labelled by TUNEL, implying an apoptotic mode of cell death. However, mice at the hypekinetic phase as well as younger animals (i.e. pre-hyperkinetic) did not show visible neuronal loss in the target regions. Thus the mouse models described here from the HD48(B,D) and HD89(A–C) lines show not only progressive behavioural deficits but also gradual loss of neurons that appears to parallel the ordered and topographic neurodegeneration seen in the HD brain.

In other transgenic mouse models of HD (see table 4.2), no clear demonstration of apoptotic neuronal loss has been shown. Mice expressing exon 1 of the HD gene with 115–156 polyglutamines (Mangiarini *et al.* 1996) develop a progressive neurological phenotype; however, the abnormalities consisted of seizures and tremors, a feature that has been associated with the less frequent form of juvenile-onset HD. Although these mice had smaller brains compared with control animals, no neuronal loss has been described in these animals, thus it remains unclear if the neurological deficits seen in these mice can be attributed to neuronal dysfunction in the striatum in the absence of clear macroscopic lesions. On the other hand, metabotrophic glutamate receptor levels in symptomatic 12-week-old mice were decreased selectively compared with control mice for mGluR1, mGluR2 and mGluR3 but not for the mGluR5 subtype of G-protein-linked metabotropic glutamate receptors (Cha *et al.* 1998). AMPA and kainate receptor levels were also decreased, while NMDA receptor levels were not different from controls. Levels of other neurotransmitter receptors, including dopamine and acetylcholine receptors, but not GABA receptors, that are known to be affected in HD were decreased in the cortex and striatum. Schilling and colleagues (1999) generated truncated version of cDNA mouse models (1–171 amino acids) under the control of the prion protein promoter. These mice have relatively low levels of mutant huntingtin (with 82 glutamines) expression but develop behavioural abnormalities, including loss of coordination, tremors and abnormal gait, and these mice have a reduced lifespan. Neuronal inclusions and neuritic aggregates were seen in animals that exhibited behavioural abnormalities. Neuronal loss was not observed in these mice (Schilling *et al.* 1999). Thus it appears that expression of the full-length protein is necessary for neuronal loss in mice.

Recently, Hodgson and colleagues (1999) developed a transgenic mouse model using a YAC spanning a 350 kb human chromosomal segment from 4p16.3 that includes the human *HD* gene which was modified to include either 18, 46 or 72 CAG repeats. Transgenic YAC46 and YAC72 mice exhibited early electrophysiological abnormalities indicating cytoplasmic dysfunction prior to nuclear inclusion formation (Hodgson *et al.* 1999). Only the line with 72 polyglutamines expressing elevated levels of mutant huntingtin showed neuronal loss in the striatum at 12 months of age. Neurodegeneration was observed irrespective of formation of macro- or micropolyglutamine aggregates.

In order to investigate the role of the protein sequence that flanks polyglutamine repeats in terms of toxicity, Ordway *et al.* (1997) developed a knock-in mouse model with 146 CAG repeats by targeting this into the mouse hypoxanthine phosphoribosyltransferase (*Hprt*) gene which is not involved in any CAG repeat disorders. These mutant mice produced a polyglutamine-expanded form of the HPRT protein and developed a late-onset neurological phenotype that progresses to premature death. Feet clasping was observed at 12 weeks of age and seizures were observed in mice older than 18 weeks although there was no difference in the weight of the brain in mutant transgenic mice compared with age-matched controls. This experiment showed that expanded polyglutamine repeats by itself has a toxic gain-in-function effect that can produce general neurological abnormalities (Ordway *et al.* 1997). However, the lack of selective neuronal loss in these mice also suggests that expression of polyglutamines alone or in the context of another protein is not sufficient to cause neurodegeneration.

Our results demonstrate that overexpression of pathogenic CAG repeat lengths in the context of the HD holoprotein correlates with and leads to region-specific neuronal loss in mice. Comparisons of other proteins capable of forming pathogenic polyglutamine repeat expansions in other triplet repeat disorders do not show any sequence or functional similarities in their protein domains other than the glutamine tract (Reddy & Housman 1997). We speculate that region-specific neuronal loss in each of these disorders, including HD, is conferred by the remaining protein sequences outside of the polyglutamine tract. Despite the widespread expression of mutant huntingtin in the transgenic mice in this study, selective neurodegeneration was observed, raising the possibility that polyglutamines within the proper context of the holoprotein can modulate aberrant protein–protein interactions in a cell-specific manner. These interactions would be dependent on the subcellular compartment in which the mutant protein is found and on the expression level of the mutant protein in any given cell type. The polyglutamine tract can lead to structural changes that can either act as a sink for proteins or prevent them from performing normal neuronal function and/or maintenance or it can result in transactivation of genes that promote apoptosis. Indeed the LANP protein has been shown to interact in a repeat-dependent manner with ataxin 1 (Matilla *et al.* 1997) and causes its translocation to the nuclear matrix, possibly disrupting normal cellular architecture.

A number of proteins have been shown to bind *in vitro* to huntingtin in a repeat-dependent manner. These polyglutamine-interacting proteins includes two novel proteins, HAP1 (Li *et al.* 1995) and HIP1 (Wanker *et al.* 1997) that are highly expressed in the brain. Other proteins that bind preferentially to mutant huntingtin, which may result in disruption of their normal cellular functions, include glyceraldehyde-3-phosphate dehydrogenase (Burke *et al.* 1996), calmodulin (Bao *et al.* 1996) and hE2-25K (Kalchman *et al.* 1996), an ubiquitin-conjugating enzyme. GAPDH is involved in the control of ATP production, and its abnormal interaction with huntingtin conceivably can result in metabolic energy impairment in the brain. Similarly, aberrant interaction of huntingtin with calmodulin may result in changes in intracellular calcium levels, leading to activation of proteases, phospholipases and endonucleases. The ubiquitination of proteins is used to mark proteins targeted for intracellular degradation by way of the proteasome complex. It is interesting that huntingtin can interact with the hE2-25K protein given that polyglutamine aggregates are highly ubiquitinated (Davies *et al.* 1997; DiFiglia *et al.* 1997) and have been speculated to be causative of HD.

The identification of polyglutamine aggregation in the nuclei of striatal neurons of HD patients (DiFiglia *et al.* 1997) and mice (Davies *et al.* 1997) suggested a causative role for these inclusion bodies in HD pathogenesis. However, we have found that NIIs are found only in a small number of striatal and cortical neurons. Moreover, we have not seen an increase in the proportion of striatal neurons that have NIIs from animals at 12 weeks compared with those that are older, including hypokinetic and akinetic mice. In addition, we have also identified inclusion bodies in Purkinje cells of the cerebellum, which are not known to be affected regions in HD. Thus our results do not support a causative role for NIIs in HD pathogenesis.

It remains unclear what the pathogenic mechanism is for HD and whether neuronal loss is a key component of HD pathogenesis or if it represents the culmination of the disease process. It is entirely possible that very mild pathological, physiological and neurochemical changes in the neurons may proceed undetected in the early stages of the disease, which may then precipitate a chain of events ultimately leading to neuronal loss. The mouse models expressing the mutated full-length huntingtin closely resemble the selective neuropathological and progressive clinical features of HD. Future studies involving these mice using more sensitive neuroanatomical techniques and cell counts may better characterize and reveal early events of the disease process. These studies may serve to elucidate the early events leading to locomotor changes and neuronal dysfunction. It will also be interesting to be able to identify emotional and cognitive changes in these animals that may be informative in managing HD symptoms. Moreover, these animals can be of considerable use in screening for potential neuroprotective compounds. Finally, the efficacy of experimental treatments can be evaluated at various times of disease progression, such as the potential benefits of administering treatments during the pre-symptomatic stage, during hyperactivity or at advanced stages when hypoactivity and neuronal loss become obvious.

We thank Lisa Pike-Buchanan, Maya Williams, Lisa Garrett, Theresa Hernandez, Amy Chen, Cecilia Rivas and Elaine Stockburger and Tracie Moss for technical assistance. We also thank Marcy Macdonald for the generous gift of the HF1 antibody. V.C. is supported by a PRAT fellowship from NIGMS and P.H.R. is supported by a Fellowship from the Huntington's Disease Society of America, Inc. This work was supported in part by PHS grant NS28236 to W.O.W. and by a grant from Cure HD Initiative–Hereditary Disease Foundation to D.A.T.

References

Andrew, S. E. (and 12 others) 1993 The relationship between trinucleotide (CAG) repeat length and clinical features of Huntington's disease. *Nature Genet.* **4**, 398–403.

Bao, J., Sharp, A. H., Wagster, M. V., Becher, M., Schilling, G., Ross, C. A., Dawson, V. L. & Dawson, T. M. 1996 Expansion of polyglutamine repeat in huntingtin leads to abnormal protein interactions involving calmodulin. *Proc. Natl Acad. Sci. USA* **93**, 5037–5042.

Beal, M. F., Kowall, N. W., Swartz, K. J., Ferrante, R. J. & Martin, J. B. 1988 Systemic approaches to modifying quinolinic acid striatal lesions in rats. *J. Neurosci.* **8**, 3901–3908.

Borlongan, C. V., Koutouzis, T. K., Freeman, T. B., Cahill, D. W. & Sanberg, P. R. 1995 Behavioral pathology induced by repeated systemic injections of 3-nitropropionic acid mimics the motoric symptoms of Huntington's disease. *Brain Res.* **697**, 254–257.

Brinkman, R. R., Mezei, M. M., Theilmann, J., Almqvist, E. & Hayden, M. R. 1997 The likelihood of being affected with Huntington disease by a particular age, for a specific CAG size. *Am. J. Hum. Genet.* **60**, 1202–1210.

Burke, J. R., Enghild, J. J., Martin, M. E., Jou, Y. S., Myers, R. M., Roses, A. D., Vance, J. M. & Strittmatter, W. J. 1996 Huntingtin and DRPLA proteins selectively interact with the enzyme GAPDH. *Nature Med.* **2**, 347–350.

Cha, J. H., Kosinski, C. M., Kerner, J. A., Alsdorf, S. A., Mangiarini, L., Davies, S. W., Penney, J. B., Bates, G. P. & Young, A. B. 1998 Altered brain neurotransmitter receptors in transgenic mice expressing a portion of an abnormal human huntington disease gene. *Proc. Natl Acad. Sci. USA* **95**, 6480–6485.

Davies, S. W., Turmaine, M., Cozens, B. A., DiFiglia, M., Sharp, A. H., Ross, C. A., Scherzinger, E., Wanker, E. E., Mangiarini, L. & Bates, G. P. 1997 Formation of neuronal intranuclear inclusions (NII) underlies the neurological dysfunction in mice transgenic for the HD mutation. *Cell* **90**, 537–5348.

DiFiglia, M., Sapp, E., Chase, K. O., Davies, S. W., Bates, G. P., Vonsattel, J. P. & Aronin, N. 1997 Aggregation of huntingtin in neuronal intranuclear inclusions and dystrophic neurites in brain. *Science* **277**, 1990–1993.

Dragunow, M., Faull, R. L., Lawlor, P., Beilharz, E. J., Singleton, K., Walker, E. B. & Mee, E. 1995 In situ evidence for DNA fragmentation in Huntington's disease striatum and Alzheimer's disease temporal lobes. *NeuroReport* **6**, 1053–1057.

Duyao, M. (and 42 others) 1993 Trinucleotide repeat length instability and age of onset in Huntington's disease. *Nature Genet* **4**, 387–392.

Duyao, M. P. (and 11 others) 1995 Inactivation of the mouse Huntington's disease gene homolog Hdh. *Science* **269**, 407–410.

Ferrante, R. J., Gutekunst, C. A., Persichetti, F., McNeil, S. M., Kowall, N. W., Gusella, J. F., MacDonald, M. E., Beal, M. F. & Hersch, S. M. 1997 Heterogeneous topographic and cellular distribution of huntingtin expression in the normal human neostriatum. *J. Neurosci.* **17**, 3052–3063.

Folstein, S. E. 1990 *Huntington's disease.* Baltimore, MD: The Johns Hopkins University Press.

Hedreen, J. C. & Folstein, S. E. 1995 Early loss of neostriatal striosome neurons in Huntington's disease. *J. Neuropathol. Exp. Neurol.* **54**, 105–120.

Hedreen, J. C., Peyser, C. E., Folstein, S. E. & Ross, C. A. 1991 Neuronal loss in layers V and VI of cerebral cortex in Huntington's disease. *Neurosci. Lett.* **133**, 257–261.

Hodgson, J. G. (and 18 others) 1999 A YAC mouse model for Huntington's disease with full-length mutant huntingtin, cytoplasmic toxicity, and selective striatal neurodegeneration. *Neuron* **23**, 181–192.

Holmberg, M. (and 10 others) 1998 Spinocerebellar ataxia type 7 (SCA7): a neurodegenerative disorder with neuronal intranuclear inclusions. *Hum. Mol. Genet.* **7**, 913–918.

Huntington's Disease Collaborative Research Group. 1993 A novel gene containing a trinucleotide repeat that is expanded and unstable on Huntington's disease chromosomes. *Cell* **72**, 971–983.

Ishiguro, H. (and 10 others) 1999 Tissue specific and age dependent occurrence of somatic mosaicism in expanded CAG repeats in mice carrying the mutated huntington's disease gene. *18th International Meeting of the World Federation of Neurology Research Group on Huntington's Disease.* Abstract, 49.

Jou, Y. S. & Myers, R. M. 1995 Evidence from antibody studies that the CAG repeat in the Huntington disease gene is expressed in the protein. *Hum. Mol. Genet.* **4**, 465–469.

Kalchman, M. A., Graham, R. K., Xia, G., Koide, H. B., Hodgson, J. G., Graham, K. C., Goldberg, Y. P., Gietz, R. D., Pickart, C. M. & Hayden, M. R. 1996 Huntingtin is ubiquitinated and interacts with a specific ubiquitin-conjugating enzyme. *J. Biol. Chem.* **271**, 19385–19394.

Kawakami, H., Maruyama, H., Nakamura, S., Kawaguchi, Y., Kakizuka, A., Doyu, M. & Sobue, G. 1995 Unique features of the CAG repeats in Machado–Joseph disease. *Nature Genet.* **9**, 344–345.

Kodsi, M. H. & Swerdlow, N. R. 1997 Mitochondrial toxin 3-nitropropionic acid produces startle reflex abnormalities and striatal damage in rats that model some features of Huntington's disease. *Neurosci. Lett.* **231**, 103–107.

Kosinski, C. M., Cha, J. H., Young, A. B., Persichetti, F., MacDonald, M., Gusella, J. F., Penney Jr, J. B. & Standaert, D. G. 1997 Huntingtin immunoreactivity in the rat neostriatum: differential accumulation in projection and interneurons. *Exp. Neurol.* **144**, 239–247.

Kurohara, K., Kuroda, Y., Maruyama, H., Kawakami, H., Yukitake, M., Matsui, M. & Nakamura, S. 1997 Homozygosity for an allele carrying intermediate CAG repeats in the dentatorubral–pallidoluysian atrophy (DRPLA) gene results in spastic paraplegia. *Neurology* 48, 1087–1090.

Li, S. H. (and 11 others) 1993 Huntington's disease gene (IT15) is widely expressed in human and rat tissues. *Neuron* **11**, 985–993.

Li, X. J., Li, S. H., Sharp, A. H., Nucifora Jr, F. C. Schilling, G., Lanahan, A., Worley, P., Snyder, S. H. & Ross, C. A. 1995 A huntingtin-associated protein enriched in brain with implications for pathology. *Nature* **378**, 398–402.

Ludolph, A. C., He, F., Spencer, P. S., Hammerstad, J. & Sabri, M. 1991 3-Nitropropionic acid—exogenous animal neurotoxin and possible human striatal toxin. *Can. J. Neurol. Sci.* **18**, 492–498.

McNeil, S. M., Novelletto, A., Srinidhi, J., Barnes, G., Kornbluth, I., Altherr, M. R., Wasmuth, J. J., Gusella, J. F., MacDonald, M. E. & Myers, R. H. 1997 Reduced penetrance of the Huntington's disease mutation. *Hum. Mol. Genet.* **6**, 775–779.

Mangiarini, L. (and 10 others) 1996 Exon 1 of the HD gene with an expanded CAG repeat is sufficient to cause a progressive neurological phenotype in transgenic mice. *Cell* **87**, 493–506.

Mason, S. T. & Fibiger, H. C. 1979 Kainic acid lesions of the striatum in rats mimic the spontaneous motor abnormalities of Huntington's disease. *Neuropharmacology* **18**, 403–407.

Matilla, A., Koshy, B. T., Cummings, C. J., Isobe, T., Orr, H. T. & Zoghbi, H. Y. 1997 The cerebellar leucine-rich acidic nuclear protein interacts with ataxin-1. *Nature* **389**, 974–978.

Miller, P. J. & Zaborszky, L. 1997 3-Nitropropionic acid neurotoxicity: visualization by silver staining and implications for use as an animal model of Huntington's disease. *Exp. Neurol.* **146**, 212–229.

Nasir, J., Floresco, S. B., O'Kusky, J. R., Diewert, V. M., Richman, J. M., Zeisler, J., Borowski, A., Marth, J. D., Phillips, A. G. & Hayden, M. R. 1995 Targeted disruption of the Huntington's disease gene results in embryonic lethality and behavioral and morphological changes in heterozygotes. *Cell* **81**, 811–823.

Ordway, J. M. (and 11 others) 1997 Ectopically expressed CAG repeats cause intranuclear inclusions and a progressive late onset neurological phenotype in the mouse. *Cell* **91**, 753–764.

Paulson, H. L., Perez, M. K., Trottier, Y., Trojanowski, J. Q., Subramony, S. H., Das, S. S., Vig, P., Mandel, J. L., Fischbeck, K. H. & Pittman, R. N. 1997 Intranuclear inclusions of expanded polyglutamine protein in spinocerebellar ataxia type 3. *Neuron* **19**, 333–344.

Persichetti, F. (and 16 others) 1995 Normal and expanded Huntington's disease gene alleles produce distinguishable proteins due to translation across the CAG repeat. *Mol. Med.* **1**, 374–383.

Portera-Cailliau, C., Hedreen, J. C., Price, D. L. & Koliatsos, V. E. 1995 Evidence for apoptotic cell death in Huntington disease and excitotoxic animal models. *J. Neurosci* .**15**, 3775–3787.

Reddy, P. S. & Housman, D. E. 1997 The complex pathology of trinucleotide repeats. *Curr. Opin. Cell Biol.* **9**, 364–372.

Reddy, P. H., Williams, M., Charles, C., Garrett, L., Pike-Buchanan, L., Whetsell Jr.,

W. O., Miller, G. & Tagle, D. A. 1998 Behavioural abnormalities and selective neuronal loss in HD transgenic mice expressing mutated full-length HD cDNA. *Nature Genet.* **20**, 198–202.

Reddy, P. H., Charles, V., Williams, M., Miller, G., Whetsell Jr, W. O. & Tagle, D. A. 1999 Transgenic mice expressing mutated full-length HD cDNA: a paradigm for locomotor changes and selective neuronal loss in Huntington's disease. *Philos. Trans. R. Soc. Biol.* **354**, 1037–1046.

Rubinsztein, D. C. (and 37 others) 1996 Phenotypic characterization of individuals with 30–40 CAG repeats in the Huntington disease (HD) gene reveals HD cases with 36 repeats and apparently normal elderly individuals with 36–39 repeats. *Am. J. Hum. Genet.* **59**, 16–22.

Sanberg, P. R., Calderon, S. F., Giordano, M., Tew, J. M. & Norman, A. B. 1989 The quinolinic acid model of Huntington's disease: locomotor abnormalities. *Exp. Neurol.* **105**, 45–53.

Sato, K., Kashihara, K., Okada, S., Ikeuchi, T., Tsuji, S., Shomori, T., Morimoto, K. & Hayabara, T. 1995 Does homozygosity advance the onset of dentatorubral–pallidoluysian atrophy? *Neurology* **45**, 1934–1936.

Schilling, G. (and 13 others) 1999 Intranuclear inclusions and neuritic aggregates in transgenic mice expressing a mutant N-terminal fragment of huntingtin. *Hum. Mol. Genet.* **8**, 397–407.

Schwarcz, R., Whetsell Jr, W. O. & Mangano, R. M. 1983 Quinolinic acid: an endogenous metabolite that produces axon-sparing lesions in rat brain. *Science* **219**, 316–318.

Sharp, A. H. (and 16 others) 1995 Widespread expression of Huntington's disease gene (IT15) protein product. *Neuron* **14**, 1065–1074.

Shelbourne, P. F. (and 11 others) 1999 A Huntington's disease CAG expansion at the murine Hdh locus is unstable and associated with behavioural abnormalities in mice. *Hum. Mol. Genet.* **8**, 763–774.

Skinner, P. J., Koshy, B. T., Cummings, C. J., Klement, I. A., Helin, K., Servadio, A., Zoghbi, H. Y. & Orr, H. T. 1997 Ataxin-1 with an expanded glutamine tract alters nuclear matrix-associated structures. *Nature* **389**, 971–974.

Spargo, E., Everall, I. P. & Lantos, P. L. 1993 Neuronal loss in the hippocampus in Huntington's disease: a comparison with HIV infection. *J. Neurol. Neurosurg. Psychiat.* **56**, 487–491.

Stine, O. C., Pleasant, N., Franz, M. L., Abbott, M. H., Folstein, S. E. & Ross, C. A. 1993 Correlation between the onset age of Huntington's disease and length of the trinucleotide repeat in IT-15. *Hum. Mol. Genet.* **2**, 1547–1549.

Strong, T. V., Tagle, D. A., Valdes, J. M., Elmer, L. W., Boehm, K., Swaroop, M., Kaatz, K. W., Collins, F. S. & Albin, R. L. 1993 Widespread expression of the human and rat Huntington's disease gene in brain and nonneural tissues. *Nature Genet.* **5**, 259–265.

Thomas, L. B., Gates, D. J., Richfield, E. K., O'Brien, T. F., Schweitzer, J. B. & Steindler, D. A. 1995 DNA end labeling (TUNEL) in Huntington's disease and other neuropathological conditions. *Exp. Neurol.* **133**, 265–272.

Vonsattel, J. P., Myers, R. H., Stevens, T. J., Ferrante, R. J., Bird, E. D. & Richardson Jr, E. P., 1985 Neuropathological classification of Huntington's disease. *J. Neuropathol. Exp. Neurol.* **44**, 559–577.

Wagster, M. V., Hedreen, J. C., Peyser, C. E., Folstein, S. E. & Ross, C. A. 1994 Selective loss of [3H]kainic acid and [3H]AMPA binding in layer VI of frontal cortex in Huntington's disease. *Exp. Neurol.* **127**, 70–75.

Wanker, E. E., Rovira, C., Scherzinger, E., Hasenbank, R., Walter, S., Tait, D., Colicelli, J. & Lehrach, H. 1997 HIP-I: a huntingtin interacting protein isolated by the yeast two-hybrid system. *Hum. Mol. Genet.* **6**, 487–495.

Wexler, N. S. (and 20 others) 1987 Homozygotes for Huntington's disease. *Nature* **326**, 194–197.

Wheeler, V. C. (and 10 others) 1999. A progressive nuclear phenotype in precise genetic mouse models of Huntington's disease. *18th International Meeting of the World Federation of Neurology Research Group on Huntington's Disease*. 28–31 August 1999, abstr, p. 44.

White, J. K., Auerbach, W., Duyao, M. P., Vonsattel, J.-P., Gusella, J. F., Joyner, A. L. & MacDonald, M. E. 1997 Huntingtin is required for neurogenesis and is not impaired by the Huntington's disease CAG expansion. *Nature Genet.* **17**, 404–410.

Wood, J. D., MacMillan, J. C., Harper, P. S., Lowenstein, P. R. & Jones, A. L. 1996 Partial characterisation of murine huntingtin and apparent variations in the subcellular localisation of huntingtin in human, mouse and rat brain. *Hum. Mol. Genet.* **5**, 481–487.

Zeitlin, S., Liu, J. P., Chapman, D. L., Papaioannou, V. E. & Efstratiadis, A. 1995 Increased apoptosis and early embryonic lethality in mice nullizygous for the Huntington's disease gene homologue. *Nature Genet.* **11**, 155–163.

5

Evidence for both the nucleus and cytoplasm as subcellular sites of pathogenesis in Huntington's disease in cell culture and in transgenic mice expressing mutant huntingtin

Abigail S. Hackam, J. Graeme Hodgson, Roshni Singaraja, Taiqi Zhang, Lu Gan, Claire-Anne Gutekunst, Steven M. Hersch and Michael R. Hayden

Introduction

The expansion of a polymorphic CAG tract encoding glutamine is the causative mutation in eight human neurodegenerative diseases, including Huntington's disease (HD), dentatorubral–pallidoluysian atrophy (DRPLA), spinobulbar muscular atrophy (SBMA) and spinocerebellar ataxia (SCA) types 1, 2, 3, 6 and 7 (Andrew *et al.* 1997; Ross 1997). Each disease affects specific populations of neurons and results in a characteristic clinical phenotype. Additionally, the mutant genes responsible for these diseases have no sequence similarity except for the CAG tracts. However, there may be a common step in the pathogenic pathways of the diseases that involve novel properties of the expanded CAG tract.

Several reports in recent years have described intracellular protein inclusions, or aggregates, within and outside the nuclei of cells expressing proteins with expanded polyglutamine tracts (reviewed in Ross 1997) (Lunkes & Mandel 1997; Hackam *et al.* 1998*b*). Several lines of evidence from patient samples, transgenic mice and cell culture models suggest that the aggregates are associated with the pathology of CAG expansion diseases. First, the inclusions are only observed in the brains of individuals carrying the disease allele, and are present predominantly in regions and neuronal populations affected by the disease (DiFiglia *et al.* 1997; Paulson *et al.* 1997; Sapp *et al.* 1997; Skinner *et al.* 1997; Becher *et al.* 1998; Gourfinkel-An *et al.* 1998; Igarashi *et al.* 1998; Li *et al.* 1998). Second, neuronal inclusions have also been observed in mice transgenic for genes with expanded CAG tracts, which develop inclusions prior to the onset of neurological symptoms (Davies *et al.* 1997; Ordway *et al.* 1997; Skinner *et al.* 1997). Third, increasing frequency of aggregates is associated with increasing toxicity in *in vitro* models of HD, DRPLA and SBMA (Hackam *et al.* 1998*a*; Igarashi *et al.* 1998; Ellerby *et al.* 1999). Further, decreasing the frequency of aggregates *in vitro* results in reduced toxicity (Igarashi *et al.* 1998; Ellerby *et al.* 1999). However, despite these data suggesting a causal relationship, none can differentiate between aggregates as being crucial to pathogenesis, from being markers of pathology.

Recent attention has focused on the nucleus as the primary site of pathogenic changes in polyglutamine expansion diseases. One line of evidence for nuclear involvement is the exclusive nuclear localization of aggregates in post-mortem brains in SCA-1 (Skinner *et al.* 1997), SCA-3 (Paulson *et al.* 1997), SCA-7 (Holmberg *et al.* 1998) and DRPLA (Becher *et al.* 1998; Igarashi *et al.* 1998). Intranuclear aggregates were also observed in mice transgenic for a hypoxanthine phosphoribosyl transferase (HPRT) gene containing an expanded polyglutamine stretch and in mice expressing exon 1 of the HD gene (Davies *et al.* 1997; Ordway *et al.* 1997). Unusual nuclear morphology, including irregular indentation of the nuclear membrane, increased pore density and chromatin condensation, has also been described in transgenic mice (Davies *et al.* 1997; Skinner *et al.* 1997) and patient brains (Tellez-Nagel *et al.* 1974; Bots & Bruyn 1981).

Aggregates were found to be exclusively nuclear in one study of adult and juvenile HD patients (Becher *et al.* 1998). However, other studies using different antibodies have identified both intranuclear and cytosolic accumulations of huntingtin in brain tissue of HD patients (DiFiglia *et al.* 1997; Gourfinkel-An *et al.* 1998). In brains of severely affected juvenile HD patients, huntingtin-containing inclusions were identified within the nuclei of neurons in the cortex and striatum (DiFiglia *et al.* 1997). Adult HD patients also displayed extranuclear accumulations of huntingtin in dystrophic neurites and perikarya (DiFiglia *et al.* 1997; Gourfinkel-An *et al.* 1998). Neurons containing cytoplasmic accumulation were more frequent in adult patients than neurons with nuclear inclusions (DiFiglia *et al.* 1997), indicating that cytoplasmic aggregates can also be cytotoxic. Therefore, whether the specific subcellular localization of the huntingtin aggregates is a contributing factor to the pathology in HD is not clear.

In vitro studies have demonstrated that nuclear localization of huntingtin aggregates is influenced by the length of the protein (Cooper *et al.* 1998; Hackam *et al.* 1998*a*). Small huntingtin proteins are both nuclear and cytoplasmic whereas larger proteins are only cytoplasmic, suggesting that passive diffusion plays a role in intracellular localization. Proteolytic cleavage would therefore be necessary to reduce the large full-length huntingtin protein to a fragment that is capable of diffusion through the nuclear pores (< 60 kDa) (Görlich & Mattaj 1996). However, we also identified a basic amino acid-rich sequence within the N-terminus of huntingtin with significant homology to a functional nuclear localization signal (NLS) (Hackam *et al.* 1998*a*), which suggested that huntingtin may also enter the nucleus by active transport. In order to determine the mode of nuclear transport, we assessed the ability of the predicted NLS to transport mutant huntingtin into the nucleus.

The predominance of nuclear aggregates has led us and others to postulate the nucleus as the site of pathology in HD (Davies *et al.* 1997; DiFiglia *et al.* 1997; Lunkes & Mandel 1997; Paulson *et al.* 1997; Ross 1997; Hackam *et al.* 1998*b*). This would suggest that a key event in the pathogenesis of HD is the translocation of truncated mutant huntingtin to the nucleus, where it exerts its toxic effect. An unanswered question is whether these aggregates may be similarly toxic outside the nucleus. Additionally, toxicity could be mediated by cell-specific vulnerability to mutant huntingtin and aggregate formation in the nucleus, or by interaction with specific nuclear proteins.

In this study we have designed experiments to specifically investigate the subcellular sites of toxicity in HD. Our results indicate that both the nucleus and cytoplasm represent sites of toxicity of huntingtin.

Results

Altering the location of aggregates formed by the 1955-128 protein does not influence toxicity

To assess whether the subcellular site of aggregation influences toxicity, we altered the localization of aggregates formed by the 1955-128 protein and compared their toxicity in different subcellular locations. The 1955-128 huntingtin fragment includes amino acids 1 to 548, corresponding in size to the fragment produced by caspase 3 cleavage (Goldberg *et al.* 1996; Wellington *et al.* 1998), and contains 128 polyglutamines. This fragment has been consistently identified in transfected cells undergoing stress (Martindale *et al.* 1998), suggesting that it may be a stable fragment produced from full-length huntingtin.

The cytoplasmic perinuclear location of 1955-128 aggregates was changed to a nuclear location by the addition of an NLS. The NLS from the SV40 large T antigen (PKKKRKV) was inserted into the N-terminus of 1955-128, forming 1955-128-NLS (figure 5.1). Immunofluorescence of cells transfected with the 1955-128 constructs was used to quantify their subcellular localization (table 5.1). In several cell types and using multiple anti-huntingtin antibodies, the 1955-128 protein is exclusively cytoplasmic (Martindale *et al.* 1998) (figure 5.2 (plate 7)). The 1955-128 protein has a predicted molecular mass of 73 kDa, which is too large to enter the nucleus by passive diffusion.

In contrast to 1955-128, 100% of cells expressing 1955-128-NLS protein had nuclear huntingtin stain (figure 5.2 (plate 7)), indicating that the ectopic NLS is functional when introduced into huntingtin. The total frequency of aggregates was similar between the 1955-128 and 1955-128-NLS proteins (table 5.1). However, the proportion of aggregates in the nucleus differed

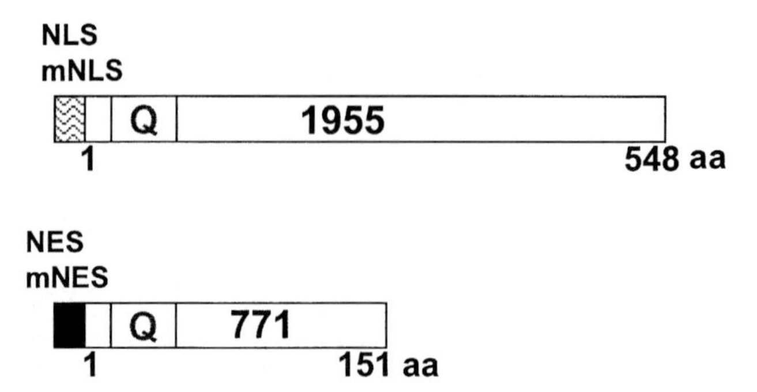

Figure 5.1 cDNA constructs used in this study. The 771 bp (1–151 amino acids) and 1955 bp (1–548 amino acids) constructs are represented. The position of the polyglutamine tract, either 15 or 128 units, is indicated by the letter Q. The position of the first huntingtin residue is indicated by the number 1. An NLS or mutant NLS (dashed box) was ligated into the 5'-end of the 1955-15 and 1955-128 constructs. An NES or mutant NES (filled box) was ligated into the 5'-end of the 771-15 and 771-128 constructs.

Table 5.1 *Frequency and subcellular localization of huntingtin aggregates*

construct	*nuclear aggregates (%)*	*cytoplasmic aggregates (%)*	*total number of cells with aggregates (%)*[a]
1955-128	0 ± 0	100 ± 0	7.8 ± 2.6
1955-128-NLS	100 ± 0	0 ± 0	5.4 ± 1.7
1955-128-mNLS	0 ± 0	100 ±0	2.7 ± 1.5
771-128	49.2 ± 15.7	50.8 ± 15.7	75.4 ± 10.7
771-128-NES	2.0 ± 2.0	98.0 ± 2.0	51.3 ± 16.5
771-128-mNES	51.6 ± 18.0	48.4 ± 18.0	68.4 ± 19.0

[a]There were no significant differences between the total aggregate frequency between the different 1955 proteins or 771 proteins, except between 1955-128 and the control protein 1955-128-mNLS ($p < 0.01$).

Figure 5.2 The localization of huntingtin is altered by the addition of an active NLS sequence. Huntingtin, detected by mAb 2166, appears as red stain, the nucleus is counter-stained in blue. Nuclear huntingtin stain is pink when the red stain is overlapped with blue. The huntingtin aggregates appear as large clumped masses, easily differentiated from normal diffuse stain. The size of the aggregates varied from cell to cell, but there was no consistent size difference between aggregates formed by the protein products of the three 1955-128 constructs. The aggregates formed by the 1955-128 protein are cytoplasmic (*a*), by the 1955-128-NLS protein, with an active NLS, are nuclear (*b*), and by the 1955-128-mNLS control protein, containing a mutant NLS, are cytoplasmic (*c*). The protein product of the 1955-15 construct is cytoplasmic (*d*), the 1955-15-NLS protein, with an active NLS, is nuclear (*e*), and the 1955-15-mNLS control protein, containing a mutant NLS, is cytoplasmic (*f*). (See also colour plate section.)

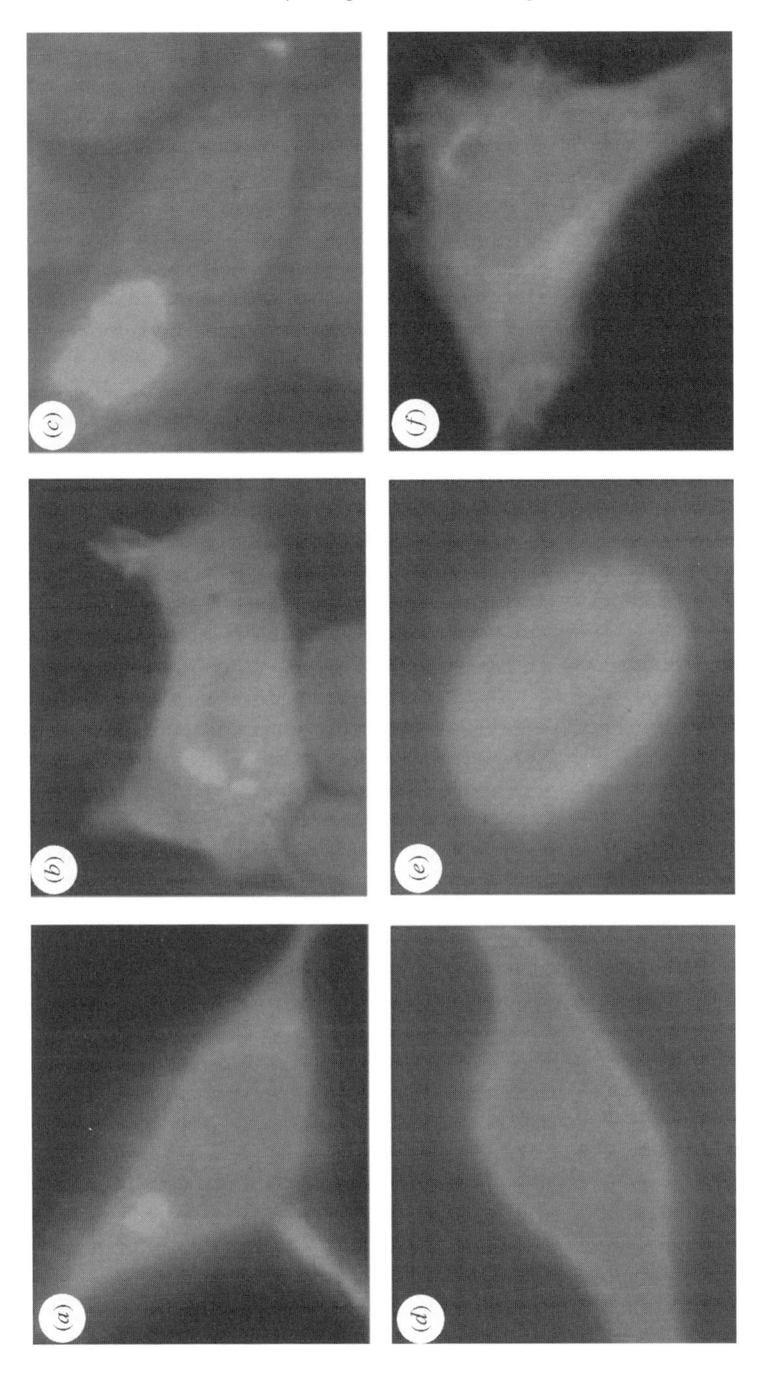
(a)
(b)
(c)
(d)
(e)
(f)

Table 5.2 *Importing huntingtin into the nucleus does not change its toxicity*

protein	*toxicity relative to LacZ (%)*
1955-128	71.88 ± 1.36
1955-128-NLS	72.66 ± 1.37
1955-128-mNLS	72.70 ± 0.67
1955-15	92.60 ± 0.91
1955-15-NLS	92.30 ± 0.53
1955-15-mNLS	92.00 ± 1.05

(Apoptosis assessed at 48 h post-transfection. The 1955-128 proteins are significantly more toxic than the 1955-15 proteins at $p < 0.001$ ($n = 5$).)

between the proteins, with 0% nuclear aggregates for 1955-128 and 100% for 1955-128-NLS, in parallel with the total nuclear stain.

A mutant NLS (PAAAAAV) was also inserted into 1955-128 (forming 1955-128-mNLS), to control for any effect of an introduced peptide on toxicity and aggregate formation. Immunofluorescence on transfected cells showed that the 1955-128-mNLS protein had 0% nuclear stain (figure 5.3 (plate 8)). Despite the differences in subcellular localization of 1955-128-mNLS and 1955-128-NLS proteins, there was no significant difference in the frequency of aggregates (table 5.1).

In the 293T cell model, expression of proteins with an expanded polyglutamine tract results in an increase in susceptibility to apoptotic stress from treatment with a sub-lethal concentration of tamoxifen (Hackam *et al.* 1998*a*; Martindale *et al.* 1998; Ellerby *et al.* 1999*b*). The resultant cell death is quantified by a monotetrazolium (MTT) assay, a standard apoptosis assay (Carmichael *et al.* 1987) that is a sensitive indicator of cell viability. Mock-transfected cells and LacZ-transfected cells, both treated with tamoxifen, are used as controls. To compare the toxicity of aggregates from the same sized huntingtin proteins in different cellular compartments, 293T cells were transfected with the 1955-128 constructs containing the functional and mutant NLS sequences. As shown in table 5.2, there was no significant difference in cell death due to expression of the 1955-128 and 1955-128-NLS proteins ($n = 5$), indicating that the toxicity of nuclear aggregates was the same as

Figure 5.3 The localization of huntingtin is altered by the addition of an active NES sequence. Huntingtin is shown in red, the nucleus is counter-stained in blue, and nuclear huntingtin stain is pink when the red stain is overlapped with blue. The size of the aggregates varied from cell to cell, but there was no consistent size difference between aggregates formed by the three 771-128 proteins. The 771-128 protein forms nuclear aggregates (*a*) 771-128-NES protein, containing an active NES, forms predominantly cytoplasmic aggregates (*b*), and the 771-128-mNES control protein, containing a mutant NES, forms nuclear aggregates (*c*). In contrast, 771-15 protein is nuclear (*d*), 771-15-NES protein, containing an active NES, is cytoplasmic (*e*), and the 771-15-mNES control protein, containing a mutant NES, is nuclear (*f*). (See also colour plate section.)

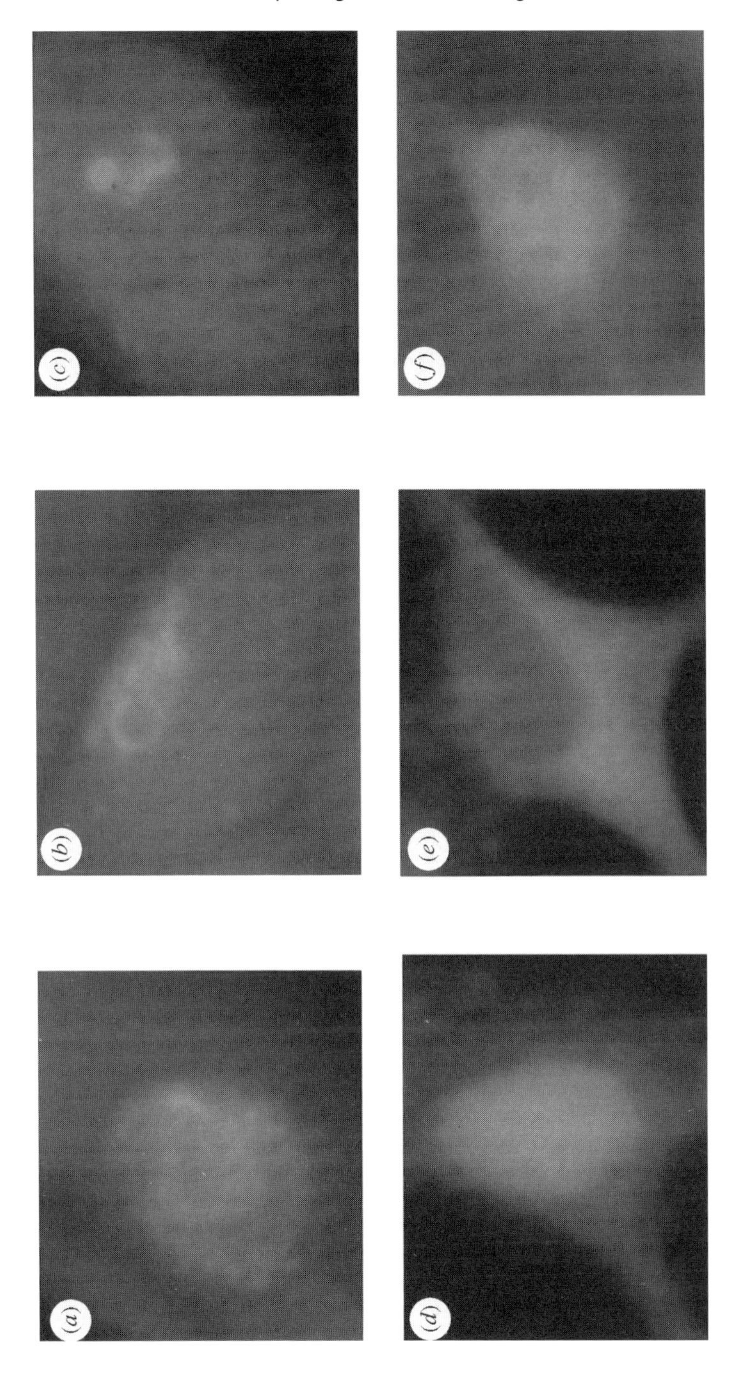
(a)
(b)
(c)
(d)
(e)
(f)

that of cytoplasmic aggregates. The control, 1955-128-mNLS, which forms cytoplasmic aggregates, also had the same toxicity as 1955-128 (table 5.2), indicating that addition of a peptide had no influence on toxicity. Western blotting demonstrated equivalent expression levels for each construct (data not shown). Therefore, aggregates generated by 1955-128 proteins have similar toxicity regardless of their localization.

Moving 771-128 aggregates out of the nucleus does not change their toxicity

The 771-128 protein has the highest frequency of aggregates in the nucleus, compared with other huntingtin fragments assessed (Hackam *et al.* 1998*a*). The 771-128 protein is also extremely toxic to cells in the presence of an apoptotic stress. To assess further the influence of subcellular localization of huntingtin on toxicity, we created a construct that brought the 771-128 protein out of the nucleus, and compared the toxicity of nuclear and cytoplasmic aggregates.

A nine-residue nuclear export sequence (NES: LALKLAGLDI) from the cAMP-dependent protein kinase inhibitor was inserted into 771-128, forming 771-128-NES (figure 5.1). Immunofluorescence studies on 293T cells expressing the 771-128 proteins are shown in figure 5.3 (plate 8). At 36 h post-transfection, the per cent of cells expressing 771-128 with nuclear aggregates was 49%. In contrast, the per cent of nuclear aggregates of the 771-128-NES protein was < 2% (table 5.1), indicating that the NES is functional when inserted into huntingtin. The total frequency of aggregates formed by 771-128 and 771-128-NES was not significantly different (table 5.1).

At 48 h post-transfection, the proportion of nuclear aggregates was higher for the 771-128-NES protein than at 36 h, most likely because passive diffusion into the nucleus occurred at a greater rate than the energy-dependent active transport out of the nucleus using the NES. A time-course assessment of aggregate formation determined that the highest proportion of cytoplasmic aggregates formed by 771-128-NES occurred at 36 h post-transfection (data not shown). Therefore, toxicity studies for this set of experiments were performed at 36 h post-transfection. Since the analysis of 771-128 was performed at 36 h, while the analysis of the 1955-128 proteins was at 48 h, the aggregate frequency and cell viabilities between these experiments cannot be compared.

To control for addition of a peptide, a 771-128-mNES construct was created with a mutant NES, AAAKAAGADA. Immunofluorescence studies demonstrated that the proportion of aggregates in the nucleus was 52% for 771-128-mNES, which is similar to that of the 771-128 protein, but differs substantially from the 771-128-NES protein. The total frequency of aggregates formed by the 771-128-mNES protein was not statistically different from either the 771-128 or 771-128-NES protein (table 5.1).

The toxicity of aggregates formed by the 771-128 proteins in different subcellular locations was compared at 36 h (table 5.3). There was no significant difference in toxicity between aggregates formed by 771-128 and 771-128-NES proteins (n = 5). In addition, there was no difference in toxicity between aggregates formed by 771-128 and 771-128-mNES, indicating that addition of a peptide had no effect (n = 5). There was also no difference in toxicity between the 771 proteins at 48 h, although overall toxicity at 48 h was greater than at 36 h due to higher protein expression (data not shown; Hackam *et al.* 1998*a*). Western blotting confirmed equivalent expression levels for each construct (data not shown). These results show that the 771-128 aggregates in the nucleus were associated with levels of toxicity similar to those of aggregates outside the nucleus. Therefore, the subcellular localization of 771-128 aggregates does not influence susceptibility to cell death.

The subcellular localization of huntingtin fragments with wild-type polyglutamine tracts does not influence toxicity

In the 293T cell model, huntingtin with longer CAG tracts (128 glutamines) has significantly higher toxicity than that with shorter tracts (15 glutamines) (Hackam *et al.* 1998*a*; Martindale *et al.* 1998). This repeat-length dependence on toxicity occurs for both full-length huntingtin and truncated huntingtin fragments (Hackam *et al.* 1998*a*). However, it has been noted previously that cells expressing truncated huntingtin containing 15 polyglutamines have an increased susceptibility to cell death (albeit less than seen with mutant huntingtin), which may represent a role for wild-type huntingtin in the regulation of cell viability (Hackam *et al.* 1998*a*; Martindale *et al.* 1998). Similar findings of toxic effects for other wild-type truncated polyglutamine-containing proteins have been observed (Ellerby *et al.* 1999*a*).

In 293T cells, the small 771-15 protein has been shown to be more toxic than the 1955-15 protein (Hackam *et al.* 1998*a*). The 771-15 is predominantly nuclear in transfected cells, whereas 1955-15 is exclusively cytoplasmic. Unlike their counterparts with 128 glutamines, the wild-type proteins do not form aggregates. Therefore, the greater toxicity of 771-15 over 1955-15 is not due to

Table 5.3 *Exporting huntingtin from the nucleus does not change its toxicity*

protein	*toxicity relative to LacZ (%)*
771-128	72.13 ± 0.95
771-128-NES	72.06 ± 1.03
771-128-mNES	71.81 ± 0.53
771-15	90.24 ± 0.85
771-15-NES	90.67 ± 0.99
771-15-mNES	90.41 ± 0.68

(Apoptosis was assessed for the 771 constructs at 36 h post-transfection instead of 48 h to maximize the proportion of 771-128-NES in the cytoplasm. The 771-128 proteins are significantly more toxic than the 771-15 proteins at $p < 0.001$ ($n = 5$).)

Table 5.4 *The proportion of cells with nuclear or cytoplasmic huntingtin is presented as a percent of the total number of cells expressing huntingtin*

construct	*nuclear localization (%)*	*cytoplasmic localization (%)*
1955-15	0 ± 0	100 ± 0
1955-15-NLS	100 ± 0	0 ± 0
1955-15-mNLS	0 ± 0	100 ± 0
771-15	76.0 ± 8.6	24.0 ± 8.6
771-15-NES	58.0 ± 12.5	42.0 ± 12.5
771-15-mNES	75.0 ± 9.2	25.3 ± 9.2

aggregate formation but could be associated with its nuclear localization. To test the influence of subcellular location on the toxicity of the wild-type huntingtin fragments, we altered the normal localization of 771-15 and 1955-15.

To alter the localization of the 1955-15 protein, the SV40 NLS sequence was inserted into 1955-15, forming 1955-15-NLS (figure 5.1). Immunofluorescence of the 1955-15 and 1955-15-NLS is shown in figure 5.2 (plate 8). At 48 h post-transfection, the percent of cells with nuclear stain was 0% for 1955-15 and 100% for 1955-15-NLS (table 5.4). The control peptide, encoding a non-functional NLS, was also inserted into 1955-15, forming 1955-15-mNLS. The 1955-15-mNLS protein had 0% nuclear stain (figure 5.2 (plate 7), table 5.4). Although the subcellular localizations were obviously different, when tested for toxicity there was no significant difference between the 1955-15, 1955-15-NLS and 1955-15-mNLS proteins (table 5.2). Therefore, for the same sized proteins, the wild-type huntingtin fragments in the cytoplasm have the same susceptibility to cell death as nuclear huntingtin.

However, the toxicity associated with the 1955-128 proteins was significantly greater than that seen with the 1955-15 proteins ($p < 0.001$, $n = 5$) (table 5.2). This result demonstrates the potent influence of increasing CAG repeat length on susceptibility to cell death, as shown previously (Hackam *et al.* 1998*a*; Martindale *et al.* 1998).

To change the predominantly nuclear 771-15 to a predominantly cytoplasmic protein, the NES sequence was inserted into 771-15, forming 771-15-NES (figure 5.1). At 36 h post-transfection, the per cent of cells with 771-15 in the nucleus was 76%. The per cent of cells with nuclear protein was reduced to 58% for 771-15-NES. The control peptide encoding a non-functional NES was also inserted into 771-15, to create 771-15-mNES. The per cent of nuclear stain for 771-15-mNES was 75%, similar to that with the parental protein 771-15 (figure 5.3 (plate 8), table 5.4). Consistent with the experiments described above, altering the localization of the 771-15 proteins also did not result in significant differences in toxicity (table 5.3). The 771-15 protein, which is predominantly nuclear, had similar toxicity to the 771-15-NES protein, which is predominantly in the cytosol. Furthermore, the 771-15-NES protein had similar toxicity to that of the product of the control construct, 771-15-mNES, with predominantly nuclear localization. Therefore, the subcellular localization of the 771-15 proteins does not influence their toxicity. In addition, there was a significant decrease in cell death associated with expression of the 771-15 proteins compared with the 771-128 proteins ($p < 0.001$, $n = 5$), as described for the 1955 proteins. Since the assessment of viabilities of the products of the 771-15 constructs was performed at 36 h, and the 1955-15 analysis was at 48 h, the values cannot be compared between experiments.

Discussion

The subcellular localization of huntingtin aggregates does not influence toxicity

Several lines of evidence have suggested that HD is a disease of the nucleus. Intranuclear inclusions are the predominant marker in affected patients (DiFiglia *et al.* 1997; Becher *et al.* 1998; Gourfinkel-An *et al.* 1998), and mice transgenic for huntingtin exon 1 develop nuclear aggregates prior to the onset of debilitating neurological symptoms (Davies *et al.* 1997). Furthermore, increasing toxicity of successively smaller huntingtin fragments in our cell culture model was associated with the formation of nuclear aggregates (Hackam *et al.* 1998*a*). Increasing toxicity *in vitro* was also associated with increased aggregate frequency (Hackam *et al.* 1998*a*). In this study, we have directly addressed the questions of whether the site of huntingtin influences its toxicity, and how huntingtin enters the nucleus.

We altered the subcellular localizations of huntingtin protein fragments with the addition of NLS and NES peptides, while the frequency of aggregates remained equivalent. The cytoplasmic protein 1955-128 was changed to a protein that forms nuclear aggregates. The 771-128 protein, which forms predominantly nuclear aggregates, was altered to form predominantly cytoplasmic aggregates. Our results in this cell culture model indicate that toxicity is not dependent on the subcellular localization of aggregates, but toxicity is associated with the frequency of aggregate formation.

The cell culture model can mimic in vivo events

Despite the fact that HD is a disease of selective neuronal death, this 293T *in vitro* model (Martindale *et al.* 1998) has previously been shown to recapitulate several features of HD. As observed *in vivo*, mutant huntingtin and other disease proteins containing expanded polyglutamine tracts form aggregates *in vitro*, whereas wild-type proteins with a normal tract do not (Skinner *et al.* 1997; Butler *et al.* 1998; Cooper *et al.* 1998; Hackam *et al.* 1998*a*; Igarashi *et al.* 1998; Li & Li 1998; Lunkes & Mandel 1998; Merry *et al.* 1998; Ellerby *et al.* 1999*a*). Aggregates formed *in vivo* and *in vitro* are frequently ubiquitinated

(Cooper *et al.* 1998; Cummings *et al.* 1998; Igarashi *et al.* 1998; Lunkes & Mandel 1998; A. S. Hackam unpublished observations). Furthermore, as observed in post-mortem neocortical tissue (Becher *et al.* 1998), increasing polyglutamine length is associated with an increased frequency of aggregates *in vitro* (Li & Li 1998; Lunkes & Mandel 1998; Martindale *et al.* 1998). More severe grades of HD have a higher frequency of cortical nuclear inclusions (Becher *et al.* 1998). Similarly, huntingtin fragments that are more toxic *in vitro* form aggregates with higher frequency (Cooper *et al.* 1998; Hackam *et al.* 1998*a*). The formation of huntingtin cleavage fragments is observed both *in vivo* and in cell culture (DiFiglia *et al.* 1997; Lunkes & Mandel 1998; Martindale *et al.* 1998). Finally, the selective vulnerability of affected neurons in HD is mimicked by increased susceptibility to apoptotic stress of cultured cells expressing mutant huntingtin (Cooper *et al.* 1998; Hackam *et al.* 1998*a*; Martindale *et al.* 1998). Therefore, it is plausible that these *in vitro* data, which recapitulate the *in vivo* situation to a significant extent, are relevant, suggesting that nuclear and cytoplasmic aggregates may also have equivalent toxicity in humans.

Aggregates are not the sole contributors to toxicity

Intracellular aggregates, regardless of their cellular compartment, are associated with toxicity. Therefore, reducing aggregate frequency remains an important therapeutic target for CAG expansion diseases. At the present time, it is unclear whether aggregates are the primary cause of neurodegeneration, or whether they are formed as an early secondary response to cell injury. The observation that the development of aggregates precedes cell death *in vivo* and *in vitro* does not distinguish between aggregate formation as a causal event, or aggregates as a by-product of other cytotoxic events that lead to death. Aggregates could theoretically even serve a protective role by sequestering toxic polyglutamine-containing fragments.

There is increasing evidence that aggregates are not the sole contributors to toxicity. Although the majority of inclusions have been identified in neuronal populations that degenerate during disease progression, several studies have shown that the concordance between nuclear inclusions and neurodegeneration is not absolute (Becher *et al.* 1998; Holmberg *et al.* 1998; Li *et al.* 1998; Warrick *et al.* 1998). There are several examples in which inclusions have been identified in cells not destined to die. Aggregates were present in the dentate nucleus of the cerebellum in HD patients (Becher *et al.* 1998), an area that does not frequently exhibit neurodegeneration. Intranuclear inclusions were also identified in SCA-7 (Holmberg *et al.* 1998) and DRPLA (Becher *et al.* 1998) patients in regions of the brain not affected by the disease. Ubiquitinated intranuclear inclusions formed by mutant androgen receptor aggregates were observed in SMBA patients in peripheral tissues (Li *et al.* 1998). In addition, mice transgenic for mutant huntingtin exon 1 (Davies *et al.* 1997) and a *Drosophila* model of SCA-3 (Warrick *et al.* 1998) contained aggregates in regions that do not exhibit cell death.

These observations suggest that aggregate formation is associated with, but is clearly not sufficient to cause cell death.

There are also reports in which nuclear inclusions were not observed in affected tissues. Neurodegeneration of Purkinje cells in SCA-7 and juvenile HD patients occurred in the absence of aggregates (Becher *et al.* 1998; Holmberg *et al.* 1998). Further, there is no apparent correlation of nuclear inclusion frequency with length of the CAG tract and Vonsattel grade in the striatum of HD patients (Becher *et al.* 1998).

These observations argue against a role for aggregates as a direct cause of neurodegeneration. However, aggregates may result in deficiencies in neuronal function (Davies *et al.* 1997; Lunkes & Mandel 1997; Ross 1997) and additional events may be required to lead to neurodegeneration subsequent to the formation of aggregates. For example, although neurons may initially form aggregates as a result of stress-induced protein cleavage, a particular cellular environment may be required for cytotoxicity, analogous to the tamoxifen-induced stress in our 293T cell model. In neurons, this 'toxic environment', possibly resulting from expression of certain glutamate receptors leading to uncontrolled excitotoxicity, may be the stimulus needed for aggregate-containing cells to die. The cells in non-affected tissues that form aggregates could be more resistant to toxicity if they have a higher threshold of injury needed for death. The threshold could be set by the particular repertoire of glutamate receptors, by the cell's ability to deal with metabolic stress or by levels of anti-apoptotic factors. Thus, aggregates alone are insufficient for cell death, but selective populations of vulnerable neurons may be more susceptible to a 'toxic environment' when their viability is compromised by the presence of aggregates.

Huntingtin is toxic in both the nucleus and cytoplasm of 293T cells

Determining the primary site of pathology of the CAG diseases is important for designing therapeutic interventions. Nuclear inclusions are associated with disease in the other CAG diseases studied so far (Skinner *et al.* 1997), whereas for HD both nuclear and extranuclear aggregates are seen *in vivo* and *in vitro* (DiFiglia *et al.* 1997).

The different localization of aggregates in HD compared with other expansion diseases may depend on several factors. First, the size of a protein influences nuclear entry (Görlich & Mattaj 1996). Thus, the ability of the protein to be cleaved into fragments small enough to enter the nucleus is important. Ataxin-3 is small enough to diffuse into the nucleus. The mutant exon 1 fragment in the HD mice forms nuclear inclusions since it is also small enough to diffuse into the nucleus. Only antibodies against N-terminal epitopes recognize nuclear inclusions in HD post-mortem tissue, suggesting proteolytic processing of full-length huntingtin (DiFiglia *et al.* 1997; Becher *et al.* 1998).

Second, the polyglutamine-containing proteins may usually reside in the nucleus as part of their normal function. For example, ataxin-1 (Skinner *et al.* 1997) is predominantly normally localized in the nucleus of neurons affected in SCA-1. Thus, nuclear entry is not a feature of this disease. In addition, ataxin-3 (Paulson *et al.* 1997; Tait *et al.* 1998), ataxin-7 (Trottier *et al.* 1995; Stevanin *et al.* 1996), the androgen receptor (Li *et al.* 1998) and atrophin-1 (Miyashita *et al.* 1997) have putative NLSs and have been identified in the nucleus. By contrast, we have not found an active NLS in the N-terminus of huntingtin. Third, there may be inherent differences in the pathogenesis of these diseases due to differences in functional properties or protein partners of the respective proteins. For example, androgen receptor toxicity is influenced by ligand concentration in 293T cells (Ellerby *et al.* 1999*a*), and huntingtin associates with several proteins that have altered interactions with increased polyglutamine length (Li *et al.* 1995; Burke *et al.* 1996; Kalchman *et al.* 1997). Ataxin-1 and ataxin-3 associate with the nuclear matrix (Matilla *et al.* 1997; Tait *et al.* 1998) and may interfere with essential nuclear events as part of their toxicity. Indeed, recent results from ataxin-1 transgenic mouse lines by Klement *et al.* (1998) indicate that nuclear localization of ataxin-1 is critical for SCA-1 pathology. Finally, detection of the localization of a particular protein could vary with tissue preparation and the antisera used.

There is clearly *in vivo* evidence for extranuclear aggregates being toxic (DiFiglia *et al.* 1997). While this paper was under review, Saudou *et al.* (1998) presented data that indicated that nuclear localization was required for cytotoxicity of huntingtin in a striatal neuronal line. There are several methodological differences that could account for the discordant conclusions between the present chapter and the work of Saudou *et al.* In this study, we have assessed the total frequency of aggregates, including cytoplasmic aggregates, which allowed direct comparison of the influence of total cellular aggregates on toxicity in 293T cells. Additionally, there are differences in sensitivities and timing of the cell viability assays used. The MTT assay quantifies mitochondrial changes, which are considered as earlier indicators of apoptosis than nuclear morphological changes (Green & Reed 1998). Mitochondrial markers of apoptosis may record early subtle alterations of cell viability caused by cytoplasmic huntingtin. In addition, further studies are needed to determine whether cell line-specific factors may contribute to the differences in these findings.

YAC transgenic mice expressing mutant huntingtin have cytoplasmic and nuclear changes

We have produced yeast artificial chromosome (YAC) transgenic mice expressing normal and mutant huntingtin that is expressed in a developmentally regulated and cell-specific manner essentially identical to that seen with endogenous huntingtin. The YAC transgenic mice expressing mutant huntingtin with 46 repeats do not have a clinical phenotype as evidenced by detailed neurological and behavioural assessment up until the latest stage of assessment at 24 months. However, mild electrophysiological abnormalities do become evident in these mice at six months of age. These abnormalities become more obvious and are readily apparent at ten months of age when diminished hippocampal long-term potentiation in the CA1 neurons is evident. This is also associated with an increase in intracellular calcium stores. At 10 months of age, detailed assessment using electron microsopy and immunogold labelling has failed to reveal any evidence for nuclear translocation of huntingtin aggregates or evidence for neurodegeneration. However, further examination of these mice at approximately two years of age has revealed evidence for labelling of N-terminal huntingtin, which is seen traversing the nuclear pore, and also seen in the nucleus of selected neurons. This indicates that nuclear translocation of huntingtin occurs long after some cellular abnormalities are apparent in these mice. Furthermore, these mice have no evidence for a clinical phenotype, also indicating that nuclear translocation of huntingtin occurs prior to evidence for a clinical phenotype.

Mice expressing huntingtin with 72 glutamines at levels less than endogenous levels showed a clinical phenotype at nine months. Detailed pathological examination has revealed evidence for increased cellular staining and medium spiny neuronal loss in the lateral striatum. The examination of a founder mouse with 72 repeats that had increased copy number of the YACs with evidence for increased expression, had an obvious clinical phenotype by six weeks of age and when examined pathologically at one year of age had obvious evidence for proteolytic processing of huntingtin with some N-terminal labelled huntingtin present in the cytoplasm, some labelling being seen in the nuclear pores, and other neurons with clear evidence of significant seeding into the nucleus of this huntingtin fragment. These N-terminal huntingtin fragments either coalesced with a few N-terminal huntingtin fragments interacting with each other, and then in some cells this was associated with evidence for large numbers of N-terminal fragments coalescing into microaggregates, and in other cells with a development of obvious aggregates seen at light microscopy (Hodgson *et al.* 1999).

These studies give some indication of the process for the development of aggregates with initial proteolytic processing of huntingtin in the cytoplasm, and liberation of an N-terminal fragment which traverses the nuclear pore and enters into the nucleus. Within the nucleus there is initial seeding of huntingtin with other N-terminal huntingtin fragments over time, and as the protein concentration within the nucleus of internal huntingtin increases, this allows N-terminal huntingtin fragments to interact with each other with eventual production of aggregates associated with neurodegeneration.

The YAC transgenic mice expressing 46 repeats demonstrated cytoplasmic toxicity with increased calcium concentration and obvious electrophysiological abnormalities. The earlier presentation of a neurological phenotype associated with neurodegeneration by one year of age of the YAC-72 mice is compatible with the accelerated disease seen in patients who manifest with juvenile HD associated with polyglutamine expansion in a similar repeat range. Polyglutamine expansion of this range is associated with presentation by ten years of age and clearly is compatible with a highly accelerated disease process. In juvenile onset, the cytoplasmic phase of this illness may be shorter, associated with the accelerated development of intranuclear aggregates occurring at a much earlier stage than seen, for example, in the mice with 46 repeats when N-terminal translocation of huntingtin was only first seen by two years of age. This would suggest that in patients with 46 repeats the cytoplasmic phase of the illness with some evidence for toxicity may occur for a longer period and that only much closer to, but prior to, onset of the clinical illness would this be associated with nuclear translocation of huntingtin as seen in the YAC transgenic mice expressing 46 glutamines. These data provide further *in vivo* evidence for the nucleus and cytoplasm as sites of pathogenesis in HD.

We are grateful to Dr Gideon Dreyfuss for use of the pyruvate kinase construct and for helpful discussion. We also thank Krista McCutcheon and Keith Fichter for their excellent technical expertise, and Dr Cheryl Wellington and the entire HD laboratory group for useful discussions. This work is supported by the Canadian Network Centres of Excellence (NCE, Genetics) and the Medical Research Council (MRC, Canada). A.S.H. is an MRC postdoctoral fellow. M.R.H is an established investigator of the British Columbia Children's Hospital. The *in vitro* work reported here is also part of an original publication in *Human Molecular Genetics* (Hackam *et al.* 1999).

References

Andrew, S. E., Goldberg, Y. P. & Hayden, M. R. 1997 Rethinking genotype and phenotype correlations in polyglutamine expansion disorders. *Hum. Mol. Genet.* **6**, 2005–2010.

Becher, M. W., Kotzuk, J. A., Sharp, A. H., Davies, S. W., Bates, G. P., Price, D. L. & Ross, C. A. 1998 Intranuclear neuronal inclusions in Huntington's disease and dentatorubral and palidoluysian atrophy: correlation between the density of inclusions and *IT15* CAG triplet repeat length. *Neurobiol. Dis.* **4**, 387–397.

Bots, G. T. & Bruyn, G. W. 1981 Neuropathological changes of the nucleus accumbens in Huntington's chorea. *Acta Neuropathol.* **55**, 21–22.

Burke, J. R., Enghild, J. J., Martin, M. E., Jou, Y. S., Myers, R. M., Roses, A. D., Vance, J. M. & Strittmatter, W. J. 1996 Huntingtin and DRPLA proteins selectively interact with the enzyme GAPDH. *Nature Med.* **2**, 347–350.

Butler, R., Leigh, P. N., McPhaul, M. J. & Gallo, J.-M. 1998 Truncated forms of the androgen receptor are associated with polyglutamine expansion in X-linked spinal and bulbar muscular atrophy. *Hum. Mol. Genet.* **7**, 121–127.

Carmichael, J., DeGraff, W. G., Gazdar, A. F., Minna, J. D. & Mitchell, J. B. 1987 Evaluation of a tetrazolium-based semiautomated colorimetric assay: assessment of chemosensitivity testing. *Cancer Res.* **47**, 936–942.

Cooper, J. K. (and 12 others) 1998 Truncated N-terminal fragments of huntingtin with expanded glutamine repeats form nuclear and cytoplasmic aggregates in cell culture. *Hum. Mol. Genet.* **7**, 783–790.

Cummings, C. J., Mancini, M. A., Antalffy, B., DeFranco, D. B., Orr, H. T. & Zoghbi, H. Y. 1998 Chaperone suppression of aggregation and altered subcellular proteasome localization imply protein misfolding in SCA1. *Nature Genet.* **19**, 148–154.
Davies, S. W., Turmaine, M., Cozens, B. A., DiFiglia, M., Sharp, A. H., Ross, C. A., Scherzinger, E., Wanker, E. E., Mangiarini, L. & Bates, G. P. 1997 Formation of neuronal intranuclear inclusions (NII) underlies the neurological dysfunction in mice transgenic for the HD mutation. *Cell* **90**, 537–548.
DiFiglia, M., Sapp, E., Chase, K. O., Davies, S. W., Bates, G. P., Vonsattel, J. P. & Aronin, N. 1997 Aggregation of huntingtin in neuronal intranuclear inclusions and dystrophic neurites in brain. *Science* **277**, 1990–1993.
Ellerby, L. M. (and 10 others) 1999*a* Kennedy's disease: caspase cleavage of the androgen receptor is a crucial event in cytotoxicity. *J. Neurochem.* **72**, 185–195.
Ellerby, L. M. (and 11 others) 1999*b* Cleavage of atrophin-1 at caspase site aspartic acid 109 modulates cytotoxicity. *J. Biol. Chem.* **274**, 8730–8736.
Goldberg, Y. P. (and 10 others) 1996 Cleavage of huntingtin by apopain, a proapoptotic cysteine protease, is modulated by the polyglutamine tract. *Nature Genet.* **13**, 442–449.
Gourfinkel-An, I., Cancel, G., Duyckaerts, C., Faucheux, B., Hauw, J.-J., Trottier, Y., Brice, A., Agid, Y. & Hirsch, E.C. 1998 Neuronal distribution of intranuclear inclusions in Huntington's disease with adult onset. *Neuroreport* **9**, 1823–1826.
Görlich, D. & Mattaj, I. W. 1996 Nucleocytoplasmic transport. *Science* **271**, 1513–1518.
Green, D. R. & Reed, J. C. 1998 Mitochondria and apoptosis. *Science* **281**, 1309–1312.
Hackam, A. S., Singaraja, R., Wellington, C. L., Metzler, M., McCutcheon, K., Zhang, T., Kalchman, M. & Hayden, M. R. 1998*a* The influence of huntingtin protein size on nuclear localization and cellular toxicity. *J. Cell Biol.* **141**, 1097–1105.
Hackam, A. S., Wellington, C. L. & Hayden, M. R. 1998*b* The fatal attraction of polyglutamine-containing proteins. *Clin. Genet.* **53**, 233–242.
Hackam, A. S., Singaraja, R , Zhang, T., Gan, L. & Hayden, M. R. 1999 *In vitro* evidence for both the nucleus and cytoplasm as subcellular sites of pathogenesis in Huntinton disease. *Hum. Mol. Genet.* **8**, 25–33.
Hodgson, J. G. (and 18 others) 1999 A YAC mouse model for Huntington's disease with full-length mutant huntingtin, cytoplasmic toxicity and striatal neurodegeneration. *Neuron* **23**, 181–192.
Holmberg, M. (and 10 others) 1998 Spinocerebellar ataxia type 7 (SCA7): a neurodegenerative disorder with neuronal intranuclear inclusions. *Hum. Mol. Genet.* **7**, 913–918.
Igarashi, S. (and 18 others) 1998 Suppression of aggregate formation and apoptosis by transglutaminase inhibitors in cells expressing truncated DRPLA protein with an expanded polyglutamine stretch. *Nature Genet.* **18**, 111–117.
Kalchman, M. A. (and 13 others) 1997 HIP1, a human homolog of *S. cerevisiae* Sla2p, interacts with membrane-associated huntingtin in the brain. *Nature Genet.* **16**, 44–53.
Klement, I. A., Skinner, P. J., Kaytor, M. D., Yi, H., Hersch, S. M., Clark, H. B., Zoghbi, H. Y. & Orr, H. T. 1998 Ataxin-1 nuclear localization and aggregation: role in polyglutamine-induced disease in *SCA1* transgenic mice. *Cell* **95**, 41–53.
Li, M., Nakagomi, Y., Kobayashi, Y., Merry, D. E., Tanaka, F., Doyu, M., Mitsuma, T., Hashizume, Y., Fischbeck, K. H. & Sobue, G. 1998 Non-neural inclusions of androgen receptor protein in spinal and bulbar muscular atrophy. *Am. J. Pathol.* **153**, 695–701.
Li S.-H. & Li, X.-J. 1998 Aggregation of N-terminal huntingtin is dependent on the length of its glutamine repeats. *Hum. Mol. Genet.* **7**, 777–782.
Li, X. J., Li, S. H., Sharp, A. H., Nucifora, F. C. Jr, Schilling, G., Lanahan, A., Worley, P., Snyder, S. H. & Ross, C. A. 1995 A huntingtin-associated protein enriched in brain with implications for pathology. *Nature* **378**, 398–402.
Lunkes, A. & Mandel, J.-L. 1997 Polyglutamines, nuclear inclusions and neurodegeneration. *Nature Med.* **8**, 1201–1202.
Lunkes, A. & Mandel, J.-L. 1998 A cellular model that recapitulates major pathogenic steps of Huntington's disease. *Hum. Mol. Genet.* **7**, 1355–1361.

Martindale, D. (and 11 others) 1998 Length of the protein and polyglutamine tract influence localization and frequency of intracellular aggregates of huntingtin. *Nature Genet.* **18**, 150–154.
Matilla, A., Koshy, B. T., Cummings, C. J., Isobe, T., Orr, H. T. & Zoghbi, H. Y. 1997 The cerebellar leucine-rich acidic nuclear protein interacts with ataxin-1. *Nature* **389**, 974–978.
Merry, D. E., Kobayashi, Y., Bailey, C. K., Taye, A. A. & Fischbeck, K. H. 1998 Cleavage, aggregation and toxicity of the expanded androgen receptor in spinal and bulbar muscular atrophy. *Hum. Mol. Genet.* **7**, 693–701.
Miyashita, T., Okamura-Oho, Y., Mito, Y., Nagafuchi, S. & Yamada, M. 1997 Dentatorubral pallidoluysian atrophy (DRPLA) protein is cleaved by caspase-3 during apoptosis. *J. Biol. Chem.* **272**, 29238–29242.
Ordway, J. M. (and 11 others) 1997 Ectopically expressed CAG repeats cause intranuclear inclusions and a progressive late onset neurological phenotype in the mouse. *Cell* **91**, 753–763.
Paulson, H. L., Perez, M. K., Trottier, Y., Trojanowski, J. Q., Subramony, S. H., Das, S. S., Vig, P., Mandel, J.-L., Fischbeck, K. H. & Pittman, R. N. 1997 Intranuclear inclusions of expanded polyglutamine protein in spinocerebellar ataxia type 3. *Neuron* **19**, 333–344.
Ross, C. 1997 Intranuclear neuronal inclusions: a common pathogenic mechanism for glutamine-repeat neurodegenerative diseases? *Neuron* **19**, 1147–1150.
Sapp, E., Schwarz, C., Chase, K., Bhide, P. G., Young, A. B., Penney, J., Vonsattel, J. P., Aronin, N. & DiFiglia, M. 1997 Huntingtin localization in brains of normal and Huntington's disease patients. *Ann. Neurol.* **42**, 604–612.
Saudou, F., Finkbeiner, S., Devys, D. & Greenberg, M. E. 1998 Huntingtin acts in the nucleus to induce apoptosis but death does not correlate with the formation of intranuclear inclusions. *Cell* **95**, 55–66.
Skinner, P. J., Koshy, B. T., Cummings, C. J., Klement, I. A., Helin, K., Servadio, A., Zoghbi, H. Y. & Orr, H. T. 1997 Ataxin-1 with an expanded glutamine tract alters nuclear matrix-associated structures. *Nature* **389**, 971–974.
Stevanin, G. (and 17 others) 1996 Screening for proteins with polyglutamine expansions in autosomal dominant cerebellar ataxias. *Hum. Mol. Genet.* **5**, 1887–1892.
Tait, D., Riccio, M., Sittler, A., Scherzinger, E., Santi, S., Ognibene, A., Maraldi, N. M., Lehrach, H. & Wanker, E. E. 1998 Ataxin-3 is transported into the nucleus and associates with the nuclear matrix. *Hum. Mol. Genet.* **7**, 991–997.
Tellez-Nagel, I., Johnson, A. B. & Terry, R. D. 1974 Studies on brain biopsies of patients with Huntington's chorea. *J. Neuropathol. Exp. Neurol.* **33**, 308–332.
Trottier, Y. (and 12 others) 1995 Polyglutamine expansion as a pathological epitope in Huntington's disease and four dominant cerebellar ataxias. *Nature* **378**, 403–406.
Warrick, J. M., Paulson, H. L., Gladys, L., Gray-Board, G. L., Bui, Q. T., Fischbeck, K. H., Pittman, R. N. & Bonini, N. M. 1998 Expanded polyglutamine protein forms nuclear inclusions and causes neural degeneration in *Drosophila*. *Cell* **93**, 939–949.
Wellington, C. L. (and 20 others) 1998 Caspase cleavage of gene products associated with triplet expansion disorders generates truncated fragments containing the polyglutamine tract. *J. Biol. Chem.* **273**, 9159–9167.

6

A genetic model for human polyglutamine-repeat disease in *Drosophila melanogaster*

Nancy M. Bonini

Introduction

We are interested in pioneering new approaches to understand the mechanisms of, and ultimately prevent, human neurodegenerative disease. To do this, we are bringing to bear the power of genetics by developing models for human brain degenerative disease in the simple system *Drosophila melanogaster* (Warrick *et al.* 1998). The polyglutamine-repeat diseases are a class of human disease that results from the expansion of a polyglutamine run within the open reading frame of the disease protein. The expanded polyglutamine run is thought to confer a dominant toxic effect on the otherwise unrelated proteins, leading to neuronal dysfunction and eventual cell loss (reviewed in Paulson & Fischbeck 1996). In order to address mechanisms by which the polyglutamine-repeat disease proteins induce neural dysfunction, as well as to identify genes that can ameliorate the effects of expanded polyglutamine proteins, we asked whether it was possible to recreate human polyglutamine-repeat disease in the fruit fly *Drosophila melanogaster*.

Drosophila has a complex nervous system, organized into neural centres much like the human brain. In addition, *Drosophila* displays complex behaviours, such as learning and memory. Genes between humans and flies are highly conserved, including entire gene pathways, such as the tyrosine kinase–ras–raf signalling pathways. *Drosophila*, however, has the advantages of a rapid reproductive cycle (ten days from egg to adult), as well as ease of growing large numbers. Thus, if it were possible to recreate polyglutamine disease in *Drosophila*, it would ultimately prove feasible to use the fly to elucidate mechanisms by which polyglutamine proteins cause neural loss, as well as to define genes or compounds that can ameliorate the effects of these proteins. A critical question is, if a phenotype is seen in *Drosophila*, to what extent does it recapitulate features of human polyglutamine-repeat disease?

Expression of a human disease gene in the fly

We initiated our studies with the polyglutamine-repeat disease gene (*MJD1*) responsible for the most common dominantly inherited ataxia, spinocerebellar ataxia type 3, also called Machado–Joseph disease (SCA-3/MJD). Since mouse transgenic models for Huntington's disease and SCA-3/MJD had shown that truncated versions of the respective disease proteins were more potent than full-length forms of the proteins (Ikeda *et al.* 1996; Mangiarini *et al.* 1996), we used a truncated version of MJD. The truncated gene is comprised of the C-terminal part of the MJD protein, which includes about 55 amino acids of the MJD protein, plus the polyglutamine stretch. The polyglutamine stretch included 27 glutamines for the non-disease protein (MJDtr-Q27) and 78 glutamines for the expanded polyglutamine form (MJDtr-Q78). The genes were expressed using a two-component system of targeted gene expression, the GAL4-UAS system (Brand & Perrimon 1993), such that transgene expression could be directed specifically to the nervous system as well as to other tissues as desired.

Human polyglutamine protein causes late onset, progressive degeneration in *Drosophila*

Our initial studies directed transgene expression to the eye using the *glass multiple reporter* (*gmr*) promoter. This promoter drives expression in all cells of the developing eye, including photoreceptor neurons and pigment cells. Expression of the MJDtr-Q27 protein had no effect on the eye when compared to normal eye morphology. However, expression of the MJDtr-Q78 protein caused progressive degeneration. When MJDtr-Q78 was strongly expressed, the eye failed to show pigmentation, indicating loss of pigment cells, and collapsed, indicating loss of photoreceptor neurons (figure 6.1). Weaker expression resulted in an eye that was only mildly disrupted; however, the eyes underwent degeneration during adult life. In this instance, degeneration was progressive, although of developmental onset. Analysis of the brain phenotype indicated that the effect occurred late in development: areas of the brain dependent upon photoreceptor neuron innervation were present, indicating that the photoreceptor neurons survived for quite a while prior to degenerating. Therefore, these studies indicate that expression of the MJDtr-Q78 protein in flies resulted in late onset, progressive cellular degeneration. This phenotype is similar to that seen in humans, and suggests that the fundamental mechanisms of human polyglutamine-repeat disease are conserved in *Drosophila*. Our findings have subsequently been confirmed by others (Jackson *et al.* 1998).

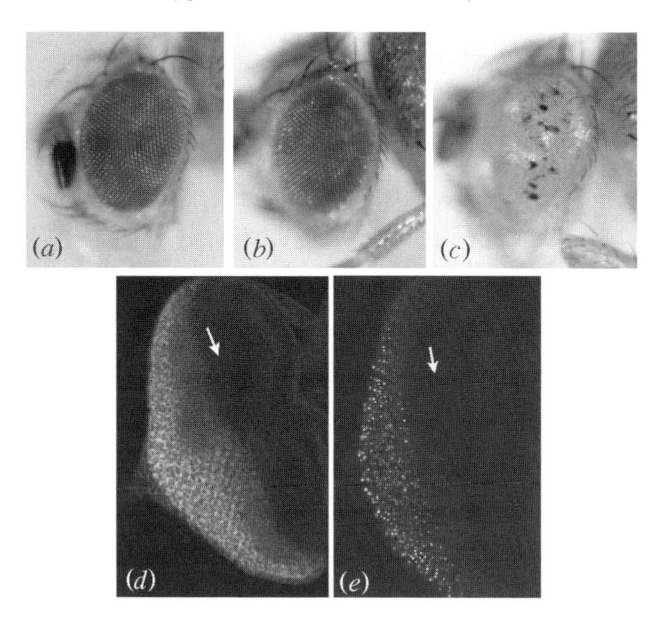

Figure 6.1 Expanded polyglutamine protein causes degeneration of the *Drosophila* eye. (*a*) Eye of a normal fly, showing a highly ordered precise neurocrystalline lattice, and red pigmentation. (*b*) Eye of a fly expressing the control protein, MJDtr-Q27. The eye morphology is identical to that of a normal fly. Fly of genotype *w*; *gmr-GAL4/UAS-MJDtr-Q27*. (*c*) Eye of a fly expressing the expanded polyglutamine protein, MJDtr-Q78. The eye shows severe degeneration compared to the normal eye. Externally, pigmentation is lost (seen here as a light coloured eye), internally (see Warrick *et al.* 1998) the eye shows complete loss of the photoreceptor neurons. (*d*) Developing eye disc of a larva expressing the MJDtr-Q27 protein. The protein is being expressed from the start of differentiation at the furrow (arrow), and shows a diffuse cytoplasmic expression pattern. Larva of genotype *w*; *gmr-GAL4/UAS-MJDtr-Q27*. (*e*) Developing eye disc of a larva expressing the MJDtr-Q78 protein. The protein is being expressed from the start of differentiation at the furrow (arrow). The protein shows eventual concentration into brightly fluorescing nuclear inclusions. Older cells are to the left. Larva of genotype *w*; *gmr-GAL4/UAS-MJDtr-Q78*.

Polyglutamine protein forms nuclear inclusions in *Drosophila*

Recently it has been noted that in the human disease tissue, as well as in transgenic mouse models and cells in culture, expanded polyglutamine proteins undergo abnormal aggregation to form nuclear inclusions (Davies *et al.* 1997; DiFiglia *et al.* 1997; Paulson *et al.* 1997; Skinner *et al.* 1997). Thus we were interested in determining whether expression of such a protein in flies would also lead to nuclear inclusion (NI) formation. We found striking evidence of NI formation. The protein, when initially expressed, was cytoplasmic, but with time accumulated in the nuclei of cells, forming brightly fluorescing inclusions (figure 6.1).

We took these studies one step further and addressed various aspects of inclusion formation, such as whether inclusions were formed in all tissues to which transgene expression was directed. Indeed, all cells expressing the mutant protein showed evidence of inclusion formation—despite the fact that the expanded polyglutamine protein was not lethal to all cells. In fact, whereas expression of the expanded polyglutamine protein was deleterious to neurons, pigment cells and muscle cells, it was not at all deleterious to dividing epithelial cells, despite the fact that NIs formed in these cells. These studies indicate that, although inclusions may be a part of the degenerative process, the mere presence of an NI does not guarantee degeneration. Additional recent evidence confirms our finding that NIs may not be necessary to disease pathogenesis (Klement *et al.* 1998; Saudou *et al.* 1998).

Mechanisms of polyglutamine-mediated neurodegeneration are conserved in *Drosophila*

We have shown that expression of an expanded polyglutamine protein shows features of polyglutamine-repeat disease that are seen in humans (table 6.1; figure 6.2; Warrick *et al.* 1998). First, expression of the protein causes a late onset, progressive degeneration. Even when the protein is strongly expressed, degeneration is not immediate but of late developmental onset, reminiscent of juvenile cases of these diseases with particularly long CAG repeats (Paulson & Fischbeck 1996). We also showed that the degeneration was of a progressive nature. Second, the protein shows abnormal protein aggregation, forming NIs in the tissues in which the protein is expressed. Irrespective of whether NIs are causal to the disease, their formation has been noted for many different polyglutamine disease proteins. Hence, with respect to this characteristic feature as well, the *Drosophila* model shows conservation.

Table 6.1 *Comparison of some of the features of polyglutamine-repeat disease between* Drosophila *and humans*[a]

(a) features of human polyglutamine repeat disease conserved in *Drosophila*
late-onset of the phenotype
progressive degeneration
abnormal protein aggregation in form of nuclear inclusions
(b) features of human polyglutamine-repeat disease not yet addressed in *Drosophila*
selective neural sensitivity of degeneration
instability of the CAG repeat from one generation to the next (known as anticipation)

[a]For a summary of the characteristic features of human polyglutamine-repeat disease, see Paulson & Fischbeck (1996).

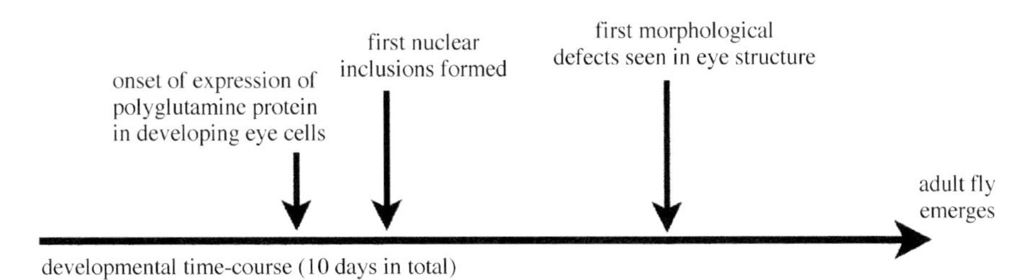

Figure 6.2 Time-course of events of polyglutamine-mediated neurodegeneration in *Drosophila*. The bottom line represents a developmental time line of *Drosophila*, from egg to the adult fly. Expression of the polyglutamine protein is initiated during the third instar larval stage, about four days after the egg is laid. About 12 h later, the first NIs are formed by the expanded polyglutamine protein. Several days later (about three days) the first morphological defects due to expression of the expanded polyglutamine protein are seen. The adult fly emerges at ten days: the adult eye undergoes further progressive degeneration, as well. For additional details, see Warrick *et al.* (1998).

Taken together, these data indicate that at least some key features of polyglutamine-repeat disease are conserved between humans and flies. This suggests that we should be able to use *Drosophila* to address mechanisms by which the disease proteins function, as well as to define genes or compounds that can ameliorate the phenotype.

Defining biological mechanisms of polyglutamine-induced degeneration

Given that we have developed a fly model, we were interested in addressing aspects of polyglutamine-repeat disease that might contribute to the biological actions of the mutant protein. We had already determined that expression of an NI *per se* is not sufficient to induce degeneration, since expression of the protein in dividing epithelial cells caused no effect. We were also interested in what other proteins might be present in the NI, and whether they might contribute to the phenotype.

A number of proteins have been tested for presence in NIs, but were not found to be present (Paulson *et al.* 1997). However, initial evidence suggested the possibility that expanded polyglutamine protein may sequester other proteins with polyglutamine repeats. A mechanism of cellular degeneration, therefore, could be depletion of other essential proteins containing polyglutamine by recruitment into aggregates. To address this, we asked whether a normal eye developmental protein that contained a polyglutamine run, *eyes absent* (*eya*), would be sequestered into the NI. To do this, we co-labelled developing eye discs for Eya protein and for NI. These data showed that indeed Eya protein was recruited and concentrated in the NI (Perez *et al.* 1998). Previous studies have shown that the Eya protein is critical to eye cells; loss of *eya* function early in eye development leads to loss of all cells of the eye (Bonini *et al.* 1993). Hence, it is possible that reduction of *eya* activity may in part lead to neuronal dysfunction. We are now in a position to generate flies that overexpress *eya*, or have reduced *eya* levels, to address whether altered *eya* activity contributes to the degenerative phenotype.

A number of studies have also suggested that altered proteolytic pathways may occur in polyglutamine-repeat diseases (Davies *et al.* 1997; Paulson *et al.* 1997; Cummings *et al.* 1998). This is suggested by the fact that the NI are typically ubiquitinated, suggesting the protein is targeted for proteolysis. In addition, studies have found that proteosomes are localized to the NI in vertebrates, as well as chaperone proteins, which aid in protein folding (Cummings *et al.* 1998; H. Paulson, unpublished observations). Hence, to address a potential role of proteolysis and chaperone activity, we asked whether heat shock protein 70 (hsp70), the major chaperone protein in *Drosophila* induced upon stress, was induced with expression of expanded polyglutamine protein. These studies indicated that hsp70 is indeed upregulated, and in fact localizes to the NI. These studies indicate that in flies, as in vertebrate cells, expanded polyglutamine protein is recognized as abnormal, inducing a stress response. Thus, we are now in a position to address whether manipulating the levels of various chaperone proteins in flies can modify the phenotype induced by the expanded polyglutamine protein. Such studies will allow us to determine whether altered chaperone activity in a living organism can have an effect on the neurodegeneration.

Screening for genes that can prevent polyglutamine-mediated degeneration

In addition to testing candidate suppressor proteins suggested by vertebrate work, it is possible in flies to define new genes that can suppress polyglutamine-induced neurodegeneration. To do this, one can screen for genes that, when reduced in dosage, will modify the phenotype induced by the expanded polyglutamine protein. Therefore, one can perform screens for dominant modifiers of the phenotype. To test this, we first addressed whether expression of P35, a potent insect viral anti-cell death gene, would have any effect on the degenerative phenotype in flies. We found that concomitant expression of P35 did partially ameliorate the phenotype. These studies indicate that anti-cell death genes may be of some benefit to these diseases. Although the effect was mild in flies, a mild effect in flies could translate in humans into a significant effect on brain function.

We are also performing genetic screens to define new genes. To do this, we are mating male flies containing various mutations to females bearing the expanded polyglutamine protein. In the progeny, we then screen for flies that have a modified phenotype—in this case, show a suppressed phenotype of the expanded polyglutamine protein. By this method, we can define mutations that dominantly modify the phenotype. The screen relies on restoration of pigmentation to the flies; subsequent studies reveal the degree to which the photoreceptor neurons are also restored. This screen is currently ongoing in our laboratory. Initial studies indicate that the screen indeed allows identification of new genes that modify degeneration induced by expanded polyglutamine protein.

Conclusions

We have shown that the mechanisms of human polyglutamine-repeat disease are conserved in *Drosophila*. This indicates that *Drosophila* can provide a powerful model system in which to address biological mechanisms of human polyglutamine-repeat disease, allowing direct tests of specific genes for their ability to modify neural degeneration, as well as to define new genes involved in the onset and progression of neuronal dysfunction. The extent to which aspects of human polyglutamine-repeat disease are conserved in flies is remarkable, showing characteristic late onset, progressive degeneration as noted in humans, as well as abnormal protein aggregation in the form of NIs (table 6.1; figure 6.2). Although clearly not all aspects of polyglutamine-repeat disease can be conserved between humans and flies, our *Drosophila* model will contribute to understanding mechanisms of human polyglutamine-repeat disease, by allowing rapid genetics to define conserved aspects. We have already shown that expression of the disease proteins is not lethal to all cells, and that *in vivo* in *Drosophila*, other polyglutamine-repeat proteins are sequestered into the abnormal protein aggregates. Our initial studies indeed indicate that it will prove possible to identify genes that can ameliorate the effect of these disease genes: we have already shown that P35 has effects, and have also defined some new suppressor mutations. Our efforts now are directed toward using this fly model to define additional mechanisms that contribute to polyglutamine disease progression, and to define genes that can slow or prevent altogether the effect of these disease proteins.

The author receives funding for this research through grants from the Alzheimer's Association (RG2-96-005), the David and Lucile Packard Foundation, the National Institutes of Health, the John Merck Fund and the Huntington's Disease Society of America Coalition for the Cure. I thank Henry Paulson for comments on the manuscript.

References

Bonini, N. M., Leiserson, W. M. & Benzer, S. 1993 The eyes absent gene: genetic control of cell survival and differentiation in the developing *Drosophila* eye. *Cell* **72**, 379–395.

7

Polyglutamine pathogenesis, potential role of protein interactions, proteolytic processing and nuclear localization

Christopher A. Ross, Jonathan D. Wood, Matthew F. Peters, Gabriele Schilling, Frederick C. Nucifora, Jr, Jillian K. Cooper, Alan H. Sharp, Russell L. Margolis, Akira Sawa and David R. Borchelt

Introduction

The number of diseases in the class in which expanding CAG repeats, coding for polyglutamine, cause progressive neuronal degeneration is itself rapidly expanding. This class currently includes Huntington's disease (HD), dentatorubral–pallidoluysian atrophy (DRPLA), several forms of spinocerebellar ataxia (SCA) and spinal and bulbar muscular atrophy (SBMA) (Ross 1995; Paulson & Fischbeck 1996; Zoghbi 1996; Gusella *et al.* 1997). For all the diseases, there is a threshold length of repeat which causes disease. Above this threshold, longer expanded repeats lead to earlier ages of onset (Andrew *et al.* 1993, 1997; Duyao *et al.* 1993; Stine *et al.* 1993; Nance 1997).

HD is the most common of these disorders. It usually affects individuals in mid-adulthood, but can begin anywhere from childhood to old age. Its symptoms include abnormal involuntary movements, such as chorea or dystonia, incoordination of voluntary movements, emotional symptoms and cognitive impairment leading to dementia. The disorder progresses gradually, leading to death about 15 to 20 years after onset of the disease. Patients who have adult-onset DRPLA have a clinical picture very similar to that of HD (Ross *et al.* 1997). SBMA presents predominantly with muscle weakness. The spinocerebellar ataxias can present with a variety of syndromes, including ataxia, dystonia, incoordination, weakness, and other signs and symptoms.

In all of these disorders there is neuronal cell death in an overlapping set of brain regions, including the basal ganglia, cerebral cortex, brainstem nuclei and the cerebellum (Ross 1995). In HD, the corpus striatum (comprising the caudate and putamen) of the basal ganglia, is most severely affected. In DRPLA, the deep cerebellar nuclei, including the dentate nucleus, show the greatest degeneration. Clinical features of the diseases are believed to result both from death of neurons and from dysfunction of affected neurons before they die. In HD, the severity of symptoms generally correlates well with the extent of striatal neuronal cell death (Vonsattel *et al.* 1985; Myers *et al.* 1988). Death of neurons may be due to a form of cell death related to apoptosis, in which cells become condensed and die without substantial inflammation or injury to surrounding cells. Activation of intracellular proteases termed caspases is a feature of apoptotic cell death, and caspase activation has been implicated in polyglutamine-mediated cell death (Kim *et al.* 1999; Ona *et al.* 1999; Sanchez *et al.* 1999). All of the polyglutamine diseases appear to involve a genetic 'gain of function' mechanism at the protein level. The protein product of the HD gene is termed huntingtin, and the product of the DRPLA gene is termed atrophin-1.

In all of the glutamine repeat diseases studied so far, inclusion bodies containing the protein product of the disease gene form in the nucleus of neurons (Davies *et al.* 1997; DiFiglia *et al.* 1997; Paulson *et al.* 1997; Ross 1997; Skinner *et al.* 1997; Becher *et al.* 1998). In addition, in some of the diseases there appears to be protein aggregation within neurons outside of the nucleus, in structures referred to as dystrophic neurites (DiFiglia *et al.* 1997). Thus, these diseases bear a resemblance to other neurodegenerative diseases in which there are inclusion bodies or other proteinaceous deposits, including Alzheimer's disease, characterized by amyloid plaques and neurofibrillary tangles, and Parkinson's disease, characterized by Lewy bodies (Lansbury 1997). However, the regions of the brain affected in these other diseases are quite different.

In this chapter, we will focus on three aspects of the pathogenesis of polyglutamine disorders: protein interactions (including aggregation), nuclear localization and proteolytic processing. We will emphasize our own studies of HD and DRPLA, but will also discuss other studies, including those of the spinocerebellar ataxias and spinal and bulbar muscular atrophy. For all of these disorders, cell and animal models are becoming increasingly important for elucidating pathogenic mechanisms.

Aggregation and inclusion bodies

In vitro *studies*

Scherzinger *et al.* (1997) were the first to demonstrate the inherent ability of polyglutamine-containing protein to aggregate. The N-terminus of huntingtin was synthesized as a fusion protein in bacteria and then released from the bacterial fusion protein partner by proteolytic cleavage to yield an N-terminal fragment (corresponding to exon 1) of *ca.* 90 amino acids. When this N-terminal fragment had a normal glutamine repeat it remained soluble. However, when the glutamine repeat length was in the range which causes

HD, the protein formed aggregates in solution which could be detected by collection on a filter membrane. Electron micrographs of these aggregates revealed a fibrillar or ribbon-like morphology. When the aggregates were stained with Congo red dye, they exhibited green birefringence characteristic of a β-pleated sheet conformation. These characteristics are typical of amyloid fibrils in Alzheimer's disease and prion diseases. Subsequent quantitative study indicated that the threshold for aggregation *in vitro* is remarkably similar to the threshold for causing disease (Scherzinger *et al.* 1999). In addition, the kinetics were similar to those predicted by a model involving seeded polymerization (Lansbury 1997). *In vivo* protein aggregates appear to contain molecules other than that harbouring the expanded polyglutamine repeat. For example, aggregates of pathological expanded polyglutamine-containing protein recruit normal length polyglutamine-containing proteins, such as normal huntingtin and TATA-binding protein (Huang *et al.* 1998; Perez *et al.* 1998).

Remarkably, formation of a β-pleated sheet had been predicted previously by Max Perutz, who hypothesized that glutamine repeats could form a so-called 'polar zipper' and result in non-covalent associations due to hydrogen bonds (Perutz 1994; Stott *et al.* 1995). This polar zipper was predicted to result when the glutamine repeat length exceeded a certain threshold. It was predicted to form either between two different molecules with glutamine repeats or within one molecule, forming a hairpin structure.

Neuropathological studies

Protein aggregation in the form of intranuclear inclusions has been found in every polyglutamine disease examined to date. The first report was in HD (DiFiglia *et al.* 1997). Huntingtin aggregation was found in intranuclear inclusions and dystrophic neurites in HD cortex and striatum. The density of the inclusions correlated with the length of the glutamine repeat (Becher *et al.* 1998). Intranuclear inclusions have also been found in DRPLA (Becher *et al.* 1998; Igarashi *et al.* 1998), SCA3 (Paulson *et al.* 1997), SCA1 (Skinner *et al.* 1997) and other polyglutamine diseases (Holmberg *et al.* 1998; M. Li *et al.* 1998). One study (Karpuj *et al.* 1999) has searched for evidence of amyloid fibrils (using techniques which can detect amyloid in Alzheimer's disease) and been unable to detect amyloid in post-mortem HD brain tissue. However, it is unknown whether these techniques can detect all forms of β-pleated sheets found in amyloid-like materials.

Animal and cell models

The first animal model of HD (Mangiarini *et al.* 1996) was made with a short N-terminal fragment of huntingtin comprising exon 1 with a very long expanded repeat. Intranuclear inclusions were seen densely in almost all regions of the brain (Davies *et al.* 1997). We have also seen intranuclear inclusions in a mouse model of HD with a slightly longer N-terminal truncation of huntingtin (the first 171 amino acids or N171) and a glutamine repeat length (82Q) within the range commonly seen in juvenile HD patients (Schilling *et al.* 1999*a*). In both models, aggregates are also seen in neurites in several brain regions. Intranuclear inclusions have been seen in other animal models of several other glutamine repeat disorders (Burright *et al.* 1995; Skinner *et al.* 1997; Jackson *et al.* 1998; Reddy *et al.* 1998; Warrick *et al.* 1998; Faber *et al.* 1999). Polyglutamine has also been inserted ectopically into an unrelated protein (HPRT), resulting in both a behavioural phenotype and the formation of intranuclear inclusions (Ordway *et al.* 1997).

In addition, polyglutamine aggregation has been seen in numerous cell models using a variety of different constructs (Ikeda *et al.* 1996; Paulson *et al.* 1997; Cooper *et al.* 1998; Igarashi *et al.* 1998; Li and Li 1998; Lunkes and Mandel 1998; Martindale *et al.* 1998; Merry *et al.* 1998; Miyashita *et al.* 1998; Hackam *et al.* 1999; Moulder *et al.* 1999). One recent study indicates that the extent of aggregation is correlated with the degree of toxicity, using an indirect measure of toxicity (Hackam *et al.* 1999).

Microaggregates and soluble protein

Two recent studies have raised the possibility that aggregation (or at least the large aggregates which are seen as inclusions bodies) may not be the primary factor leading to cell toxicity. In one study, an animal model of SCA1 was generated using a construct similar to that used for the previous model (Burright *et al.* 1995; Klement *et al.* 1998). However, in this new model a deletion was made within a region of ataxin-1 which had been found to be a self-association domain (Burright *et al.* 1997; Klement *et al.* 1998). These mice developed ataxia and Purkinje cell pathology similar to the original SCA1 mice, at least in the early stages of the illness. However, no ataxin-1 aggregates were found (Klement *et al.* 1998). A second study involved a cell model using transient transfection of primary neurons in culture (Saudou *et al.* 1998). In this model, intranuclear aggregates were formed. However, the presence of huntingtin aggregates did not correlate with cell toxicity. Under some circumstances, there was an increase in aggregation but a decrease in toxicity. Thus, both of these experiments raise the possibility that polyglutamine protein aggregation, at least in large inclusions, may not be necessary for cellular toxicity.

Protein interactions

In addition to the self-interaction which leads to aggregation, polyglutamine-containing proteins can interact with a variety of other proteins within the cell. Some of these protein interaction partners provide information regarding the normal function of polyglutamine-containing proteins. In addition, there is indirect evidence for a role for some of the interactions in polyglutamine pathogenesis, though there is no direct evidence to date that any of the interactions so far described are necessary for pathogenesis.

Interactions of huntingtin

The first protein to be described which can interact with huntingtin was huntingtin-associated protein 1 (HAP1) (X.-J. Li *et al.* 1995). The interaction was modestly stronger when huntingtin had an expanded polyglutamine repeat, suggesting a possible role for HAP1 in the pathogenesis of HD. However, subsequent studies have been unable to produce clear evidence of a role for HAP1 in HD pathogenesis. The distribution of HAP1 expression does not correspond to the distribution of HD pathology (X.-J. Li *et al.* 1996; S.-H. Li *et al.* 1998*a*; Page *et al.* 1998; Martin *et al.* 1999; Sharp *et al.*, 2000). In cell models, there is some evidence for the presence for HAP1 in aggregates containing huntingtin (S.-H. Li *et al.* 1998*a*). However, there has not been evidence to date for the presence of HAP1 in aggregates or altered distribution of HAP1 in mouse models or human pathological material. Furthermore, the exon 1 fragment of huntingtin used in the first mouse model of HD (Mangiarini *et al.* 1996) does not appear to interact with HAP1 (the region of huntingtin which interacts with HAP1 appears to be between amino acids 171 and 223), suggesting that HAP1 is not involved in the behavioural and pathological features of this model (Bertaux *et al.* 1998). Similarly, the fragment of huntingtin used in the mouse model of Schilling *et al.* (1999*a*) does not appear to interact with HAP1, suggesting that HAP1 is not involved in the pathology in this model either, and this model is thought to show some striatal cell death (Schilling *et al.* 1999*a*). The most direct way to test the possibility that HAP1 (or any other interactor) is involved in HD pathogenesis is to make a knock-out mouse and cross it with the available HD mouse models.

Even if HAP1 is not involved in the pathogenesis of HD, it may help to shed light on the normal functions of huntingtin. HAP1 itself has no striking homology to other proteins in the databases and has no known function. However, several interactors of HAP1 have been described, including a protein termed duo or kalirin (Colomer *et al.* 1997). This protein is believed to be able to activate the rac1 signalling pathway and to be involved in regulation of cytoskeletal function and vesiclular transport. Another protein with which HAP1 interacts is the p150^{glued} component of the dynactin complex (Engelender *et al.* 1997; S.-H. Li *et al.* 1998*b*). Thus huntingtin and HAP1 may be involved in regulation of cellular motor transport proteins. Both huntingtin and HAP1 can undergo axonal transport in neurons (Block-Galarza *et al.* 1997), consistent with this possibility.

Another protein with which huntingtin can interact is huntingtin-interacting protein 1 (HIP1). In this case, the strength of the interaction decreases when huntingtin has an expanded polyglutamine repeat (Kalchman *et al.* 1997;

Wanker *et al.* 1997). Overexpression of HIP1 recently has be shown to be toxic to cells (A. S. Hackam and M. R. Hayden, unpublished results), raising the possibility that expansion of the polyglutamine repeat might decrease binding of HIP1, releasing it into the cell and contributing to cell death in HD. How this would fit with HD involving a genetic gain of function is unclear. However, recent data suggest that huntingtin may normally have an anti-apoptotic function (D. Rigamonti, E. Cattaneo *et al.*, submitted), suggesting that partial loss of function could also contribute to disease progression. HIP1 also appears to be involved in cytoskeletal and membrane interactions, consistent with the role of huntingtin in regulation of cytoskeletal and membrane functions postulated above.

Another huntingtin interactor is glyceraldehyde-3-phosphate dehydrogenase (GAPDH). This interaction originally was proposed to be modulated by the length of the polyglutamine tract, though this may not now be the case (Burke *et al.* 1996; Koshy *et al.* 1996). The interaction appears to be mediated directly by the polyglutamine tract, potentially making it relevant to all of the polyglutamine diseases. GAPDH has itself been implicated in cellular death processes, possibly via nuclear translocation (Sawa *et al.* 1997; Ishitani *et al.* 1998; Shashidharan *et al.* 1999; Saunders *et al.* 1999).

Huntingtin can interact with an SH3 domain-containing protein termed SH3GL3 (Sittler *et al.* 1998). This interaction is mediated via one of the proline-rich regions in huntingtin and is of potential interest since it is stronger with an expanded polyglutamine tract. SH3GL3 has been detected in huntingtin aggregates, suggesting that it may be involved in the pathogenic process in some way. The proline-rich region of huntingtin can also interact with a number of proteins containing WW domains (Faber *et al.* 1998). Interestingly, a number of these proteins are localized to the nucleus. Huntingtin can also interact with the E2-25k ubiquitin-conjugating enzyme, also known as HIP2 (Kalchman *et al.* 1996), and with cystathione β-synthetase (Boutell *et al.* 1998). Any significance of these interactions to pathogenesis is unclear.

Huntingtin can also interact directly with a component of nuclear co-repressor complexes termed N-CoR (Boutell *et al.* 1999). This interaction is of potential interest since it is strengthened by expansion of the polyglutamine tract. In addition, it raises the question of a role for huntingtin in the regulation of gene transcription, which may be of interest in view of the possible involvement of nuclear processes in huntingtin pathogenesis (see below).

Atrophin-1 interactions

Like huntingtin, atrophin-1 can interact with a number of proteins containing WW domains (Wood *et al.* 1998), though the WW domain-containing proteins with which atrophin-1 and huntingtin interact belong to different subfamilies of WW domain proteins. Some of the atrophin-1-interacting proteins may be involved in interactions with the cytoskeleton or membranes, suggesting a role for atrophin-1 in such processes. The finding that atrophin-1 can interact with the SH3 domain of an insulin receptor tyrosine kinase substrate termed IRSp53 (Okamura-Oho *et al.* 1999) is consistent with this idea.

Atrophin-1 interacts with a protein called MTG8 or ETO (Wood *et al.*, 2000). This protein can in turn interact with N-CoR in nuclear co-repressor complexes and is believed to be involved in transcriptional repression. The role of this interaction in pathogenesis of DRPLA is unclear since the interaction does not appear to require the polyglutamine tract or to be modulated by changes in length of the polyglutamine tract. Atrophin-1 and ETO/MTG8 redistribute and co-localize with histone deacetylases (HDACs) and mSin3, two components of nuclear co-repressor complexes, in cell transfection experiments. Atrophin-1 can repress transcription in cell transfection-based assays, suggesting that it may regulate the activity of nuclear co-repressor complexes in some way.

Ataxin-1 interactions

Ataxin-1 can interact in a repeat length-dependent fashion with a nuclear protein called leucine-rich acidic nuclear protein (LANP) (Matilla *et al.* 1997). This interaction may be of pathogenic relevance since the distribution of expression of LANP appears to match the pattern of pathology in SCA1, with the highest levels of expression in Purkinje cells of the cerebellum and lower expression in other brainstem regions. However, no direct evidence has yet implicated LANP in SCA1 pathogenesis. Ataxin-1 also interacts with GAPDH. This interaction is not modulated by the length of the polyglutamine tract (Koshy *et al.* 1996).

Androgen receptor interactions

The androgen receptor is the only polyglutamine-containing protein with a known function. It is involved in androgen-dependent regulation of gene transcription via nuclear co-activator and nuclear co-repressor complexes. The DNA-binding and ligand-binding domains of the androgen receptor are well defined (for a review, see MacLean *et al.* 1997). The N-terminal region of the protein, which contains the polyglutamine tract, is a transactivation domain and interacts with other nuclear proteins, including receptor co-activators. In a transient transfection cell culture system, polyglutamine-expanded androgen receptors form aggregates that sequester molecular chaperones, proteasome components, the steroid receptor co-activator SRC-1 and the ubiquitin-like molecule NEDD8 (Stenoien *et al.* 1999).

Nuclear localization

Several studies suggest that the localization of the protein may be more important than whether it forms macroaggregates or not. In the ataxin-1 study of Klement *et al.* (1998), a construct in which the nuclear localization signal was mutated was also used to generate transgenic mice. These mice did not develop disease within the same time period as controls, suggesting that nuclear localization of the protein is critical for cell toxicity. Similar results were obtained in the study of Saudou *et al.* (1998). Experiments in this study involved the use of an exogenous nuclear export signal on their huntingtin construct. Again, when the protein did not localize to the nucleus, there was dramatically less cellular toxicity.

These latter experiments may be complicated by the possibility that huntingtin may have endogenous nuclear localization and nuclear export signals within its N-terminal portion. Our group has performed additional experiments with a shorter fragment of huntingtin (N63) which does not have any putative localization signals (Peters *et al.* 1999). Addition of an exogenous nuclear localization signal led to greater toxicity, while addition of an exogenous nuclear export signal led to less toxicity. By contrast, one cell model study (Hackam *et al.* 1999) suggested that nuclear localization does not increase toxicity, though toxicity was measured as a response to an exogenous stimulus, rather than directly. Animal model studies have not yet addressed in detail the issue of nuclear versus cytoplasmic localization of soluble polyglutamine proteins. Our own experience is that in animal models of both HD and DRPLA (see below), there is prominent diffuse-appearing nuclear label for the relevant protein (in addition to the intranuclear inclusions). Thus, while more work needs to be done, these results, taken together, suggest the possibility that nuclear localization is an important component of polyglutamine pathogenesis.

These studies raise the issue of the normal localization of polyglutamine-containing proteins. Several polyglutamine-containing proteins are normally present, at least in part, in the nucleus, including the androgen receptor, atrophin-1, ataxin-1, ataxin-3 and ataxin-7 (Brooks and Fischbeck 1995; Servadio *et al.* 1995; Brooks *et al.* 1997; Miyashita *et al.* 1998; Tait *et al.* 1998; Trottier *et al.* 1998; Kaytor *et al.* 1999). The HPRT gene product used for the mouse model with ectopic polyglutamine expansion is small enough to diffuse into the nucleus without active transport (Ordway *et al.* 1997).

The normal cellular localization of huntingtin is controversial. Most studies have shown cytoplasmic localization (Aronin *et al.* 1995; DiFiglia *et al.* 1995; Sharp *et al.* 1995; Ferrante *et al.* 1997; Sapp *et al.* 1997; Wilkinson *et al.* 1999). Interestingly, one recent study has suggested that the cellular expression profile of huntingtin in striatal and cortical neurons does not correlate with neuronal vulnerability in HD (Fusco *et al.* 1999). Other studies have suggested that huntingtin may have a nuclear localization, at least in some cells (Hoogeveen *et al.* 1993; de Rooij *et al.* 1996). In addition, the presence of putative nuclear export and nuclear localization signals raises the possibility that huntingtin normally may shuttle through the nucleus, perhaps in a regulated fashion.

What nuclear structures could the polyglutamine proteins be associated with? Early evidence suggests the possibility of association with structures in the nuclear matrix. The ataxin-1-associated protein LANP (Matilla *et al.* 1997) is present in the nuclear matrix. In addition, co-transfection of cells with ataxin-1 alters the distribution of the nuclear matrix-associated PML protein (Matilla *et al.* 1997; Skinner *et al.* 1997). The interaction of atrophin-1 with the ETO/MTG8 protein also takes place within the nuclear matrix. Co-transfection of atrophin-1 with ETO/MTG8 yields matrix localization for both proteins, whereas transfection of atrophin-1 by itself yields mainly soluble nuclear protein (Wood *et al.*, 2000). The localization of the huntingtin interaction with N-CoR has not been clearly documented but could also take place in the nuclear matrix. Ataxin-3 has also been reported to associate with the nuclear matrix (Tait *et al.* 1998).

Truncation of polyglutamine proteins

Numerous cell model studies have indicated that truncated portions of polyglutamine proteins are more toxic than the full-length protein. In several different cell transfection paradigms, truncated N-terminal portions of huntingtin protein form more aggregates than longer or full-length huntingtin polypeptides in cells (Cooper *et al.* 1998; Hackam *et al.* 1998; Li and Li 1998; Martindale *et al.* 1998; Saudou *et al.* 1998). Furthermore, small fragments of huntingtin are more likely to aggregate in the nucleus and appear to be more likely to cause cell toxicity (Cooper *et al.* 1998; Hackam *et al.* 1998; Martindale *et al.* 1998). Similarly, short portions of atrophin-1, ataxin-3 and the androgen receptor, containing the polyglutamine tract, are more toxic to cells than full-length protein (Ikeda *et al.* 1996; Igarashi *et al.* 1998; Merry *et al.* 1998).

It is too early to be sure about the situation with mouse models, but initial studies suggest that truncated fragments of huntingtin may be more toxic to mice. The very short N-terminal fragment used in the study of Mangiarini *et al.* (1996) produced a dramatic behavioural phenotype, though substantial neuronal cell death has not yet been documented in this model. Progressive motor deficits in these mice have now been fully characterized (Carter *et al.* 1999), although phenotypic analysis of these mice is complicated by the demonstration that they are diabetic (Hurlbert *et al.* 1999). Expression of full-length huntingtin using different methods led to relatively delayed expression of a behavioural phenotype, and neuronal toxicity (Reddy *et al.* 1998; Hodgson *et al.* 1999). By contrast, expression of very low levels of an N-terminal fragment (N171) caused a relatively early and dramatic behavioural phenotype and cell toxicity (Schilling *et al.* 1999*a*). However, because the promoters, mouse strains and repeat lengths are not exactly comparable in these model systems, the issue of the role of truncation remains to be resolved.

The neuropathological data for huntingtin, however, do favour the idea that truncation of huntingtin protein is involved in the disease process. Post-mortem study of HD brains revealed intranuclear inclusions only by using antibodies directed at the N-terminus or epitopes close to the N-terminus.

Epitopes several hundred amino acids away showed no reactivity (DiFiglia *et al.* 1997; Becher *et al.* 1998).

The neuropathological studies do not reveal anything about the mechanism of truncation of the protein. *In vitro* studies have indicated that several of the polyglutamine proteins can be cleaved by the pro-apoptotic cysteine proteases termed caspases. The initial report for huntingtin suggested that this cleavage might be repeat length dependent (Goldberg *et al.* 1996). However, subsequent experiments indicate that the full-length protein with an expanded repeat or a normal repeat are equally well cleaved. Caspase cleavage sites have been defined for huntingtin, atrophin-1 and the androgen receptor, though cleavage of huntingtin and atrophin-1 was much more efficient (Wellington *et al.* 1998). Huntingtin is cleaved by caspase 3 most efficiently at amino acid 513 (Wellington *et al.* 1998). However, other sites are present within a cluster C-terminal to this site, including a caspase 6 site. Atrophin-1 is cleaved by caspase 3 at position 109 (Miyashita *et al.* 1997; Wellington *et al.* 1998).

The relationship of these *in vitro* data to *in vivo* events is still uncertain. Identification of the mechanism of truncation *in vivo* has proved difficult. Cleavage of huntingtin was not described in the full-length cDNA transgenic mouse model of Reddy *et al.* (1998). Truncation has not been described in the SCA1 model which utilizes full-length ataxin-1 protein (Burright *et al.* 1995).

A recent study has suggested that caspase activation may be involved in the pathogenesis of the exon 1 transgenic mouse model (Mangiarini *et al.* 1996). Two experiments were done in this study (Ona *et al.* 1999). In the first experiment, the exon 1 transgenic mouse was crossed with a mouse overexpressing a dominant-negative caspase 1 mutant. This led to a delay in the appearance, and a slowing of the progression, of both the behavioural and pathologic aspects of the phenotype. In the second experiment, a caspase 1 inhibitor was injected intraventricularly into the exon 1 transgenic mice, again leading to amelioration of the phenotype. In both experiments, the magnitude of the effect was *ca.* 20–30%. The mechanism of the effect is still rather unclear. This mouse model does not have prominent neuronal cell death, so it is not likely to be due simply to inhibition of caspases involved in neuronal cell death. One possibility was suggested by the observation that endogenous huntingtin appeared to be preserved in a full-length form rather than truncated. This suggests that cleavage of endogenous huntingtin might somehow be involved in pathogenesis, conceivably through loss of a neuroprotective function. However, this remains speculative.

Our studies with mice transgenic for full-length atrophin-1 indicate that there is repeat length-dependent accumulation of truncated atrophin-1 products in the nuclei of neurons (Schilling *et al.* 1999*b*). The mice express atrophin-1 with 65 glutamines under the control of the mouse prion promoter. These animals develop a dramatic progressive behavioural phenotype with incoordination, seizures, involuntary movements and weight loss, progressing to early death. Post-mortem examination reveals both intranuclear inclusions and diffuse nuclear label. Mice expressing atrophin-1 with a normal length repeat have no phenotype.

Western blot analysis of nuclear fractions from these mice indicated that truncated atrophin-1 products accumulate preferentially only in neuronal nuclei of mice expressing atrophin-1 with the expanded repeat. Full-length atrophin-1 is 1184 amino acids long and has an apparent molecular mass of *ca.* 190 kDa. In the cytoplasm, this is the predominant form of atrophin-1 present. By contrast, in the nucleus of the mutant animal, there is very little of the full-length protein, but there is a doublet of bands at around 120 kDa. The levels of these lower molecular mass products increase with age. These data suggest that truncated products of atrophin-1 accumulate in the nucleus in a repeat length-dependent fashion, though the mechanism is still uncertain.

The truncated atrophin-1 fragments are reactive with the 1C2 antibody (Trottier *et al.* 1995), indicating that the fragments contain the polyglutamine tract. In addition, they react with antibodies to the N-terminus of atrophin-1, indicating that the N-terminus of the protein is intact. Therefore, the fragments cannot arise from the known caspase 3 cleavage site at amino acid 109. However, it is not known whether they arise from caspase cleavage at some other site distal to the polyglutamine repeat, via processive degradation from the C-terminus, or from endoproteolytic cleavage by some other enzyme or enzymes.

Other issues in pathogenesis

The preceding discussion has emphasized processes occurring within the nucleus. However, other effects may be relevant. The cytoplasmic aggregates observed in both mouse models and patients may suggest that non-nuclear processes could contribute to neuronal dysfunction and perhaps neuronal toxicity (DiFiglia *et al.* 1997; Gutekunst *et al.* 1999; H. Li *et al.* 1999). In fact, in post-mortem brain tissue from most adult-onset cases there appear to be few nuclear inclusions, and more cytoplasmic inclusions and dystrophic neurites (DiFiglia *et al.* 1997). Long before the identification of the intranuclear inclusions, neurons from HD patients had been shown to have altered morphology using Golgi techniques (Graveland *et al.*, 1985). These included not just the expected atrophic changes, but also changes suggestive of abnormal growth and plasticity of dendrites. How these fit in with current models has yet to be determined.

It has been proposed that huntingtin toxicity may relate to cellular phosphorylation pathways (Liu *et al.* 1997; Liu 1998), but this remains to be demonstrated in animal models or human pathological material.

Recent data suggest that polyglutamine can activate caspase 8 (perhaps via dimerization), triggering cell death pathways (Sanchez *et al.* 1999). A number of lines of evidence indicated that caspase 8 was critically involved. Transfection of the Q78 fragment of ataxin-3 (Ikeda *et al.* 1996) led to activation of caspase 8. Aggregates containing polyglutamine also contained caspase 8. Cell death could be blocked by caspase inhibitors or by transfecting cells deficient in caspase 8 but not other caspases. The authors suggested the possibility that dimerization of huntingtin might cause dimerization of death domain-containing caspase activator proteins such as FADD. This would provide an attractive mechanism which would directly link the formation of microaggregates to activation of cell death enzymes, though it is unknown whether the effects are direct or indirect. In this model, cells die via a pathway which has some similarities to apoptosis, but also some differences. For instance, synthesis of new proteins would not be necessary.

A recent study has taken advantage of the availability of lymphoplastoid cell lines from HD patients (Sawa *et al.* 1999). Cells from juvenile-onset HD patients are more susceptible to inhibition of mitochondria and redox enzymes, and cell death initiators such as staurosporine. In this system, mitochondria become depolarized and the cells die via an apoptotic pathway. These data raise the possibility that polyglutamine may be directly or indirectly toxic to mitochondria. Data suggesting that there are biochemical abnormalities and evidence for excitotoxicity in HD continue to accumulate (Tabrizi *et al.* 1999). Somewhat paradoxically, huntingtin exon 1 transgenic mice (Mangiarini *et al.* 1996) appear to be resistant to quinolinic acid-induced striatal excitotoxicity (Hansson *et al.* 1999).

Conclusions

The field of polyglutamine pathogenesis is making rapid progress. A number of mechanisms have been identified, but how they all fit together is uncertain. The *in vitro* data and the repeat length dependence of aggregation are highly suggestive that aggregation is involved in some way. The visible aggregates may be a marker for an initial conformational change. Microaggregates may be more relevant to toxicity than macroaggregates or inclusion bodies.

Proteolytic processing may be involved, though the mechanisms are unknown. They may involve cleavage of polyglutamine-containing proteins by caspases or other proteases. Nuclear mechanisms are now being implicated. Whether these involve interaction with transcription factors or other molecules such as splicing factors is unknown. Polyglutamine microaggregates may also be able to activate caspases, perhaps via intermediary death domain-containing proteins. It is provocative that both huntingtin and atrophin-1 can interact with proteins in nuclear co-repressor complexes (N-CoR and ETO/MTG8, respectively). The androgen receptor, which undergoes polyglutamine expansion in SBMA, is itself a hormone receptor which can functionally interact with both nuclear co-activator and co-repressor complexes. The mice made in the Bates' laboratory have dramatic alterations in levels of a number of receptor proteins, suggestive of alterations in gene transcription, though the mechanism is still unknown (Cha *et al.* 1998). PC12

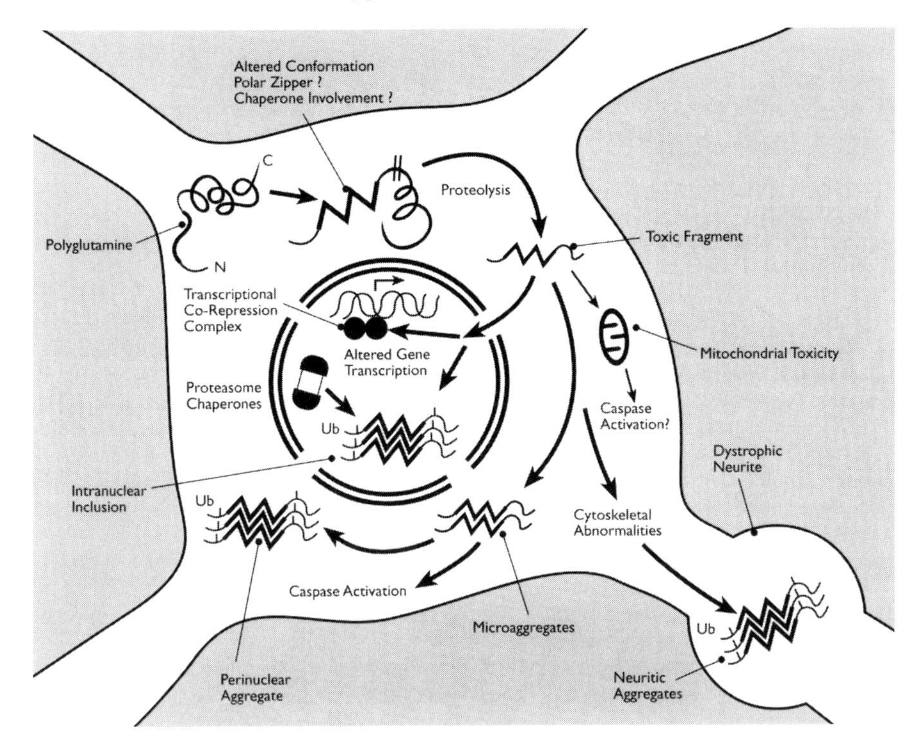

Figure 7.1 Model for polyglutamine pathogenesis.

cells stably expressing mutant huntingtin also show altered gene expression (S.-H. Li *et al.* 1999), although this could represent an adaptive response to the selection procedure during the generation of the cell lines. The relationship between ataxin-1 and PML-oncogenic domains (PODs), which may be related to transcription and/or cell death, is provocative and needs to be followed up.

A model of some of the mechanisms discussed in the text is shown in figure 7.1. No doubt the model will need to be modified as further data are generated. A better understanding of pathogenic mechanisms will be critical for the development of rational therapeutics. The *in vitro*, cellular and animal models described above will be critical for screening and then for further testing of possible therapeutic compounds which can then be tried in HD patients.

We thank the Hereditary Disease Foundation 'Cure HD Initiative', the Huntington's Disease Society of America 'Coalition for the Cure' and the NINDS P01 NS16375, R01 NS 34172 and R01 38144 for support.

References

Andrew, S. E. (and 11 others) 1993 The relationship between trinucleotide (CAG) repeat length and clinical features of Huntington's disease. *Nature Genet.* **4**, 398–403.

Andrew, S. E., Goldbery, Y. P. & Hayden, M. R. 1997 Rethinking genotype and phenotype correlations in polyglutamine expansion disorders. *Hum. Mol. Genet.* **6**, 2005–2010.

Aronin, N. (and 16 others) 1995 CAG expansion affects the expression of mutant huntingtin in the Huntington's disease brain. *Neuron* **15**, 1193–1201.

Becher, M. W., Kotzuk, J. A., Sharp, A. H., Davies, S. W., Bates, G. P., Price, D. L. & Ross, C. A. 1998 Intranuclear neuronal inclusions in Huntington's disease and dentatorubral and pallidoluysian atrophy: correlation between the density of inclusions and *IT15* triplet repeat length. *Neurobiol. Dis.* **4**, 387–397.

Bertaux F., Sharp, A. H., Ross, C. A., Lehrach, H., Bates, G. P. & Wanker, E. 1998 HAP1–huntingtin interactions do not contribute to the molecular pathology in Huntington's disease transgenic mice. *FEBS Lett.* **426**, 229–232.

Block-Galarza, J., Chase, K. O., Sapp, E., Vaughn, K. T., Vallee, R. B., DiFiglia, M. & Aronin, N. 1997 Fast transport and retrograde movement of huntingtin and HAP1 in axons. *NeuroReport* **8**, 2247–2251.

Boutell, J. M., Wood, J. D., Harper, P. S. & Jones, A, L. 1998 Huntingtin interacts with cystathionine beta-synthase. *Hum. Mol. Genet.* **7**, 371–378.

Boutell, J. M., Thomas, P., Neal, J. W., Weston, V. J., Duce, J., Harper, P. S. & Jones, A. L. 1999 Aberrant interactions of transcriptional repressor proteins with the Huntington's disease gene product, huntingtin. *Hum. Mol. Genet.* **8**, 1647–1655.

Brooks, B. P. & Fischbeck, K. H. 1995 Spinal and bulbar muscular atrophy: a trinucleotide-repeat expansion nuerodegenerative disease. *Trends Neurosci.* **18**, 459–461.

Brooks, B. P., Paulson, H. L., Merry, D. E., Salazar-Grueso, E. F., Brinkmann, A. O., Wilson, E. M. & Fischbeck, K. H. 1997 Characterization of an expanded glutamine repeat androgen receptor in a neuronal cell culture system. *Neurobiol. Dis.* **4**, 313–323.

Burke, J. R., Enghild, J. J., Martin, M. E., Jou, Y.-S., Myers, R. M., Roses, A. D., Vance, J. M. & Strittmatter, W. J. 1996 Huntingtin and DRPLA proteins selectively interact with the enzyme GAPDH. *Nature Med.* **2**, 347–350.

Burright, E. N., Clark, H. B., Servadio, A., Matilla, T., Feddersen, R. M., Yunis, W. S., Duvick, L. A., Zoghbi, H. Y. & Orr, H. T. 1995 SCA1 transgenic mice: a model for neurodegeneration caused by an expanded CAG trinucleotide repeat. *Cell* **82**, 937–948.

Burright, E. N., Davidson, J. D., Duvick, L. A., Koshy, B., Zoghbi, H. Y. & Orr, H. T. 1997 Identification of a self-association region within the SCA1 gene product, ataxin-1. *Hum. Mol. Genet.* **6**, 513–518.

Carter, R. J., Lione, L. A., Humby, T., Mangiarini, L., Mahal, A., Bates, G. P., Dunnett, S. B. & Morton, A. J. 1999 Characterization of progressive motor deficits in mice transgenic for the human Huntington's disease mutation. *J. Neurosci.* **19**, 3248–3257.

Cha, J.-H. J., Kosinski, C. M., Kerner, J. A., Alsdorf, S. A., Mangiarini, L., Davies, S. W., Penney, J. B., Bates, G. P. & Young, A. B. 1998 Altered brain neurotransmitter receptors in transgenic mice expressing a portion of an abnormal human Huntington's disease gene. *Proc. Natl Acad. Sci. USA* **95**, 6480–6485.

Colomer, V., Engelender, S., Sharp, A. H., Duan, K., Cooper, J. K., Lanahan, A., Lyford, G., Worley, P. & Ross, C. A. 1997 Huntingtin-associated protein 1 (HAP1) binds to a tri-like polypeptide, with a rac1 guanine nucleotide exchange factor domain. *Hum. Mol. Genet.* **6**, 1519–1525.

Cooper, J. K. (and 12 others) 1998 Truncated N-terminal fragments of huntingtin with expanded glutamine repeats form nuclear and cytoplasmic aggregates in cell culture. *Hum. Mol. Genet.* **7**, 783–790.
Davies, S. W., Turmaine, M., Cozens, B. A., DiFiglia, M., Sharp, A. H., Ross, C. A., Scherzinger, E., Wanker, E. E., Mangiarini, L. & Bates, G. P. 1997 Formation of neuronal intranuclear inclusions underlies the neurological dysfunction in mice transgenic for the HD mutation. *Cell* **90**, 537–548.
de Rooij, K. E., Dorsman, J. C., Smoor, M. A., Den Dunnen, J. T. & Van Ommen, G. J. 1996 Subcellular localization of the Huntington's disease gene product in cell lines by immunofluorescence and biochemical subcellular fractionation. *Hum. Mol. Genet.* **5**, 1093–1099.
DiFiglia, M. (and 11 others) 1995 Huntingtin is a cytoplasmic protein associated with vesicles in human and rat brain neurons. *Neuron* **14**, 1075–1081.
DiFiglia, M., Sapp, E., Chase, K. O., Davies, S. W., Bates, G. P., Vonsattel, J. P. & Aronin, N. 1997 Aggregation of huntingtin in neuronal intranuclear inclusions and dystrophic neurites in brain. *Science* **277**, 1990–1993.
Duyao, M. (and 41 others) 1993 Trinucleotide repeat length instability and age of onset in Huntington's disease. *Nature Genet.* **4**, 387–392.
Engelender, S., Sharp, A. H., Colomer, V., Tokito, M. K., Lanahan, A., Worley, P., Holzbaur, E. L. F. & Ross, C. A. 1997 Huntingtin associated protein 1 (HAP1) interacts with dynactin $p150^{Glued}$ and other cytoskeletal related proteins. *Hum. Mol. Genet.* **6**, 2205–2212.
Faber, P. W., Barnes, G. T., Srinidhi, J., Chen, J., Gusella, J. F. & MacDonald, M. E. 1998 Huntingtin interacts with a family of WW proteins. *Hum. Mol. Genet.* **7**, 1463–1474.
Faber, P. W., Alter, J. R., MacDonald, M. E. & Hart, C. 1999 Polyglutamine-mediated dysfunction and apoptotic death of a *Caenorhabditis elegans* sensory neuron. *Proc. Natl Acad. Sci. USA* **96**, 179–184.
Ferrante, R. J., Gutekunst, C.-A., Persichetti, F., McNeil, S. M., Kowall, N. W., Gusella, J. F., MacDonald, M. E., Beal, M. F. & Hersch, S. M. 1997 Heterogeneous topographic and cellular distribution of huntingtin expression in the normal human neostriatum. *J. Neurosci.* **17**, 3052–3063.
Fusco, F. R., Chen, Q., Lamoreaux, W. J., Figueredo-Cardenas, G., Jiao, Y., Coffman, J. A., Surmeier, D. J., Honig, M. G., Carlock, L. R. and Reiner, A. 1999 Cellular localization of huntingtin in striatal and cortical neurons in rats: lack of correlation with neuronal vulnerability in Huntington's disease. *J. Neurosci.* **19**, 1189–1202.
Goldberg, Y. P. (and 10 others) 1996 Cleavage of huntingtin by apopain, a proapoptotic cysteine protease, is modulated by the polyglutamine tract. *Nature Genet.* **13**, 442–449.
Graveland, G. A., Williams, R. S. & DiFiglia, M. 1985 Evidence for degenerative and regenerative changes in neostriatal spiny neurons in Huntington's disease. *Science* **227**, 770–773.
Gusella, J. F., Persichetti, F. & MacDonald, M. E. 1997 The genetic defect causing Huntington's disease: repeated in other contexts? *Mol. Med.* **3**, 238–246.
Gutekunst, C. A., Li, S.-H., Yi, H., Mulroy, J. S., Kuemmerle, S., Jones, R., Rye, D., Ferrante, R. J., Hersch, S. M. & Li, X.-J. 1999 Nuclear and neuropil aggregates in Huntington's disease: relationship to neuropathology. *J. Neurosci.* **19**, 2522–2534.
Hackam, A. S., Singaraja, R., Wellington, C. L., Metzler, M., McCutcheon, K., Zhang, T., Kalchman, M. & Hayden, M. R. 1998 The influence of huntingtin protein size on nuclear localization and cellular toxicity. *J. Cell Biol.* **141**, 1097–1105.
Hackam, A. S., Singaraja, R., Zhang, T., Gan, L. & Hayden, M. R. 1999 In vitro evidence for both the nucleus and cytoplasmas subcellular sites of pathogenesis in Huntington's disease. *Hum. Mol. Genet.* **8**, 25–33.

Hansson, O., Peterson, A., Leist, M., Nicotera, P., Castilho, R. F. & Brundin, P. 1999 Transgenic mice expressing a Huntington's disease mutation are resistant to quinolinic acid-induced striatal excitotoxicity. *Proc. Natl Acad. Sci. USA* **96**, 8727–8732.

Hodgson, J. G. (and 18 others) 1999 A YAC mouse model for Huntington's disease with full-length mutant huntingtin, cytoplasmic toxicity, and selective striatal neurodegeneration. *Neuron* **23**, 181–192.

Holmberg, M. (and 10 others) 1998 Spinocerebellar ataxia type 7 (SCA7): a neurodegenerative disorder with neuronal intranuclear inclusions. *Hum. Mol. Genet.* **7**, 913–918.

Hoogeveen, A. T., Willemsen, R., Meyer, N., de Rooij, K. E., Roos, R. A., van Ommen, G. J. & Galjaard, H. 1993 Characterization and localization of the Huntington disease gene product. *Hum. Mol. Genet.* **2**, 2069–2073.

Huang, C. C., Faber, P. W., Persichetti, F., Mittal, V., Vonsattel, J. P., MacDonald, M. E. & Gusella, J. F. 1998 Amyloid formation by mutant huntingtin: threshold, progressivity and recruitment of normal polyglutamine proteins. *Somat. Cell Mol. Genet.* **24**, 217–233.

Hurlbert, M. S., Zhou, W., Wasmeier, C., Kaddis, F. G., Hutton, J. C. & Freed, C. R. 1999 Mice transgenic for an expanded CAG repeat in the Huntington's disease gene develop diabetes. *Diabetes* **48**, 649–651.

Igarashi, S. (and 18 others) 1998 Suppression of aggregate formation and apoptosis by transglutaminase inhibitors in cells expressing truncated DRPLA protein with an expanded polyglutamine stretch. *Nature Genet.* **18**, 111–117.

Ikeda, H., Yamaguchi, M., Sugai, S., Aze, Y., Narumiya, S. & Kakizuka, A. 1996 Expanded polyglutamine in the Machado–Joseph disease protein induces cell death in vitro and in vivo. *Nature Genet.* **13**, 196–202.

Ishitani, R., Tanaka, M., Sunaga, K., Katsube, N. & Chuang, D. M. 1998 Nuclear localization of overexpressed glyceraldehyde-3-phosphate dehydrogenase in cultured cerebellar neurons undergoing apoptosis. *Mol. Pharmacol.* **53**, 701–707.

Jackson, G. R., Salecker, I., Dong, X., Yao, X., Arnheim, N., Faber, P. W., MacDonald, M. E. & Zipursky, S. L. 1998 Polyglutamine-expanded human huntingtin transgenes induce degeneration of *Drosophila* photoreceptor neurons. *Neuron* **21**, 633–642.

Kalchman, M. A., Graham, R. K., Xia, G., Koide, H. B., Hodgson, J. G., Graham, K. C., Goldberg, Y. K., Gietz, R. D., Pickart, C. M. & Hayden, M. R. 1996 Huntingtin is ubiquitinated and interacts with a specific ubiquitin conjugated enzyme. *J. Biol. Chem.* **271**, 19385–19394.

Kalchman, M. A. (and 13 others) 1997 *HIP1*, a human homologue of *S. cerevisiae Sla2p*, interacts with membane-associated huntingtin in the brain. *Nature Genet.* **16**, 44–53.

Karpuj, M. V., Garren, H., Slunt, H., Price, D. L., Gusella, J., Becher, M. W. & Steinman, L. 1999 Transglutaminase aggregates huntingtin into nonamyloidogenic polymers, and its enzymatic activity increases in Huntington's disease brain nuclei. *Proc. Natl Acad. Sci. USA* **96**, 7388–7393.

Kaytor, M. D., Duvick, L. A., Skinner, P. J., Koob, M. D., Ranum, L.P. & Orr, H. T. 1999 Nuclear localization of the spinocerebellar ataxia type 7 protein, ataxin-7. *Hum. Mol. Genet.* **8**, 1657–1664.

Kim, M. (and 14 others) 1999 Mutant huntingtin expression in clonal striatal cells: dissociation of inclusion formation and neuronal survival by caspase inhibition. *J. Neurosci.* **19**, 964–973.

Klement, I. A., Skinner, P. J., Kaytor, M. D., Yi, H., Hersch, S. M., Clark, H. B., Zoghbi, H. Y. & Orr, H. T. 1998 Ataxin-1 nuclear localization and aggregation: role in polyglutamine-induced disease in *SCA1* transgenic mice. *Cell* **95**, 41–53.

Koshy, B., Matilla, T., Burright, E. N., Merry, D. E., Fischbeck, K. H., Orr, H. D. & Zoghbi, H. Y. 1996 Spino-cerebellar ataxia type-1 and spinobulbar muscular

atrophy gene products interact with glyceraldehyde-3-phosphate dehydrogenase. *Hum. Mol. Genet.* **5**, 1311–1318.
Lansbury Jr, P. T. 1997 Structural neurology: are seeds at the root of neuronal degeneration? *Neuron* **19**, 1151–1154.
Li, H., Li, S.-H., Cheng, A. L., Mangiarini, L., Bates, G. P. & Li, X.-J. 1999 Ultrastructural localization and progressive formation of neuropil aggregates in Huntington's disease transgenic mice. *Hum. Mol. Genet.* 8, 1227–1236.
Li, M., Miwa, S., Kobayashi, Y., Merry, D. E., Yamamoto, M., Tanaka, F., Doyu, M., Hashizume, Y., Fischbeck, K. H. & Sobue, G. 1998 Nuclear inclusions of the androgen receptor protein in spinal and bulbar muscular atrophy. *Ann. Neurol.* **44**, 249–254.
Li, S.-H. & Li, X.-J. 1998 Aggregation of N-terminal huntingtin is dependent on the length of its glutamine repeats. *Hum. Mol. Genet.* **7**, 777–782.
Li, S.-H., Hosseini, S. H., Gutekunst, C.-A., Hersch, S. M., Ferrante & Li, X.-J. 1998*a* A human HAP1 homologue. *J. Biol. Chem.* **273**, 19220–19227.
Li, S.-H., Gutekunst, C.-A., Hersch, S. M. & Li, X.-J. 1998*b* Interaction of huntingtin-associated protein with dynactic P150^{Glued}. *J. Neurosci.* **18**, 1261–1269.
Li, S.-H., Cheng, A.L., Li, H. and Li, X.-J. 1999 Cellular defects and altered gene expression in PC12 cells stably expressing mutant huntingtin. *J. Neurosci.* **19**, 5159–5172.
Li, X.-J., Li, S.-H., Sharp, A. H., Nucifora Jr, F. C., Schilling, G., Lanahan, A., Worley, P., Snyder, S. H. & Ross, C. A. 1995 A huntingtin-associated protein enriched in brain with implications for pathology. *Nature* **378**, 398–402.
Liu, Y. F. 1998 Expression of polyglutamine-expanded huntingtin activates the SEK1–JNK pathway and induces apoptosis in a hippocampal neuronal cell line. *J. Biol. Chem.* **273**, 28873–28877.
Liu, Y. F., Deth, R. C. & Devys, D. 1997 SH3 domain-dependent association of huntingtin with epidermal growth factor receptor signaling complexes. *J. Biol. Chem.* **272**, 8121–8124.
Lunkes, A. & Mandel, J.-L. 1998 A cellular model that recapitulates major pathogenic steps of Huntington's disease. *Hum. Mol. Genet.* **7**, 1355–1361.
MacLean, H. E., Warne, G. L. & Zajac, J. D. 1997 Localization of functional domains in the androgen receptor. *J. Steroid Biochem. Mol. Biol.* **62**, 233–242.
Mangiarini, L. (and 10 others) 1996 Exon 1 of the HD gene with an expanded CAG repeat is sufficient to cause a progressive neurological phenotype in transgenic mice. *Cell* **87**, 493–506.
Martin, E. J. (and 11 others) 1999 Analysis of huntingtin-associated protein 1 in mouse brain and immortalized striatal neurons. *J. Comp. Neurol.* **403**, 421–430.
Martindale, D. (and 12 others) 1998 Length of huntingtin and its polyglutamine tract influences localization and frequency of intracellular aggregates. *Nature Genet.* **18**, 150–154.
Matilla, A., Koshy, B., Cummings, C. J., Isobe, T., Orr, H. T. & Zoghbi, H. Y. 1997 The cellular leucine rich acidic nuclear protein (LANP) interacts with ataxin-1: further insight into the pathogenesis of SCA1. *Nature* **389**, 974–978.
Merry, D. E., Kobayashi, Y., Bailey, C. K., Taye, A. A. & Fischbeck, K. H. 1998 Cleavage, aggregation and toxicity of the expanded androgen receptor in spinal and bulbar muscular atrophy. *Hum. Mol. Genet.* **7**, 693–701.
Miyashita, T., Okamura-Oho, Y., Mito, Y., Nagafuchi, S. & Yamada, M. 1997 Dentatorubral pallidoluysian atrophy (DRPLA) protein is cleaved by caspase-3 during apoptosis. *J. Biol. Chem.* **272**, 29238–29242.
Miyashita, T., Nagao, K., Ohmi, K., Yanagisawa, H., Okamura-Oho, Y. & Yamada, M. 1998 Intracellular aggregate formation of dentatorubral–pallidoluysian atrophy (DRPLA) protein with the extended polyglutamine. *Biochem. Biophys. Res. Commun.* **249**, 96–102.

Moulder, K. L., Onodera, O., Burke, J. R., Strittmatter, W. J. & Johnson Jr, E. M. 1999 Generation of neuronal intranuclear inclusions by polyglutamine-GFP: analysis of inclusion clearance and toxicity as a function of polyglutamine length. *J. Neurosci.* **19**, 705–715.

Myers, R. H., Vonsattel, J. P., Stevens, T. J., Cupples, L. A., Richardson, E. P., Martin, J. B. & Bird, E. D. 1988 Clinical and neuropathologic assessment of severity in Huntington's disease. *Neurology* **38**, 341–347.

Nance, M. A. 1997 Clinical aspects of CAG repeat diseases. *Brain Pathol.* **7**, 881–900.

Okamura-Oho, Y., Miyashita, T., Ohmi, K. & Yamada, M. 1999 Dentatorubral–pallidoluysian atrophy protein interacts through a proline-rich region near polyglutamine with the SH3 domain of an insulin receptor tyrosine kinase substrate. *Hum. Mol. Genet.* **8**, 947–957.

Ona, V. O. (and 14 others) 1999 Inhibition of caspase-1 slows disease progression in a mouse model of Huntington's disease. *Nature* **399**, 263–267.

Ordway, J. M. (and 11 others) 1997 Ectopically expressed CAG repeats cause intranuclear inclusions and a progressive late onset neurological phenotype in the mouse. *Cell* **91**, 753–763.

Page, K. J., Potter, L., Aronni, S., Everitt, B. J. & Dunnett, S. B. 1998 The expression of huntingtin-associated protein (HAP1) mRNA in developing, adult and ageing rat CNS: implications for Huntington's disease neuropathology. *Eur. J. Neurosci.* **10**, 1835–1845.

Paulson, H. L. & Fischbeck, K. H. 1996 Trinucleotide repeats in neurogenetic disorders. *Annu. Rev. Neurosci.* **19**, 79–107.

Paulson, H. L., Perez, M. K., Trottier, Y., Trojanowski, J. Q., Subramony, S. H., Das, S. S., Vig, P., Mandel, J.-L., Fischbeck, K. H. & Pittman, R. N. 1997 Intranuclear inclusions of expanded polyglutamine protein in spinocerebellar ataxia type 3. *Neuron* **19**, 333–344.

Perez, M. K., Paulson, H. L., Pendse, S. J., Saionz, S. J., Bonini, N. M. & Pittman, R. N. 1998 Recruitment and the role of nuclear localization in polyglutamine-mediated aggregation. *J. Cell Biol.* **143**, 1457–1470.

Perutz, M. 1994 Polar zippers: their role in human disease. *Protein Sci.* **3**, 1629–1637.

Peters, M. F., Nucifora Jr, F. C., Kushi, J., Seaman, H. J., Cooper, J. K., Herring, W. J., Dawson, V. L., Dawson, T. M. & Ross, C. A. 1999 Nuclear targeting of mutant huntingtin increases toxicity. *Mol. Cell Neurosci.* **14**, 121–128.

Reddy, P. H., Williams, M., Charles, V., Garrett, L., Pike-Buchanan, L., Whetsell, W. O., Miller, G. & Tagle, D. A. 1998 Behavioural abnormalities and selective neuronal loss in HD transgenic mice expressing mutated full-length HD cDNA. *Nature Genet.* **20**, 198–202.

Rigamonti, D., Cattaneo, E. *et al.*, submitted

Ross, C. A. 1995 When more is less: pathogenesis of glutamine repeat neurodegenerative diseases. *Neuron* **15**, 493–496.

Ross, C. A. 1997 Intranuclear neuronal inclusions: a common pathogenic mechanism for glutamine-repeat neurodegenerative diseases? *Neuron* **19**, 1147–1150.

Ross, C. A., Margolis, R. L., Rosenblatt, A., Ranen, N. G., Becher, M. W. & Aylward, E. A. 1997 Reviews in molecular medicine: Huntington's disease and a related disorder, dentatorubral–pallidoluysian atrophy (DRPLA). *Medicine* **76**, 305–338.

Sanchez, I., Xu, C. J., Juo, P., Kakizaka, A., Blenis, J. & Yuan, J. 1999 Caspase-8 is required for cell death induced by expanded polyglutamine repeats. *Neuron* **22**, 623–633.

Sapp, E., Schwarz, C., Chase, K., Bide, P. G., Young, A. B., Penney, J., Vonsattel, J. P., Aronin, N. & DiFiglia, M. 1997 Huntingtin localization in brains of normal and Huntington's disease patients. *Ann. Neurol.* **42**, 604–612.

Saudou, F., Finkbeiner, S., Devys, D. & Greenberg, M. E. 1998 Huntingtin acts in the nucleus to induce apoptosis but death does not correlate with the formation of intranuclear inclusions. *Cell* **95**, 55–66.

Saunders, P. A., Chen, R. W. & Chuang, D. M. 1999 Nuclear translocation of glyceraldehyde-3-phosphate dehydrogenase isoforms during neuronal apoptosis. *J. Neurochem.* **72**, 925–932.

Sawa, A., Khan, A. A., Hester, L. D. & Snyder, S. H. 1997 Glyceraldehyde-3-phosphate dehydrogenase: nuclear translocation participates in neuronal and nonneuronal cell death. *Proc. Natl Acad. Sci. USA* **94**, 11669–11674.

Sawa, A., Wiegand, G. W., Cooper, J., Margolis, R. L., Sharp, A. H., Lawler Jr J. F., Greenamyre, J. T., Snyder, S. H. & Ross, C. A. 1999 Increased apoptosis of Huntington's disease lymphoblasts associated with repeat length-dependent mitochrondrial depolarization. *Nature Med.* (In the press.)

Scherzinger, E., Lurz, R., Turmaine, M., Mangiarini, L., Hollenbach, B., Hasenbank, R., Bates, G. P., Davies, S. W., Lehrach, H. & Wanker, E. E. 1997 Huntingtin-encoded polyglutamine expansions form amyloid-like protein aggregates *in vitro* and *in vivo*. *Cell* **90**, 549–558.

Scherzinger, E., Sittler, A., Schweiger, K., Heiser, V., Lurz, R., Hasenbank, R., Bates, G. P., Lehrach, H. & Wanker, E. E. 1999 Self-assembly of polyglutamine-containing huntingtin fragments into amyloid-like fibrils: implications for Huntington's disease pathology. *Proc. Natl Acad. Sci. USA* **96**, 4604–4609.

Schilling, G. (and 13 others) 1999*a* Intranuclear inclusions and neuritic pathology in transgenic mice expressing a mutant N-terminal fragment of huntingtin. *Hum. Mol. Genet.* 8, 397–407.

Schilling, G. (and 15 others) 1999*b* Nuclear accumulation of truncated atrophin-1 fragments in a transgenic mouse model of DRPLA. *Neuron* 24, 1–20.

Servadio, A., Koshy, B., Armstrong, D., Antalffy, B., Orr, H. T. & Zoghbi, H. Y. 1995 Expression analysis of the ataxin-1 protein in tissues from normal and spinocerebellar ataxia type 1 individuals. *Nature Genet.* **10**, 94–98.

Sharp, A. H. (and 15 others) 1995 Widespread expression of Huntington's disease gene (IT15) protein product. *Neuron* **14**, 1065–1074.

Sharp, A. H., Zhang, C., Ward, C., Ryugo, D. K. & Ross, C. A. 2000 Cellular and subcellular localization of huntingtin-associated protein-1 (HAP1): potential role in vesicle transport. Submitted.

Shashidharan, P., Chalmers-Redman, R. M., Carlile, G. W., Rodic, V., Gurvich, N., Yuen, T., Tatton, W. G. & Sealfon, S. C. 1999 Nuclear translocation of GAPDH–GFP fusion protein during apoptosis. *NeuroReport* **10**, 1149–1153.

Sittler, A., Walter, S., Wedemeyer, N., Hasenbank, R., Scherzinger, E., Eickhoff, H., Bates, G. P., Lehrach, H. & Wanker, E. E. 1998 SH3GL3 associates with the huntingtin exon 1 protein and promotes the formation of polygln-containing protein aggregates. *Mol. Cell* **2**, 427–436.

Skinner, P. J., Koshy, B. T., Cummings, C. J., Klement, I. A., Helin, K., Servadio, A., Zoghbi, H. Y. & Orr, H. T. 1997 Ataxin-1 with an expanded glutamine tract alters nuclear matrix-associated structures. *Nature* **389**, 971–978.

Stenoien, D. L., Cummings, C. J., Adams, H. P., Mancini, M. G., Patel, K., DeMartino, G. N., Marcelli, M., Weigel, N. L. & Mancini, M. A. 1999 Polyglutamine-expanded androgen receptors form aggregates that sequester heat shock proteins, proteasome components and SRC-1, and are suppressed by the HDJ-2 chaperone. *Hum. Mol. Genet.* **8**, 731–741.

Stine, O. C., Pleasant, N., Franz, M. L., Abbott, M. H., Folstein, S. E. & Ross, C. A. 1993 Correlations between the onset of Huntington's disease and length of the trinucleotide repeat in IT-15. *Hum. Mol. Genet.* **2**, 1547–1549.

Stott, K., Blackburn, J. M., Butler, P. J. G. & Perutz, M. 1995 Incorporation of glutamine repeats make protein oligomerize: implications for neurodegenerative diseases. *Proc. Natl Acad. Sci. USA* **92**, 6509–6513.

Tabrizi, S. J., Cleeter, M. W., Xuereb, J., Taanman, J. W., Cooper, J. M. & Schapira, A. H. 1999 Biochemical abnormalities and excitotoxicity in Huntington's disease brain. *Ann. Neurol.* **45**, 25–32.

Tait, D., Riccio, M., Sittler, A., Scherzinger, E., Santi, S., Ognibene, A., Maraldi, N. M., Lehrach, H. & Wanker, E. E. 1998 Ataxin-3 is transported into the nucleus and associates with the nuclear matrix. *Hum. Mol. Genet.* **7**, 991–997.

Trottier, Y. (and 10 others) 1995 Polyglutamine expansion as a pathological epitope in Huntington's disease and four dominant cerebellar ataxias. *Nature* **378**, 403–406.

Trottier, Y., Cancel, G., An-Gourfinkel, I., Lutz, Y., Weber, C., Brice, A., Hirsch, E. & Mandel, J.-L. 1998 Heterogeneous intracellular localization and expression of ataxin-3. *Neurobiol. Dis.* **5**, 335–347.

Vonsattel, J. P., Myers, R. H., Stevens, T. J., Ferrante, R. J., Bird, E. D. & Richardson, E. P. 1985 Neuropathological classification of Huntington's disease. *J. Neuropathol. Exp. Neurol.* **44**, 559–577.

Wanker, E. E., Rovira, C., Scherzinger, E., Hasenbank, R., Wälter, S., Tait, D., Colicelli, J. & Lehrach, H. 1997 HIP-1: a huntingtin interacting protein isolated by the yeast two-hybrid system. *Hum. Mol. Genet.* **6**, 487–495.

Warrick, J. M., Paulson, H. L., Gray-Board, G. L., Bui, Q. T., Fischbeck, K. H., Pittman, R. N. & Bonini, N. M. 1998 Expanded polyglutamine protein forms nuclear inclusions and causes neural degeneration in *Drosophila*. *Cell* **93**, 939–949.

Wellington, C. L. (and 20 others) 1998 Caspase cleavage of gene products associated with triplet expansion disorders generates truncated fragments containing the polyglutamine tract. *J. Biol. Chem.* **273**, 9158–9167.

Wilkinson, F. L., Nguyen, T. M., Manilal, S. B., Thomas, P., Neal, J. W., Harper, P. S., Jones, A. L. & Morris, G. E. 1999 Localization of rabbit huntingtin using a new panel of monoclonal antibodies. *Brain Res. Mol. Brain Res.* **69**, 10–20.

Wood, J. D., Yuan, J., Margolis, R. L., Colomer, V., Duan, K., Kushi, J., Kaminsky, Z., Kleiderlein, J. J., Sharp, A. H. & Ross, C. A. 1998 Atrophin-1, the DRPLA gene product, interacts with two families of WW domain-containing proteins. *Mol. Cell. Neurosci.* **11**, 149–160.

Wood, J. D., Nucifora Jr, F. C., Duan, K., Wang, J., Kim, Y., Schilling, G., Sacchi, N., Liu, J. M. & Ross, C. A. 2000 Atrophin-1, the DRPLA gene product, interacts with ETO/MTG8 in the nuclear matrix. Submitted.

Zoghbi, H. Y. 1996 The expanding world of ataxins. *Nature Genet.* **14**, 237–238.

8

Properties of polyglutamine expansion *in vitro* and in a cellular model for Huntington's disease

Astrid Lunkes, Yvon Trottier, Jerôme Fagart, Patrick Schultz, Gabrielle Zeder-Lutz, Dino Moras and Jean-Louis Mandel

Introduction

Huntington's disease (HD) and seven other neurodegenerative diseases are caused by CAG codon/polyglutamine (polyQ) expansions that lead to an increase in toxic properties in target proteins (reviewed by Paulson & Fischbeck (1996) and Wells & Warren (1998)). In five of these diseases the pathological threshold is between 35 and 40 glutamine residues, whereas it is higher for dentatorubral–pallidoluysian atrophy (DRPLA) (49 glutamine residues) and for spinocerebellar ataxia 3 (SCA3) (61 glutamine residues) (Wells & Warren 1998). In SCA6, the pathological polyQ is much smaller (21 glutamine residues); the overlap of clinical manifestations with those caused by the conventional loss of function mutations within the same gene suggests that in this case the polyQ expansion could instead cause an alteration of the normal function (Zoghbi 1997). Recent studies reviewed in this volume have shown that proteins containing a pathological polyQ stretch can form ubiquitinated aggregates within neurons of patients and of some transgenic mouse models (Davies *et al.* 1997; DiFiglia *et al.* 1997; Paulson *et al.* 1997; Skinner *et al.* 1997; Becher *et al.* 1998; Igarashi *et al.* 1998). In many cases the aggregates were found in the nucleus, despite the fact that the cognate normal protein is cytoplasmic (as for huntingtin or ataxin-3). For huntingtin, the protein that is found in nuclear inclusions or in dystrophic neurites corresponds to an N-terminal fragment of huntingtin that contains the polyQ expansion (DiFiglia *et al.* 1997). There is also evidence that in SCA3 only a subfragment containing the polyQ stretch is present in nuclear inclusions (Paulson *et al.* 1997). Studies of transgenic mice and transfected cellular models have also shown that truncated proteins containing a pathological polyQ stretch are much more toxic, or have increased aggregation properties, than the full-length mutated proteins (Davies *et al.* 1997; Paulson *et al.* 1997; Butler *et al.* 1998; Cooper *et al.* 1998; Igarashi *et al.* 1998; Klement *et al.* 1998; Li & Li 1998; Lunkes & Mandel 1998; Martindale *et al.* 1998; Merry *et al.* 1998; Saudou *et al.* 1998). This suggests that proteolysis of polyQ-containing proteins and their aggregation might be important steps in pathogenesis.

Here we present work from our laboratory on two major aspects. We summarize the properties of the 1C2 antibody that recognizes specifically, in a Western blot, proteins carrying a pathological polyQ expansion (Trottier *et al.* 1995). This indicates that in the range of 35–40 glutamine residues, which corresponds to the pathological threshold in five of the polyQ diseases, a change in conformation occurs that strengthens the interaction with 1C2 antibody. This conformational epitope is most probably involved in the aggregation properties demonstrated *in vitro* or *in vivo*, notably by mutated huntingtin (Davies *et al.* 1997; Scherzinger *et al.* 1997). We observed that even the normal stretch of 38 glutamine residues carried by the human transcription factor TBP, which is recognized by the 1C2 antibody, is able to aggregate into regular fibrils similar to those observed by Scherzinger *et al.* (1997) for the N-terminal subfragment of mutated huntingtin.

To decipher some of the cellular events that link the processing of a target protein with expanded polyQ and its aggregation, leading to neuronal dysfunction and ultimately to death, cellular models are highly desirable. Here we present a cellular model in which the expression of full-length huntingtin with a polyQ expansion can lead to the formation of nuclear inclusions that seem similar to those found in patients, notably in that they contain only a short N-terminal fragment (Lunkes & Mandel 1998). This model should therefore be useful in analysing the processing step necessary for building up the nuclear inclusion.

Conformational epitope and aggregation of polyQ tracts

Recognition properties of 1C2 monoclonal antibody reveal a pathological epitope

The 1C2 monoclonal antibody was initially developed using as antigen the human TATA-binding protein (TBP), a general transcription factor that contains a polymorphic polyQ stretch with the most common allelic form of 38 glutamine residues, and rare variants having 42 glutamine residues (Gostout *et al.* 1993; Imbert *et al.* 1994). Epitope mapping of the antibody indicated that it recognizes the polyQ domain of TBP (Lescure *et al.* 1995). We found that the 1C2 antibody shows the unexpected property of selectively recognizing pathological polyQ expansions in the proteins involved in HD, SCA1 and SCA3 (Trottier *et al.* 1995). A Western blot analysis of allelic forms of huntingtin showed that the intensity of detection by the 1C2 antibody is dependent on the size of the expansion, increasing from 39 to 85 glutamine residues. In contrast, huntingtin with a normal polyQ stretch was not detected, whereas allelic forms with polyQ at the upper normal range (*ca.* 28) were faintly detected after a very long exposure of the blot. Thus, the efficiency of detection seemed to parallel the severity of the disease.

The recognition properties of the 1C2 antibody could be explained by a higher affinity for a pathological expansion, owing to the presence of a specific conformational epitope that appears when the polyQ is more than 35 residues long. Alternatively, or synergistically, the repetition of a polyQ epitope could facilitate a bivalent binding of the antibody and thus stabilize the interaction.

To study the binding properties between 1C2 antibody and polyQ stretches, we have performed kinetic analyses of the interaction under native conditions with a surface plasmon resonance biosensor (BIAcore™, Pharmacia) (Myszka 1997). The Fab (which has a single antigen-binding site) showed a much stronger affinity for an expansion of 73 glutamine residues than for a normal polyQ stretch (15 glutamine residues). The complex formed between the Fab and the expansion to 73 glutamine residues is remarkably stable, and dissociates at 1/100 of the rate found with the stretch of 15 glutamine residues (Trottier *et al.* 1998). Similar kinetics of interaction were obtained with the whole antibody, indicating that bivalent binding is not a major factor in the detection process. The results indicate the formation of a novel conformation by an expanded polyQ stretch.

The 1C2 antibody was used to test the hypothesis of polyQ expansion as the cause of several neurological diseases that show an anticipation phenomenon. On Western blot analysis, the 1C2 antibody was able to detect novel proteins containing a polyQ expansion in patients with autosomal dominant spinocerebellar ataxia type 2 (linked to chromosome 12) and type 7 (linked to chromosome 3) (Trottier *et al.* 1995). Another protein has been recently detected by the 1C2 antibody in a single patient with autosomal dominant familial spastic paraplegia (AD-FSP) linked to chromosome 2, which also showed a specific RED product (Nielsen *et al.* 1998). However, in the AD-FSP patients that we studied (some were linked to chromosome 2), we were unable to detect novel protein with the 1C2 antibody (Trottier *et al.* 1995). Thus, the implication of polyQ expansion in this disease needs to be confirmed. Several groups have tried without success to identify new types of polyQ expansion in patients with sporadic or unassigned autosomal dominant cerebellar ataxia (Lopes-Cendes *et al.* 1996; Stevanin *et al.* 1996). One should, however, note that although a positive result proves the mechanism, false negative findings can be obtained if the pathological protein is poorly expressed in, or absent from, the cells analysed, or if the polyQ stretch is too small to be detected with high sensitivity. In the search for a putative polyQ expansion protein, the analysis of early-onset cases is warranted, given the strong polyQ length-dependent properties of the antibody (Trottier *et al.* 1998).

Anticipation was also reported to occur in some families with bipolar affective illness or schizophrenia, suggesting a possible implication of polyQ expansion (Ross *et al.* 1993). The detection of CAG repeats by using the

RED method has not given a clear indication for such an involvement. Several groups have undertaken studies with the use of 1C2 antibody to search for pathological polyQ expansion in lymphoblasts from schizophrenia patients.

The 1C2 antibody was used to screen two cDNA expression libraries from SCA2 and SCA7 lymphoblasts, respectively. No clones corresponding to the expected mutated alleles were recovered in these libraries, possibly owing to a toxic effect of expanded polyQ in *Escherichia coli*, as reported by Onodera *et al.* (1996). We did, however, observe that 1C2 was able to detect smaller polyQ repeats (more than 11 glutamine residues) in the bacteriophage plaque screening conditions, which are less stringent than for Western blotting (no SDS or reducing agent). This allowed us to identify five new polyQ-coding cDNA species, including one corresponding to a normal allele of the *SCA2* gene (Imbert *et al.* 1996) and another corresponding to a Ca^{2+}-regulated K^+ channel gene (*hSKCa3*) (Wittekindt *et al.* 1998).

The striking parallel between the effect of increased polyQ length in disease severity and in recognition by 1C2 antibody suggests that the novel conformational epitope recognized by this antibody has a key role in pathogenicity, possibly by favouring aggregation and/or by altering interaction with other cellular proteins. Determination of the crystal structure of expanded polyQ is an important goal, but is rendered difficult by the great tendency of polyQ-carrying polypeptides to aggregate *in vitro*.

Aggregation of a normal fragment of TBP containing 38 glutamine residues

The alleles of TBP carrying 38–42 glutamine residues are larger than the pathological threshold of 35–37 glutamine residues observed in HD, SCA2 and SCA7. These allelic TBP proteins are well recognized by 1C2 antibody; however, they are not pathogenic proteins. We intended to use an N-terminal fragment of TBP with a stretch of 38 glutamine residues (N-TBP-38Q) for X-ray crystallography. After overexpression in *E. coli*, the purified N-TBP-38Q showed a tendency to aggregate as seen on non-denaturing gel electrophoresis (results not shown). When freshly purified N-TBP-38Q fractions were investigated by electron microscopy, we observed long filaments (0.1–0.5 μm) that were 10–12 nm in diameter (figure 8.1*a*), very similar to those reported by Scherzinger *et al.* (1997) with a subfragment of huntingtin carrying 51 glutamine residues. The filaments that we observed were not uniformly distributed over the entire surface but were organized in clusters of three to a few tens, suggesting that they interact or that their mechanism of growth promoted the formation of discrete branching points, thus resulting in a loose network. The number of such structures increased with time during the two days after purification, whereas their number decreased after three days, a time that corresponded to the appearance of large three-dimensional aggregates composed of a dense irregular filamenteous network (figure 8.1*b*). The aggregation of N-TBP-38Q *in vitro* precludes the formation of crystals.

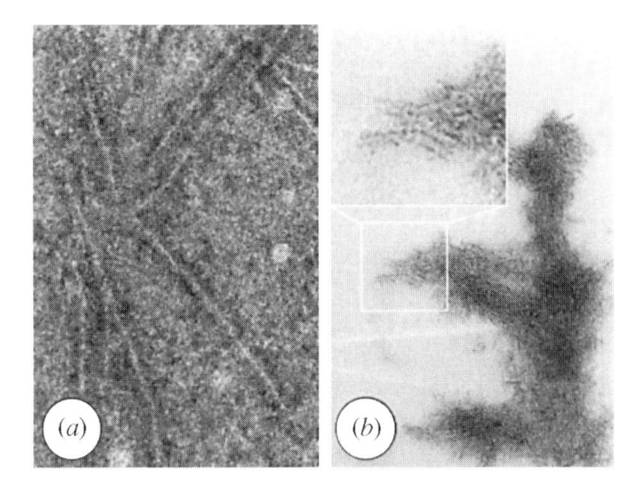

Figure 8.1 Aggregation of an N-terminal fragment of TBP with 38 glutamine residues. (*a*) The N-terminal domain of the human TBP (154 residues) with 38 glutamine residues in fusion with a polyhistidine tag was expressed in *E. coli* and purified on a nickel chelating column and by gel filtration. Aliquots of freshly purified N-TBP fragments, kept in 100 mM Hepes, pH 7.2, at a concentration of 1 mg/ml, were diluted to a final concentration of 30 μg/ml, deposited on a glow-discharged electron microscopy grid covered with a thin carbon film and negatively stained with a 20 g l^{-1} uranyl acetate solution. Filaments *ca.* 0.1–0.5 μm in length and 10–12 nm wide were clearly observed. Filaments similar in size were also observed when the polyhistidine tag was removed enzymically, thus demonstrating that the tag does not influence the aggregation process (results not shown). The bar represents 100 nm. (*b*) On longer storage (three or four days), the N-TBP fraction formed large aggregates composed of filaments similar to those described in (*a*). The scale bar represents 200 nm and the insert is enlarged twice to show the filamentous organization.

Interestingly, the preincubation of N-TBP-38Q with an Fab fragment of 1C2 antibody prevents filament formation and aggregation. Instead we observed small complexes formed of two or three stain-excluding domains whose size was consistent with that of an Fab fragment (results not shown). The observation that interaction with 1C2 antibody prevents such an aggregation led us to initiate attempts to co-crystallize N-TBP-38 glutamine residues or a pathological ataxin-2 fragment (42 glutamine residues) with the Fab fragment of the 1C2 antibody.

The stretch of 38 glutamine residues present in TBP can lead to the formation of fibrils of very similar appearance to those found with the N-terminal fragment of huntingtin (Scherzinger *et al.* 1997). In our case, the fact that an anti-polyQ antibody blocked the aggregation indicated the importance of the polyQ stretch for the formation of supramolecular assemblies. It is likely that the folding of full-length TBP prevents this aggregation from occurring *in vivo* (especially because the nuclear localization of this protein would probably favour toxicity). Thus, the variability in pathological threshold could be dependent both on the folding of each full-length target protein and on the proteolytic activities that lead to the production of more toxic subfragments.

A cellular model for Huntington's disease

Formation of inclusions in cells expressing mutant huntingtin

In none of the previous cellular models designed for Huntington's disease could the situation *in vivo* be simulated by the expression of full-length mutant protein. We have now established a regulated system for the expression of huntingtin under the control of the reverse tetracycline-inducible transactivator (Gossen *et al.* 1995), in the mouse–rat neuroblastoma–glioma cell line NG108-15. In this system, the addition of doxicycline to the culture medium induces the expression of huntingtin, thus preventing potential toxic effects of overproduced mutant huntingtin when growing the cells. The NG108-15 cells display neuronal-like properties after differentiation (Nelson *et al.* 1976), and they allow the study of the expression of mutant huntingtin over much longer periods (up to 18 days) than in transient transfection systems described previously (Cooper *et al.* 1998; Li & Li 1998; Martindale *et al.* 1998).

In a first step, the reverse tetracycline-inducible transactivator (rtTA) was stably integrated in the cell line NG108-15. To study the fate of full-length huntingtin compared with truncated forms, as well as the effect of polyQ length, double-transformed cell lines expressing the following proteins under the control of the tetracycline transactivator promoter were then generated: full-length (FL) huntingtin with 15, 73 and 116 repeats; truncated (T) huntingtin (502 residues) with 15 and 73 repeats, and a very truncated (VT) huntingtin protein (80 residues) with 15, 73 and 120 repeats (figure 8.2). All these proteins carry the Flag-tag epitope at their N-terminus. The T form is similar to the N-terminal fragment generated by cleavage at the caspase-3 site (Goldberg *et al.* 1996), whereas the VT form is similar to that expressed in transgenic mice studied by Mangiarini *et al.* (1996) and Davies *et al.* (1997).

Huntingtin expression was assessed by immunofluorescence after differentiation and induction with doxicycline, by using the anti-Flag monoclonal antibody (mAb). Homogeneous cytoplasmic staining was observed in all lines expressing the normal huntingtin constructs. In contrast, distinct intracellular inclusions (cytoplasmic inclusion (CI) or nuclear inclusion (NI)) were seen in the mutant lines T-hd73, FL-hd73 and FL-hd116, co-existing in most instances with homogeneous cytoplasmic staining (Lunkes & Mandel 1998). Both types of inclusion (NIs and CIs) also reacted with an antibody against ubiquitin, whereas the NIs were also detected with an antibody against the nuclear protein SUG1, which is a subunit of the PA700 complex of the proteasome (Fraser *et al.* 1997) (figure 8.3 (plate 8)). Those observations

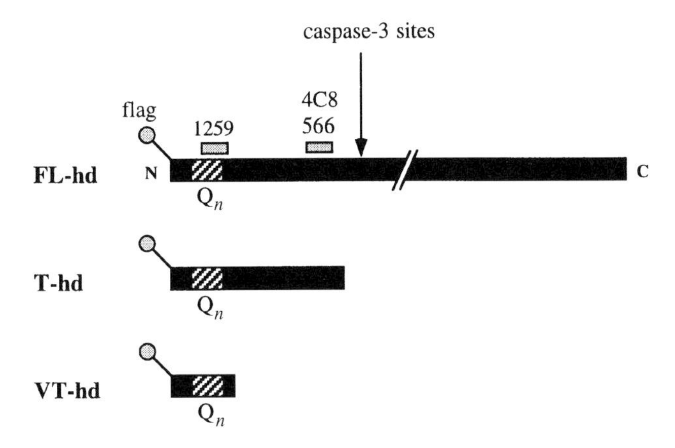

Figure 8.2 Features of the huntingtin-derived proteins. The full-length and truncated huntingtin cDNA constructs with different lengths of CAG repeats were Flag-tagged at the N-terminus and cloned into the pUHD172-1neo vector (Gossen & Bujard 1992; Gossen *et al.* 1995). The hatched boxes correspond to a polyQ repeat with n =15, 73 or 116. The localization of the epitopes (grey boxes) of the antibodies used, and the predicted caspase-3 cleavage sites, are indicated.

most probably reflect failed attempts of the cell to remove misfolded or aggregated proteins. In line VT-hd73, expressing the shortest huntingtin protein, no homogeneous cytoplasmic staining was observed. Instead the protein seemed to aggregate as dense inclusions, either in the cytoplasm (CI) or in the nucleus (NI). In a time-course experiment, the frequency of CIs decreased from 49% (day 4) to 4% (day 16), whereas the frequency of NIs increased from 20% (day 4) to 84% (day 16) (figure 8.4), suggesting that the CIs might be precursors of the NIs. Cells expressing the T-hd73 truncated construct showed a delayed and less efficient formation of dense NIs (up to 17%), and for FL-hd73 only 1% of expressing cells formed NIs at late time points (16 days) (Lunkes & Mandel 1998). A much greater formation of small dense NIs was observed in line FL-hd116, starting at around 12 days of induction, whereas CIs predominated at earlier time points (figure 8.5). The NIs were detected with the anti-Flag mAb but not with the 4C8 mAb or the 566 polyclonal antibody (whose epitopes are just proximal to the caspase-3 sites) (Lunkes & Mandel 1998). This suggests that these NIs are composed of a fragment cleaved N-terminal to the 4C8 epitope, which is in agreement with the finding that inclusions in the brain of HD patients are detected only with N-terminal antibodies close to the polyQ stretch (DiFiglia *et al.* 1997).

The dense nuclear inclusions were not stained by the 1C2 mAb that is specific for expanded polyQ tracts (Trottier *et al.* 1995). Most probably, the pathological epitope is masked in the dense aggregates, possibly owing to the interaction of polyQ tracts with each other to form β-pleated sheets (Perutz *et al.* 1994). This is consistent with the lack of staining of inclusions

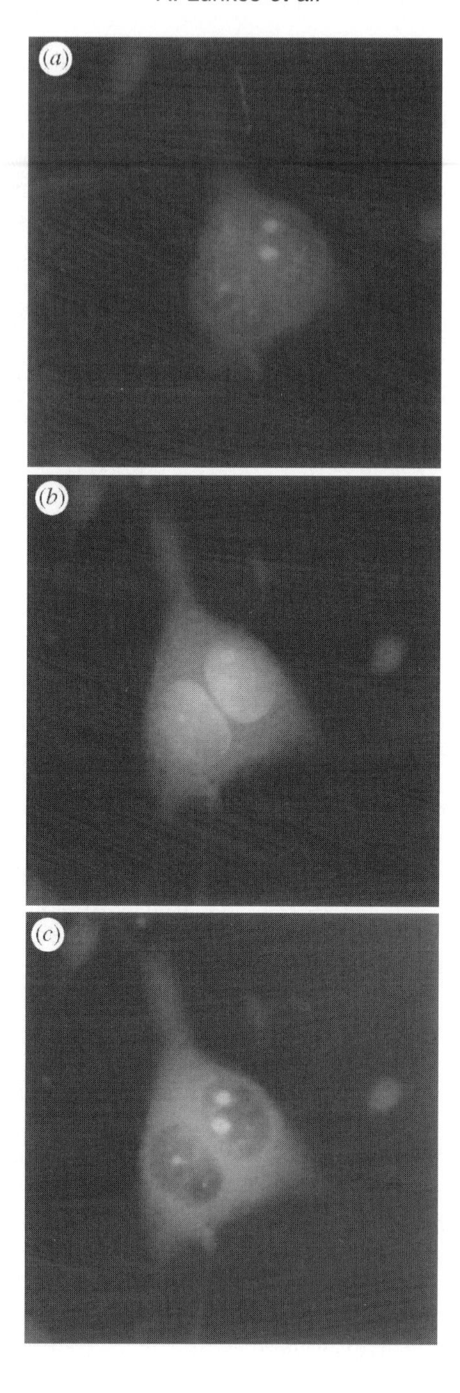
(a)
(b)
(c)

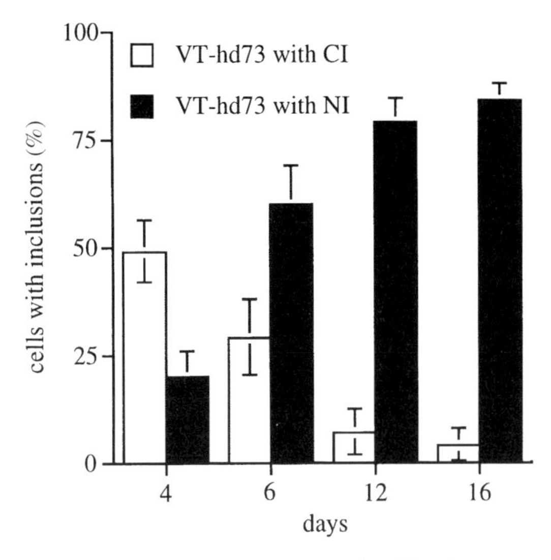

Figure 8.4 Fate of the VT-hd73 transgene protein. The frequency of cells bearing cytoplasmic inclusions (CIs) decreased over time, whereas the inclusions in the nucleus (NIs) increased. The results are from one experiment. Similar results were obtained with cells expressing the VT transgene with 120 glutamine residues. Bars represent the s.e.m. of the proportion of cells having cytoplasmic or nuclear inclusions.

by 1C2 in transgenic HD mice (Davies *et al.* 1997), with its very weak reaction with nuclear inclusions seen in the brain of SCA3 patients (Paulson *et al.* 1997), and with its failure to detect aggregates of the mutant androgen receptor on a Western blot (Merry *et al.* 1998).

Processing of mutant full-length huntingtin

To study the processing of the huntingtin protein, we performed a Western blot analysis of cells expressing mutant full-length huntingtin. A breakdown product of *ca.* 98 kDa for FL-hd116 was detected from day 10 onwards by both the anti-Flag mAb (not shown) and the 4C8 mAb (figure 8.6). The size of the fragment corresponded well to that of the predicted N-terminal

Figure 8.3 Co-localization of N-terminal huntingtin and SUG1, a subunit of the PA700 complex of the proteasome, in inclusions observed in NG108-15 cells expressing FL-hd116. Cytoplasmic huntingtin and NIs are detected with the N-terminal polyclonal antibody 1259 (produced against a glutathione *S*-transferase fusion protein containing residues 1–171 of huntingtin). (*b*). The anti-SUG1 mAb (Baur *et al.* 1996) reacts strongly with inclusions in the nucleus (*a*), which co-localize with the huntingtin-positive NIs (*c*). The co-localization has been confirmed with antibodies raised against known (such as Mss1 (Fraser *et al.* 1997)) and unknown proteins found enriched in a PA700-containing complex of high molecular mass (Y. Lutz, personal communication) (A. Lunkes, unpublished data). (See also colour plate section.)

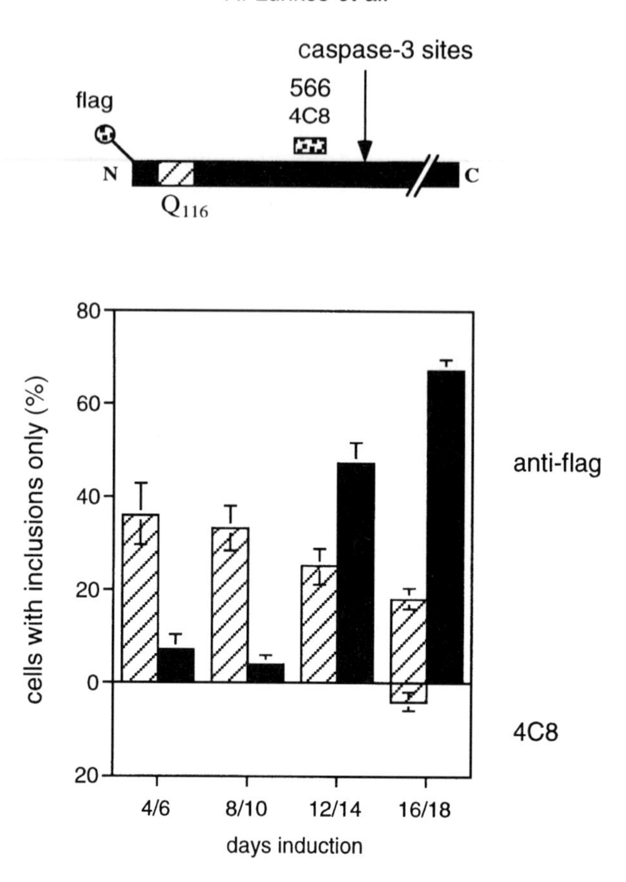

Figure 8.5 Formation of inclusions in a cell line expressing full-length huntingtin (FL-hd116). Cells expressing full-length huntingtin with tracts of 116 glutamine residues were differentiated and induced with doxicycline. Immunofluorescence analysis was performed with the anti-Flag and 4C8 mAbs. The 4C8 mAb detected only a low frequency of CIs at late stages only, and no NIs. The results of one representative experiment are shown.

caspase-3 cleavage product of huntingin (Goldberg *et al.* 1996), which includes the 4C8 epitope. Although endogenous huntingtin is produced at levels similar to that of mutant huntingtin (as detected by the 4C8 mAb), the corresponding caspase-3 fragment was not observed in these lines, which is consistent with the hypothesis that an elongated polyQ repeat enhances the cleavage of full-length huntingtin by caspase-3 (Goldberg *et al.* 1996). However, although inclusions were detectable at day 4–6 (figure 8.5), the potential caspase-3 fragment appeared rather later (day 10). This suggests that this fragment is not involved in the initial formation of inclusions.

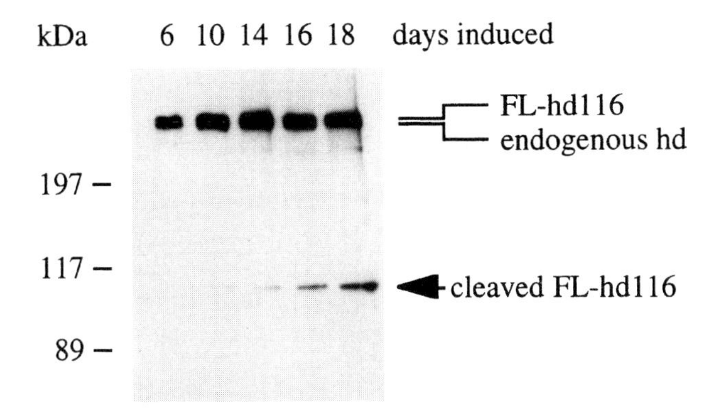

Figure 8.6 Cleavage of the mutant full-length huntingtin protein. Cells expressing FL-hd116 were differentiated and induced with doxicycline over 18 days; samples were taken at the indicated time points for Western blotting. Immunoprobing with the huntingtin-specific mAb 4C8 revealed, from day 10 onwards, a breakdown product corresponding to the size of the putative caspase-3 fragment. No cleavage product was detected for the endogenous huntingtin (also detected by the 4C8 mAb), which was expressed at about the same level as the mutant transgene protein.

However, we could not observe in our Western blot analysis the shorter fragment predicted from the immunofluorescence studies of inclusions, most probably owing to the high aggregation potential of such a fragment.

Conclusion

Cells expressing full-length mutant huntingtin show the formation of cytoplasmic and nuclear inclusions in a manner dependent on time and polyQ length. These inclusions are ubiquitinated and contain only a short N-terminal part of mutant huntingtin, thus appearing similar to those observed in the brain of HD patients (DiFiglia *et al.* 1997). This indicates that a processing step generating fragments with high aggregation potential operates in these cells. We are currently mapping the exact site of cleavage as a step towards the identification of the protease involved, which could constitute an important factor in the pathogenesis of HD, and thus a potential therapeutic target.

The cellular models presented so far for polyQ expansion diseases were based on transient transfections (Paulson *et al.* 1997; Butler *et al.* 1998; Cooper *et al.* 1998; Igarashi *et al.* 1998; Li & Li 1998; Martindale *et al.* 1998). In those short-duration assays the formation of cytoplasmic inclusions was observed only in cells expressing truncated mutant forms, or at low level in cells expressing a full-length huntingtin with 128 glutamine residues, whereas NIs were detected only in cells expressing a very short form of mutant huntingtin (exon-1 with 128 CAG repeats) (Martindale *et al.* 1998).

The model presented here provides a tool for the elucidation of the mechanisms leading to the cleavage of full-length huntingtin, its role in the formation of intracellular aggregates, and the effects of the latter on cellular physiology. It will be especially interesting to see whether in this system protease inhibition or transglutaminase inhibition might interfere with the formation of inclusions and their consequences on cell function and survival.

We thank Dr D. Devys, G. Yvert, Y. Lutz and Dr L. Tora for providing original constructs and antibodies, Dr M. H. V. Van Regenmortel for support and advice on the BIAcore experiments, and C. Weber for excellent technical assistance. This work was supported by funds from INSERM, CNRS, HUS and EEC (contract BMH4-CT96-0244). A.L. and J.F. are supported by fellowships from the DFG (Deutsche Forschungsgemeinschaft) and from l'Association Huntingtin France, respectively. Y.T. was supported by a fellowship from the Hereditary Disease Foundation.

References

Baur, E. vom, Zechel, C., Heery, D., Heine, M. J. S., Garnier, J. M., Vivat, V., Le Douarin, B., Gronemeyer, H., Chambon, P. & Losson, R. 1996 Differential ligand-dependent interactions between the AF-2 activating domain of nuclear receptors and the putative transcriptional intermediary factors mSUG1 and TIF1. *EMBO J.* **15**, 110–124.

Becher, M. W., Kotzuk, J. A., Sharp, A. H., Davies, S. W., Bates, G. P., Price, D. L. & Ross, C. A. 1998 Intranuclear neuronal inclusions in Huntington's disease and dentatorubral and pallidoluysian atrophy: correlation between the density of inclusions and *IT15* CAG triplet repeat length. *Neurobiol. Dis.* **4**, 387–397.

Butler, R., Leigh, P. N., McPhaul, J. & Gallo, J.-M. 1998 Truncated forms of the androgen receptor are associated with polyglutamine expansion in X-linked spinal and bulbar muscular atrophy. *Hum. Mol. Genet.* **7**, 121–127.

Cooper, J. K. (and 12 others) 1998 Truncated N-terminal fragments of huntingtin with expanded glutamine repeats form nuclear and cytoplasmic aggregates in cell culture. *Hum. Mol. Genet.* **7**, 783–790.

Davies, S. W., Turmaine, M., Cozens, B. A., DiFiglia, M., Sharp, A. H., Ross, C. A., Scherzinger, E., Wanker, E. E., Mangiarini, L. & Bates, G. P. 1997 Formation of neuronal intranuclear inclusions underlies the neurological dysfunction in mice transgenic for the HD mutation. *Cell* **90**, 537–548.

DiFiglia, M., Sapp, E., Chase, K. O., Davies, S. W., Bates, G. P., Vonsattel, J. P. & Aronin, N. 1997 Aggregation of huntingtin in neuronal intranuclear inclusions and dystrophic neurites in brain. *Science* **277**, 1990–1993.

Fraser, R. A., Rossignol, M., Heard, D. H., Egly, J.-M. & Chambon, P. 1997 SUG1, a putative transcriptional mediator and subunit of the PA700 proteasome regulatory complex, is a DNA helicase. *J. Biol. Chem.* **272**, 7122–7126.

Goldberg, Y. P. (and 10 others) 1996 Cleavage of huntingtin by apopain, a proapoptotic cysteine protease, is modulated by the polyglutamine tract. *Nature Genet.* **13**, 442–449.

Gossen, M. & Bujard, H. 1992 Tight control of gene expression in mammalian cells by tetracycline responsive promoters. *Proc. Natl Acad. Sci. USA* **89**, 5547–5551.

Gossen, M., Freundlieb, S., Bender, G., Mueller, G., Hillen, W. & Bujard, H. 1995 Transcriptional activation by tetracycline in mammalian cells. *Science* **268**, 1766–1769.

Gostout, B., Liu, Q. & Sommer, S. S. 1993 'Cryptic' repeating triplets of purines and pyrimidines (cRRY(i)) are frequent and polymorphic: analysis of coding cRRY(i) in the proopiomelanocortin (POMC) and TATA-binding protein (TBP) genes. *Am. J. Hum. Genet.* **52**, 1182–1190.

Igarashi, S. (and 18 others) 1998 Suppression of aggregate formation and apoptosis by transglutaminase inhibitors in cells expressing truncated DRPLA protein with an expanded polyglutamine stretch. *Nature Genet.* **18**, 111–117.

Imbert, G., Trottier, Y., Beckman, J. & Mandel, J.-L. 1994 The gene for the TATA-binding protein (TBP) that contains a highly polymorphic protein coding CAG repeat maps to 6q27. *Genomics* **21**, 667–668.

Imbert, G. (and 14 others) 1996 Cloning of the gene for spinocerebellar ataxia 2 reveals a locus with high sensitivity to expanded CAG/glutamine repeats. *Nature Genet.* **13**, 285–291.

Klement, I. A., Skinner, P. J., Kaytor, M. D., Yi, H., Hersch, S. M., Clark, H. B., Zoghbi, H. Y. & Orr, H. T. 1998 Ataxin-1 nuclear localization and aggregation: role in polyglutamine-induced disease in SCA1 transgenic mice. *Cell* **95**, 41–53.

Lescure, A., Lutz, Y., Eberhard, D., Jacq, X., Krol, A., Grummt, I., Davidson, I., Chambon, P. & Tora, L. 1995 The N-terminal domain of the human TATA-binding protein plays a role in transcription from TATA-containing RNA polymerase II and III promoters. *EMBO J.* **13**, 1166–1175.

Li, S.-H. & Li, X.-J. 1998 Aggregation of N-terminal huntingtin is dependent on the length of its glutamine repeats. *Hum. Mol. Genet.* **7**, 777–782.

Lopes-Cendes, I., Gaspar, C., Trottier, Y., Mandel, J.-L. & Rouleau, G. A. 1996 Searching for proteins containing polyglutamine expansions in a large group of spinocerebellar ataxia patients. *Am. Soc. Hum. Genet.* A**1559** (Abstract).

Lunkes, A. & Mandel, J.-L. 1998 A cellular model that recapitulates major pathogenic steps of Huntington's disease. *Hum. Mol. Genet.* **7**, 1355–1361.

Mangiarini, L. (and 10 others) 1996 Exon 1 of the *HD* gene with an expanded CAG repeat is sufficient to cause a progressive neurological phenotype in transgenic mice. *Cell* **87**, 493–506.

Martindale, D. (and 12 others) 1998 Length of huntingtin and its polyglutamine tract influences localization and frequency of intracellular aggregates. *Nature Genet.* **18**, 150–154.

Merry, D. E., Kobayashi, Y., Bailey, C. K., Taye, A. A. & Fischbeck, K. H. 1998 Cleavage, aggregation and toxicity of the expanded androgen receptor in spinal and bulbar muscular atrophy. *Hum. Mol. Genet.* **7**, 693–701.

Myszka, D. G. 1997 Kinetic analysis of macromolecular interactions using surface plasmon resonance biosensors. *Curr. Opin. Biotechnol.* **8**, 50–57.

Nelson, P., Clifford, C. & Nirenberg, M. 1976 Synapse formation between clonal neuroblastoma x glioma hybrid cells and striated muscle cells. *Proc. Natl Acad. Sci. USA* **73**, 123–127.

Nielsen, J. E, Koefoed, P., Abell, K., Hasholt, L., Eiberg, H., Fenger, K, Niebuhr, E. & Sorensen, A. S. A. 1998 CAG repeat expansion in autosomal dominant pure spastic paraplegia linked to chromosome 2p21–p24. *Hum. Mol. Genet.* **6**, 1811–1816.

Onodera, O., Roses, A. D., Tsuji, S., Vance, J. M., Strittmatter, W. J. & Burke, J. R. 1996 Toxicity of expanded polyglutamine-domain proteins in *Escherichia coli*. *FEBS Lett.* **399**, 135–139.

Paulson, H. L. & Fischbeck, K. H. 1996 Trinucleotide repeats in neurogenetic disorders. *Annu. Rev. Neurosci.* **19**, 79–107.

Paulson, H. L., Perez, M. K., Trottier, Y., Trojanowski, J. Q., Subramony, S. H., Das, S. S., Vig, P., Mandel, J. L., Fischbeck, K. H. & Pittman, R. N. 1997 Intranuclear inclusions of expanded polyglutamine protein in spinocerebellar ataxia type 3. *Neuron* **19**, 333–344.

Perutz, M. F., Johnson, T., Suzuki, M. & Finch, J. T. 1994 Glutamine repeats as polar zippers: their possible role in inherited neurodegenerative diseases. *Proc. Natl Acad. Sci. USA* **91**, 5355–5358.

Ross, C. A., McInnis, M. G., Margolis, R. L. & Li, S.-H. 1993 Genes with triplets: candidate mediators of neuropsychiatric disorders. *Trends Neurosci.* **16**, 254–260.

Saudou, F., Finkbeiner, S., Devys, D. & Greenberg, M. E. 1998 Huntingtin acts in the nucleus to induce apoptosis but death does not correlate with the formation of intranuclear inclusions. *Cell* **95**, 55–66.

Scherzinger, E., Lurz, R., Turmaine, M., Mangiarini, L., Hollenbach, B., Hasenbank, R., Bates, G. P., Davies, S. W., Lehrach, H. & Wanker, E. E. 1997 Huntingtin-encoded polyglutamine expansions form amyloid-like protein aggregates *in vitro* and *in vivo*. *Cell* **90**, 549–558.

Skinner, P. J., Koshy, B., Cummings, C. J., Klement, I. A., Helin, K., Servadio, A., Zoghbi, H. Y. & Orr, H. T. 1997 SCA1 pathogenesis involves alterations in nuclear matrix associated structures. *Nature* **389**, 971–974.

Stevanin, G. (and 17 others) 1996 Screening for proteins with polyglutamine expansions in autosomal dominant cerebellar ataxias. *Hum. Mol. Genet.* **5**, 1887–1892.

Trottier, Y. (and 12 others) 1995 Polyglutamine expansion as a pathological epitope in Huntington's disease and four dominant cerebellar ataxias. *Nature* **378**, 403–406.

Trottier, Y., Zeder-Lutz, G. & Mandel, J.-L. 1998 Selective recognition of proteins with pathological polyglutamine tracts by a monoclonal antibody. In *Genetic instabilities and hereditary neurological diseases* (ed. R. D. Wells & S. T. Warren), pp. 447–453. San Diego: Academic Press.

Wells, R. D. & Warren, S. T. 1998 *Genetic instabilities and hereditary neurological diseases*. San Diego: Academic Press.

Wittekindt, O. (and 13 others) 1998 The human small conductance calcium-regulated potassium channel gene (hSKCa3) contains two CAG repeats in exon 1, is on chromosome 1q21.3, and shows a possible association with schizophrenia. *Neurogenetics* **1**, 259–265.

Zoghbi, H. Y. 1997 CAG repeats in SCA6: anticipating new clues. *Neurology* **49**, 1–5.

9

Evidence for a recruitment and sequestration mechanism in Huntington's disease

Elizabeth Preisinger, Barbara M. Jordan, Aleksey Kazantsev and David Housman

Introduction

Eight dominantly inherited neurodegenerative diseases, including Huntington's disease, are caused by the expansion of a CAG trinucleotide repeat within protein-coding sequences (Reddy & Housman 1997). If translated, these mutant expanded trinucleotide repeat domains can encode up to 100 glutamine residues embedded within the affected proteins. Each of these polyglutamine (polyQ) expansions occurs in one of a diverse set of widely expressed and seemingly unrelated proteins. Late-onset progressive neurodegeneration is a common feature in these diseases, yet they differ in important respects. Specific regions of the brain are affected in the various diseases, with correspondingly distinct clinical presentations. Thus Huntington's disease is characterized by late-onset degeneration and profound neuronal loss in a specific subset of striatal neurons, whereas the spinocerebellar ataxias manifest themselves as degeneration of the Purkinje cell layer of the cerebellum. The molecular basis for selective vulnerability within neuronal populations in the polyQ diseases is not known, although differences in expression levels of the affected proteins between different neuronal subpopulations might have a role (Kosinski *et al.* 1997).

Perutz and colleagues (Perutz *et al.* 1994) proposed several years ago a 'polar zipper' model of polyQ interaction, in which homopolymeric polyQ domains can interact in a non-covalent manner through the formation of intermolecular hydrogen bonds. It has now been shown that extended polyQ tracts aggregate as amyloid-like protein *in vitro* and form neuronal intranuclear inclusions in mice transgenic for the Huntington's disease mutation (Davies *et al.* 1997; Scherzinger *et al.* 1997). In Huntington's disease, aggregation of extended polyQ tracts causes formation of neuronal intranuclear inclusions as well (DiFiglia *et al.* 1997).

At the cellular level, the precise mechanisms of polyQ toxicity as a cause of neuronal dysfunction and neuron loss in the trinucleotide repeat diseases are unknown. Although it is clear that a toxic 'gain of function' is attributable to the polyQ tracts themselves, the role of large polyQ aggregates in neuronal dysfunction and in neuronal loss in these progressive diseases remains to be elucidated. Moreover, at least in Huntington's disease, neuronal dysfunction as manifested by clinical abnormalities precedes the development of discernible histopathologic changes such as neuronal degeneration and neuronal loss (Vonsattel *et al.* 1985). There might well be distinct polyQ-dependent pathogenetic mechanisms that precede the formation of frank aggregates and that might be associated with neuronal dysfunction at early stages of disease. One such potential mechanism is suggested by the work of Cha *et al.* (1998), who show a decrease in levels of certain neurotransmitter receptors at the transcriptional level in mice transgenic for an expanded CAG repeat. This altered expression of specific subsets of neurotransmitter receptors precedes clinical symptoms and the formation of aggregates in these mice.

Here we describe certain anomalous intrinsic biochemical properties of polyQ tracts that might contribute to the ability to self-aggregate. We demonstrate that the subcellular localization of expanded polyQ aggregates within cells is determined by the properties of the molecules within which the polyQ stretch is contained. We clearly demonstrate polyQ-dependent interactions between different polyQ-containing molecules in transient transfection experiments. We show not only that expanded polyQ molecules aggregate on their own in cells but that they can recruit polypeptides with short polyQ domains into cytoplasmic and nuclear aggregates. These results provide evidence for a recruitment mechanism for pathogenesis in the polyQ neurodegenerative disorders. In susceptible cells, extended polyQ tracts in mutant proteins might interact with and sequester or deplete certain endogenous polyQ-containing cellular proteins.

Generating polyQ of lengths ranging from 25 to 300 glutamine residues

To study polyQ aggregation, we generated alternating CAG/CAA repeats encoding 25 glutamine residues (normal in Huntington's disease), extended 47 and 104 glutamine residues (pathological in Huntington's disease), and extended 191, 230 and 300 glutamine residues (elongated beyond the pathological range). All lengths of alternating CAG/CAA repeats were quite stable in bacteria, in contrast to native extended CAG repeats. A short N-terminus Huntington's disease complementary DNA (cDNA) fragment including the Kozak box and the first 17 residues was ligated to polyQ repeats of different lengths (25–300 glutamine residues). To monitor the formation of aggregates in cells, we fused polyQ tracts at the C-terminus with either a 28 residue c-Myc tag or a 230 residue enhanced green fluorescent protein (EGFP) tag. PolyQ fusions were subcloned into the pcDNA 3.1 expression vector, which includes the human cytomegalovirus immediate-early promoter–enhancer for high-level expression in mammalian cells.

Anomalous physical properties of polyQ on SDS–PAGE

Initially we wished to characterize the products of each of our constructs in a translation system *in vitro* that would permit an analysis of the properties of the encoded polypeptides in the absence of cell metabolism. This characterization demonstrated that polypeptides predicted by the sequence of the synthetic open reading frame were indeed produced, but it also gave an insight into an unusual property of polyQ-containing polypeptides. It has previously been observed that polypeptides containing extended polyQ segments show an atypical mobility on SDS–PAGE. The polypeptides produced by our constructs exhibited a mobility in SDS–PAGE that was proportional to predicted molecular mass but demonstrated a systematic relative decrease in electrophoretic mobility consistent with these previous reports (figure 9.1). The ability to extend the length of polyQ to 300 residues allowed a systematic evaluation of this altered mobility. We found that the observed mobility of a polypeptide composed almost exclusively of polyQ was 40% of the expected mobility of a polypeptide of identical length with a random amino acid sequence (figure 9.2). Atypical mobility of normal and extended polyQ peptides was seen on SDS–PAGE gels at various concentrations of acrylamide.

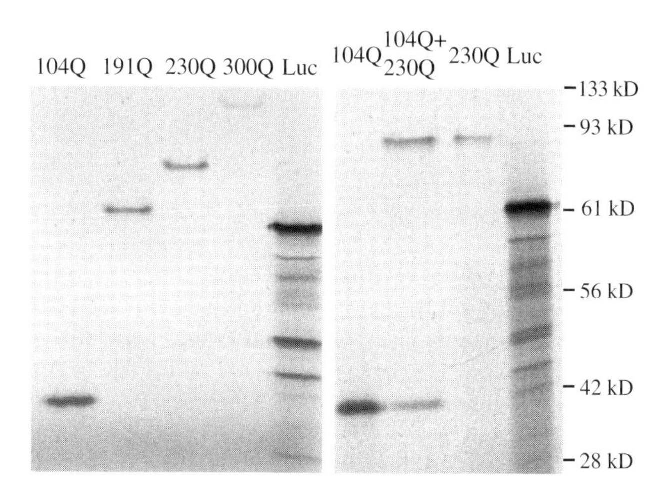

Figure 9.1 Translation and co-translation *in vitro* of polyQ expression constructs. PolyQ expression constructs were translated *in vitro* using T7 RNA polymerase and rabbit reticulocyte lysates in the presence of [^{35}S]methionine. Left, SDS–PAGE analysis of the alternating CAG/CAA repeats translated *in vitro*. Right, no detectable intermediates between co-translated 104Q and 230Q constructs. Luc, 61 kDa luciferase translated *in vitro*.

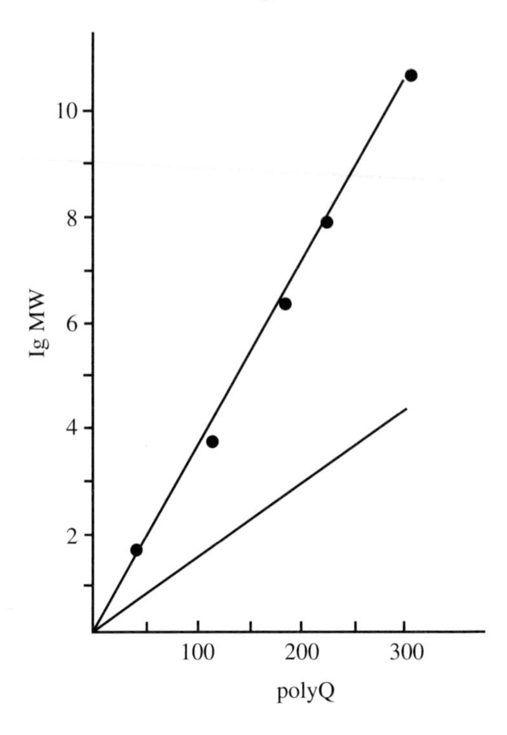

Figure 9.2 Retarded mobility in SDS–PAGE of polyQ tracts translated *in vitro*. The solid line represents the expected mobility of products translated *in vitro* on the basis of predicted molecular mass. In contrast, filled circles show the observed mobilities of these products in SDS–PAGE. The observed mobility of these products, composed almost exclusively of polyQ, was consistently 40% of that expected across a range of 25–300 glutamine repeats.

Two possible explanations for the anomalous behaviour of polyQ in SDS–PAGE were considered: first, multimers of polyQ might be stable in SDS–PAGE; second, polyQ might form an intrinsic structure with atypical physical properties. To discriminate between these hypotheses, we performed an experiment that would allow the association of polyQ tracts of two different lengths after translation *in vitro* by co-translating constructs with 104 and 230 glutamine residues respectively. The outcome of this experiment is shown in figure 9.1. Polypeptides migrating with mobilities of 104 and 230 glutamine residues are observed, but not polypeptides with an intermediate mobility. These findings led us to conclude that polyQ translated *in vitro* has an atypical intrinsic structure in SDS–PAGE, because we failed to detect stable polyQ multimers.

Naked extended polyQ peptides form cytoplasmic and perinuclear aggregates, whereas nuclear localization of aggregates depends on the flanking sequence of extended polyQ

We next tested the ability of these naked extended polyQ tracts to form aggregates in cells. In transient transfection experiments in Cos-1 cells, followed by fluorescence microscopy, we found that polypeptides of 25 glutamine residues were expressed diffusely throughout the cytoplasm of transfected cells. When we tested our highly expanded polyQ constructs, we found that 104, 191, 230 and 300 glutamine residues all formed aggregates in most transfected cells. Aggregate formation began as early as 16 h after transfection; it appeared typically at this time point as multiple small crystalline star-like structures distributed throughout the cytoplasm. At this early time point, a background of diffuse EGFP fluorescence could be seen in these cells as well. By 24 h after transfection, however, aggregate-forming cells had lost this diffuse background fluorescence, which presumably represented expressed polyQ tracts in soluble form. Instead, aggregates within a cell seemed to coalesce into a single, dense, brilliantly fluorescent spherical structure that could be 5 mm or larger (figure 9.3*a* (plate 9)). These single aggregates were clearly located in the cytoplasm but were often perinuclear or closely apposed to the nuclear membrane. In many instances the nuclear membrane seemed indented or distorted by the presence of the aggregate. Although others have reported a proportional polyQ length-dependent effect on rate of aggregate formation (Li & Li 1998), we found no appreciable difference in the rate of aggregate formation under these conditions among the constructs expressing 104, 191, 230 or 300 glutamine residues. However, we might have been unable to detect such a relationship owing to the very rapid formation of aggregates with any of our long polyQ constructs.

Because neuronal intranuclear inclusions have been reported in the human CAG-repeat diseases as well as in models in the mouse and fly (DiFiglia *et al.* 1997; Paulson *et al.* 1997; Warrick *et al.* 1998) we sought to direct our polyQ peptides to the nucleus by fusing them with a strong nuclear localization signal. To accomplish this we chose the 650 residue nucleolin protein, which has strong nuclear and nucleolar localization signals. We generated expression constructs (polyQ–nucleolin–EGFP) by inserting nucleolin cDNA between polyQ (25 (25Q), 104 (104Q) and 300 glutamine residues) and EGFP sequences. When we expressed normal-length 25Q–nucleolin–EGFP and extended polyQ–nucleolin–EGFP fusion proteins in cells, we found that all polypeptides were located in the nucleus and particularly in the nucleoli. No fluorescent signal was seen in the cytoplasm.

When we transfected cells with extended polyQ–nucleolin–EGFP constructs, we saw nucleolar aggregates formed by extended polyQ tracts (figure 9.3*b* (plate 9)). These nucleolar aggregates were quite distinct morphologically from the cytoplasmic aggregates generated by the naked polyQ

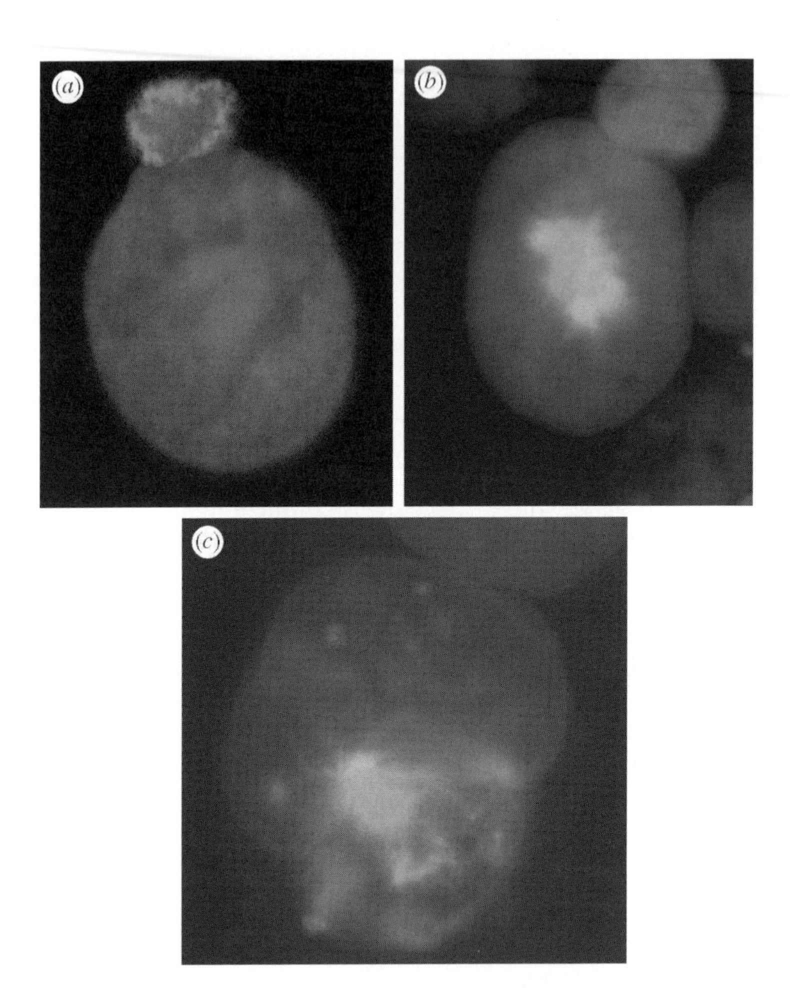

Figure 9.3 Extended polyQ constructs expressed by transient transfection in Cos-1 cells, 48 h after transfection. (*a*) 104Q–c-Myc fusion construct forms a single dense brilliantly fluorescent spherical structure located in the cytoplasm but closely apposed to the nuclear membrane. Note the poor penetration of the anti-Myc antibody, causing the appearance of a rim around the polyQ aggregate. (*b*) 104Q–nucleolin–EGFP construct. Nucleolar aggregates displacing the nucleoplasm and showing a fibrillary, almost spiculated, appearance. (*c*) 47Q–EGFP construct, showing multiple sites for initiation of aggregation within the cell and a less compact aggregate structure. (See also colour plate section.)

constructs. Cytoplasmic aggregates appeared as dense and compact bodies, whereas nucleolar aggregates showed a fibrillary, almost spiculated, appearance. These results show that the subcellular localization of polyQ aggregates can be determined by polyQ flanking sequences. The basis for the morphological distinction that we observed between cytoplasmic and nucleolar aggregates has not been thoroughly investigated, but it is possible that there are fundamental differences between subcellular compartments in the cellular machinery for processing and degrading these aberrant proteins.

Extended polyQ with a nuclear localization signal can translocate polyQ-containing proteins to the nucleus through polyQ interactions

The naked extended polyQ constructs lack a nuclear localization signal and when transfected alone are found aggregated exclusively in the cytoplasm. To test directly the intermolecular interactions between polyQ-containing molecules, we co-expressed 104Q–c-Myc with 104Q–nucleolin–EGFP. Strikingly, we now found 104Q–c-Myc in heterogeneous aggregates in the nucleus (figure 9.4*b*, plate 9)). Nuclear localization was strictly dependent on co-aggregation with 104Q–nucleoli–EGFP fusion protein. A nucleolin–EGFP construct without polyQ was unable to pull 104Q–c-Myc into the nucleus, demonstrating the polyQ-dependent nature of the interaction. Despite the presence of the strong nuclear localization signal *in cis*, we found that polyQ–nucleolin–EGFP also aggregated, in some cells, with extended polyQ in the cytoplasm, and thus was excluded from the nucleus. Thus the subcellular localization of aggregation depends in general on the functional characteristics of the protein in which the polyQ is contained. Nonetheless, strong intermolecular interactions mediated by polyQ domains can in some cases be sufficient to override the effects of such intrinsic localization signals.

Extended polyQ peptides can trap normal-length polyQ peptides in aggregates

A number of cellular proteins have been identified with naturally occurring homopolymeric polyQ segments ranging in length from six to 38 glutamine residues. We sought to establish in an experimental system whether normal-length polyQ peptides of no more than 40 glutamine residues could interact with and perhaps aggregate with extended polyQ tracts when co-expressed in cells.

To test this hypothesis, we co-transfected normal-length 25Q–EGFP and extended 104Q–c-Myc. Normal-length polyQ tracts showed a diffuse pattern of expression when transfected alone. Remarkably, these same normal-length polyQ tracts were recruited into cellular aggregates when they

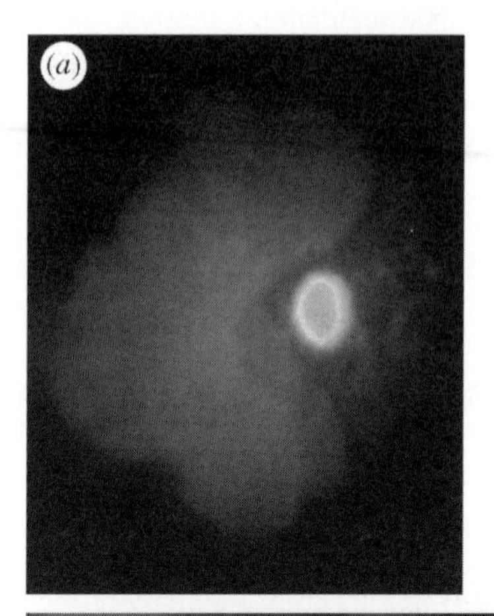
(a)

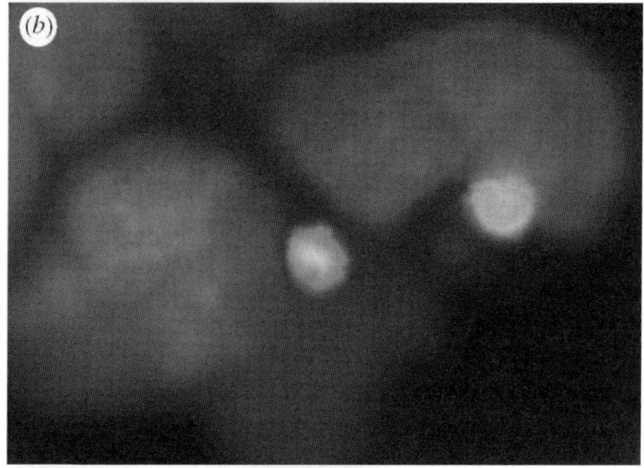
(b)

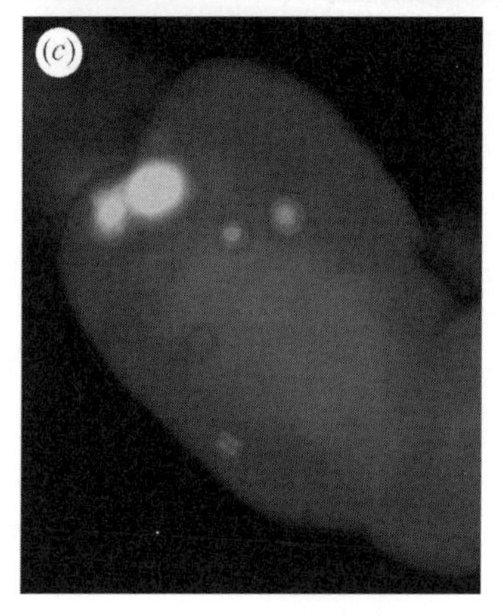
(c)

were co-expressed with extended polyQ constructs (figure 9.4*a* (plate 11)). In contrast, when EGFP lacking a polyQ segment was co-expressed with 104Q–c-Myc, EGFP fluorescence was not detected in aggregates. Co-expression experiments with 25Q–nucleolin–EGFP and 104Q–c-Myc yielded co-aggregates in nucleoli in a small fraction of co-transfected cells (figure 9.4*c* (plate 11)). Control experiments in which EGFP–nucleolin was co-transfected with 104Q–c-Myc gave cytoplasmic aggregates that had no EGFP signal. These results demonstrate the strict polyQ-dependent nature of the co-aggregation phenomenon.

Discussion

We synthesized alternating CAG/CAA triplet repeats, ranging from 25 to 300 glutamine residues. All repeats were highly stable in bacteria and were easy to manipulate by using conventional methods of molecular cloning. We generated DNA constructs expressing essentially naked polyQ tracts of different lengths. To direct extended polyQ tracts to the nucleus and to model more closely the neuronal intranuclear aggregates found in CAG repeat neurodegenerative diseases, we fused alternating CAG/CAA triplet repeats with nucleolin cDNA.

Our results highlight a critical property of polyQ aggregate formation. We show that relatively short, soluble peptides with short polyQ tracts can co-precipitate in cells in the presence of extended polyQ tracts. Once the process is initiated, an extended length of polyQ is not required for joining the aggregate. Moreover, we have demonstrated clearly in our experiments that co-aggregation is strictly dependent on the presence of a polyQ segment in both molecules. Thus, as suggested by Perutz *et al.* (1994), this interaction is mediated directly through the polyQ domains.

We propose that extended mutant polyQ has the potential to interact with any normal polyQ-containing protein in the cell. A possible pathogenic mechanism in the trinucleotide neurodegenerative diseases might involve the

Figure 9.4 Co-expression and co-aggregation of normal length and extended polyQ tracts. (*a*) Co-expressed 25Q–EGFP and 104Q–c-Myc detected in a single cytoplasmic perinuclear aggregate. Because the short polyQ (green) is tagged with the intrisically fluorescent EGFP, and the long polyQ (red) is detected by poorly penetrating fluorescent antibody, there is an artefact of the appearance of a red rim around the aggregate. We conclude that both short and long polyQ molecules are present throughout the aggregate, because experiments with reciprocally tagged constructs gave an identical appearance (results not shown). (*b*) Co-expression of 104Q–nucleolin–EGFP (green) and 104Q–c-Myc (red), showing the translocation of 104Q–c-Myc into the nucleus by 104Q–nucleolin. This figure illustrates the heterogeneous composition of the co-aggregates. (*c*) Co-expression of 25Q–nucleolin–EGFP (green) with 104Q–c-Myc (red). Short polyQ fused to nucleolin aggregates in the nucleus when co-expressed with extended polyQ. (See also colour plate section.)

Table 9.1 *Some examples of normal cellular proteins containing significant polyQ stretches, as revealed by database searching*

GenBank accession no.	*gene*	*function*	*polyQ stretch (single-letter codes)*
L37868	N-Oct 3	nervous system-specific POU domain transcription factor; homeodomain protein	QQQHQQQQQQQQQQQQQQQQQQQQQ
D26155, X72889	hSNF2a	transcriptional co-activator for glucocorticoid, oestrogen and retinoic acid receptors	QQQQQQQQQQQQQQQQQQQQQQQP QQQPPQPQTQQQQQ
AF071309	OPA-containing protein	unknown	QQQQQQQQQQQQQQQQQQQQQQQQ QQYHIRQQQQQQILRQQQQQQQQQQ QQQQQQQQQQQQQQQQHQQQQQQQ AAPPQPQPQSQPQFQRQGLQQTQQQQQ
M55654	TATA-binding protein	transcriptional activator	QQRQQQQQQQQQQQQQQQQQQQQQ QQQQQQQQQQQQQQQQQ
AF012108	amplified in breast cancer (AIB1)	steroid receptor co-activator	QQQQQQQQQQQQQQQQQQQQQQQQ QQQQQTQ
AF016031	TRAM-1	thyroid hormone receptor activator	QQQQQQQQQQQQQQQQQQQQQQQQ QQQQQTQ
AF010227	RAC3	transcriptional co-activator with intrinsic histone acetyl-transferase activity	QQQQQQQQQQQQQQQQQQQQQQQQ QQTQ
AF036892	nuclear receptor co-activator (ACTR)	histone acetyltransferase; transcriptional cofactor	QQQQQQQQQQQQQQQQQQQQQQQQ QQTQ

GenBank accession no.	*gene*	*function*	*polyQ stretch (single-letter codes)*
L32832	ATBF1	expresses in a neuronal differentiation-dependent manner; zinc-finger homeodomain protein	QQQQQQQQQQQQQQQQQQQAQ
U47741	CREB-binding protein (CBP)	transcriptional adaptor; histone acetyltransferase	QLLQQQQQQQQQQQQQQQQQQ
AF010403	ALR	unknown	QQLQQQQQLQQQQQLQQQQQQQLQ QQQQLQQQQLQQQQQQQQLQQQQQ QQLQQQQQLQQQQQQQQQQQFQQQ QQQQQMGLLNQSRTLLSPQQQQQQQ

depletion or sequestration of normal cellular proteins that contain short homopolymeric polyQ domains. Protein database searches reveal that hundreds of polyQ-containing proteins have been identified so far (see table 9.1). An interesting class of nuclear proteins that contain glutamine-rich regions and often homopolymeric glutamine stretches are transcription factors and transcriptional co-activators. The expression of many transcriptional activators (or repressors) can and must be exquisitely specific with respect to cell type and developmental stage. Most probably the vulnerability to polyQ expansion of specific neuronal populations in these diseases is governed not simply by the properties of the mutant proteins themselves. In fact, it has been shown that an expanded polyQ repeat inserted into hypoxanthine phosphoribosyltransferase caused a neurological phenotype and intranuclear neuronal inclusions in transgenic mice (Ordway *et al.* 1997). There are certain to be cell-type-specific cofactors for pathogenesis that remain to be elucidated. Among such cell-type-specific cofactors could be normal polyQ-containing molecules that might interact with expanded mutant polyQ tracts within specific cellular compartments or at particular developmental stages. An intriguing candidate for such a role in the neurodegenerative diseases, for example, is the brain-specific Pou-domain-containing homeobox protein N-Oct-3, which contains a nearly homopolymeric polyQ stretch of 27 glutamine residues.

Nuclear localization of mutant polyQ has been shown to be important in the pathogenesis of neurological disease. Klement *et al.* (1998) recently reported that nuclear localization of mutant ataxin-1, but not nuclear aggregation, was necessary for development of ataxia in a transgenic mouse model. We have shown that neither nuclear localization nor nuclear aggregation in our transfected cell lines takes place in the absence of a nuclear localization signal. However, we have also shown that interactions between heterogeneous polyQ-containing molecules can mediate the subcellular localization of polyQ aggregates. Thus, although the nuclear localization of mutant extended polyQ might indeed be critical for pathogenesis, a nuclear localization signal *in cis* might not be required for mutant huntingtin, for example, to be found in the nucleus.

References

Cha, J. H., Kosinski, C. M., Kerner, J. A., Alsdorf, S. A., Mangiarini, L., Davies, S. W., Penney, J. B., Bates, G. P. & Young, A. B. 1998 Altered brain neurotransmitter receptors in transgenic mice expressing a portion of an abnormal human Huntington disease gene. *Proc. Natl Acad. Sci. USA* **95**, 6480–6485.

Davies, S. W., Turmaine, M., Cozens, B. A., DiFiglia, M., Sharp, A. H., Ross, C. A., Scherzinger, E., Wanker, E. E., Mangiarini, L. & Bates, G. P. 1997 Formation of neuronal intranuclear inclusions underlies the neurological dysfunction in mice transgenic for the HD mutation. *Cell* **90**, 537–548.

DiFiglia, M., Sapp, E., Chase, K. O., Davies, S. W., Bates, G. P., Vonsattel, J. P. & Aronin, N. 1997 Aggregation of huntingtin in neuronal intranuclear inclusions and dystrophic neurites in brain. *Science* **277**, 1990–1993.

Klement, I. A., Skinner, P. J., Kaytor, M. D., Yi, H., Hersch, S. M., Clark, H. B., Zoghbi, H. Y. & Orr, H. T. 1998 Ataxin-1 nuclear localization and aggregation: role in polyglutamine-induced disease in SCA1 transgenic mice. *Cell* **95**, 41–53.
Kosinski, C. M., Cha, J. H., Young, A. B., Persichetti, F., MacDonald, M., Gusella, J. F., Penney Jr, J. B. & Standaert, D. G. 1997 Huntingtin immunoreactivity in the rat neostriatum: differential accumulation in projection and interneurons. *Exp. Neurol.* **144**, 239–247.
Li, S. H. & Li, X. J. 1998 Aggregation of N-terminal huntingtin is dependent on the length of its glutamine repeats. *Hum. Mol. Genet.* **7**, 777–782.
Ordway, J. M. (and 11 others) 1997 Ectopically expressed CAG repeats cause intranuclear inclusions and a progressive late onset neurological phenotype in the mouse. *Cell* **91**, 753–763.
Paulson, H. L., Perez, M. K., Trottier, Y., Trojanowski, J. Q., Subramony, S. H., Das, S. S., Vig, P., Mandel, J. L., Fischbeck, K. H. & Pittman, R. N. 1997 Intranuclear inclusions of expanded polyglutamine protein in spinocerebellar ataxia type 3. *Neuron* **19**, 333–344.
Perutz, M. F., Johnson, T., Suzuki, M. & Finch, J. T. 1994 Glutamine repeats as polar zippers: their possible role in inherited neurodegenerative diseases. *Proc. Natl Acad. Sci. USA* **91**, 5355–5358.
Reddy, P. S. & Housman, D. E. 1997 The complex pathology of trinucleotide repeats. *Curr. Opin. Cell Biol.* **9**, 364–372.
Scherzinger, E., Lurz, R., Turmaine, M., Mangiarini, L., Hollenbach, B., Hasenbank, R., Bates, G. P., Davies, S. W., Lehrach, H. & Wanker, E. E. 1997 Huntingtin-encoded polyglutamine expansions form amyloid-like protein aggregates *in vitro* and *in vivo*. *Cell* **90**, 549–558.
Vonsattel, J. P., Myers, R. H., Stevens, T. J., Ferrante, R. J., Bird, E. D. & Richardson Jr, E. P. 1985 Neuropathological classification of Huntington's disease. *J. Neuropathol. Exp. Neurol.* **44**, 559–577.
Warrick, J. M., Paulson, H. L., Gray-Board, G. L., Bui, Q. T., Fischbeck, K. H., Pittman, R. N. & Bonini, N. M. 1998 Expanded polyglutamine protein forms nuclear inclusions and causes neural degeneration in *Drosophila*. *Cell* **93**, 939–949.

Biochemistry of huntingtin in cell cultures and *in vitro*

10

Aggregation of truncated GST–HD exon 1 fusion proteins containing normal range and expanded glutamine repeats

B. Hollenbach, E. Scherzinger, K. Schweiger, R. Lurz, H. Lehrach and E. E. Wanker

Introduction

Huntington's disease (HD) is a progressive neurodegenerative disorder with autosomal dominant inheritance (Harper 1996). The disorder is characterized by selective neuronal cell death, primarily in the cortex and striatum, leading to psychiatric symptoms, choreatic movement disturbances and cognitive decline (Vonsattel *et al.* 1985). Onset of the disease is generally in midlife but can vary from early childhood until well into old age. The mutation causing HD is an expansion of polymorphic CAG repeats located within exon 1 of the *IT-15* gene (Huntington's Disease Collaborative Research Group 1993). The CAG repeat is translated into a polyglutamine (polyQ) stretch. Thus in HD patients, huntingtin proteins with 38–182 glutamine residues are expressed, whereas in healthy individuals huntingtin proteins with 8–41 glutamine residues are synthesized (Rubinsztein *et al.* 1996; Sathasivam *et al.* 1997). Furthermore, there is a strong correlation between, on the one hand, the age of onset and the severity of symptoms and, on the other hand, the length of the polyQ repeat expansion. However, the molecular mechanism by which an elongated polyQ stretch leads to selective neurodegeneration is still unknown. Several lines of mice transgenic for the HD mutation have been generated to study the pathomechanism of HD. Mangiarini *et al.* (1996) showed that transgenic mice expressing exon 1 of the human HD gene carrying CAG repeat expansions of 115–156 units exhibited a progressive neurological phenotype with similarities to HD. Electron microscopy revealed that the formation of characteristic neuronal nuclear inclusions (NIIs) containing aggregated huntingtin protein underlies the neurological dysfunction in these mice (Davies *et al.* 1997). In addition, we have shown that a recombinant glutathione *S*-transferase (GST)–HD exon 1 fusion protein with a polyQ expansion in the pathological range (51 glutamine residues) forms high-molecular-mass protein aggregates *in vitro* after site-specific proteolysis of the fusion protein. These aggregates, which were not detected after proteolysis of shorter fusion proteins (20 or 30 glutamine residues), revealed a fibrillar or ribbon-like morphology similar to that of scrapie prions and β-amyloid fibrils in Alzheimer's disease (Scherzinger *et al.* 1997). Recently, high-molecular-mass polyQ-containing protein aggregates were also detected in the brains of patients with HD (Becher *et al.* 1998; DiFiglia *et al.* 1997), SCA1 (Matilla *et al.* 1997; Skinner *et al.* 1997), SCA3 (Paulson *et al.* 1997), SCA7 (Holmberg *et al.* 1998) and DRPLA (Becher *et al.* 1997; Igarashi *et al.* 1998), suggesting that all these diseases are the result of a toxic amyloid fibrillogenesis, as has been proposed for Alzheimer´s disease and the prion diseases.

Here we show that a truncated GST–HD exon 1 fusion protein with 51 glutamine residues (GST–HD51ΔP), which lacks the proline-rich region located immediately downstream of the glutamine repeat, self-assembles into more highly ordered structures with a fibrillar morphology. Under the same conditions, fusion proteins containing only 20 or 30 glutamine residues do not self-aggregate.

Materials and methods

Strains and plasmids

Escherichia coli DH10B (BRL) was used for plasmid construction; *E. coli* SCS1 (Stratagene) was used for the expression of GST–HD fusion proteins. Standard protocols for DNA manipulations were followed. The construction of the plasmids pCAG20ΔP and pCAG51ΔP encoding the fusion proteins GST–HD20ΔP and GST–HD51ΔP, respectively, has been described elsewhere (Wanker *et al.* 1999). *IT-15* cDNA sequences encoding the N-terminal portion of huntingtin, including the CAG repeats, were amplified by PCR with the oligonucleotides ES25 (5'-TGGGATCCGCATGGCGACCCTGGAAA-AGCTGATGAAGG-3') and ES27 (3'-CTCCTCGAGCGGCGGTGGCG-GCTGTTGCTGCTGCTGCTG-5') as primers and the plasmids pCAG30 and pCAG83 as templates (Scherzinger *et al.* 1997). Conditions for PCR were as described (Mangiarini *et al.* 1996). The resulting cDNA fragments were gel-purified, digested with *Bam*HI and *Xho*I and then inserted into the *Bam*HI–*Xho*I site of the expression vector pGEX-5X-1 (Pharmacia), yielding pCAG30ΔP and pCAG83ΔP, respectively. The plasmids pCAG30ΔP and pCAG83ΔP were used for the expression of the truncated HD exon 1 fusion proteins containing 30 and 83 glutamine residues, respectively.

Purification and proteolytic cleavage of GST–HD fusion proteins

The procedure for the purification of GST–HD fusion proteins has been described elsewhere (Wanker *et al.* 1999). The purified proteins were stored at –70°C at a concentration of 1 mg ml^{-1}. The GST–HD fusion proteins (2 μg) were digested with modified trypsin (Boehringer Mannheim, sequencing grade) at an enzyme/substrate ratio of 1:20 (w/w). The reaction was performed in 20 μl of 20 mM Tris–HCl (pH 8), 150 mM NaCl and 2 mM $CaCl_2$ for 3–16 h at 37°C.

Microscopic analysis of GST–HD fusion proteins

For electron microscopic observation, the native or protease-digested GST–HD fusion proteins were adjusted to a final concentration of 50 μg ml^{-1} in 40 mM Tris–HCl (pH 8.0), 150 mM NaCl, 0.1 mM EDTA and 5% (v/v) glycerol. Samples were negatively stained with 1% (w/v) uranyl acetate and viewed in a Philips CM100 electron microscope.

Results

Purification of GST–HD fusion proteins lacking the proline-rich region

Previously we have produced GST–HD exon 1 fusion proteins containing polyQ tracts of various lengths and the proline-rich region located C-terminal to the polyQ tract in HD exon 1 (Scherzinger *et al.* 1997). In this study, truncated GST–HD exon 1 fusion proteins with 20 (GST–HD20ΔP), 30 (GST–HD30ΔP), 51(GST–HD51ΔP) and 83 (GST–HD83ΔP) glutamine residues, lacking most of the proline-rich region, were used. These proteins were purified by affinity chromatography on glutathione–agarose (Smith & Johnson 1988). The structures of the various GST–HD fusion proteins are shown in figure 10.1. SDS–PAGE analysis of the purified proteins GST–HD20ΔP, GST–HD30ΔP, GST–HD51ΔP and GST–HD83ΔP revealed major bands migrating at about 30, 33, 39 and 45 kDa, respectively (figure 10.2). These bands were also detected by immunoblot analysis with the anti-huntingtin antibody HD1 (results not shown). All recombinant proteins migrated at a size corresponding nearly to that predicted from their amino acid sequence. Interestingly, in the protein preparations of GST–HD51ΔP and GST–HD83ΔP, an additional band was detected on top of the gel, indicating the presence of insoluble high-molecular-mass protein aggregates in these preparations (Scherzinger *et al.* 1997). Such high-molecular-mass protein bands were never detected in the protein preparations of GST–HD20ΔP or GST–HD30ΔP.

Self-assembly of GST–HD51ΔP into more highly ordered structures

To examine the morphology of the aggregated GST–HD51ΔP fusion protein, the high-molecular-mass aggregates were separated from the soluble protein

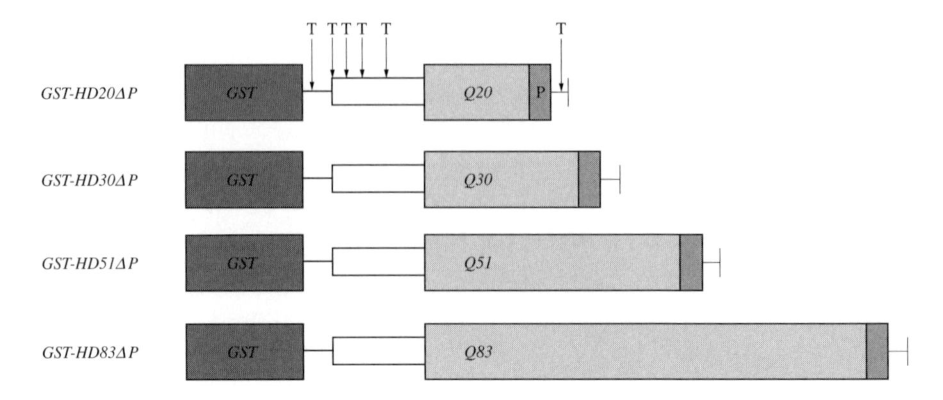

Figure 10.1 Schematic representation of the primary structure of GST–HD fusion proteins. The amino acids corresponding to the N-terminal portion of huntingtin are boxed. Q and P stand for polyglutamine and polyproline extensions, respectively. Arrows labelled T indicate cleavage sites for trypsin.

by centrifugation and then analysed by electron microscopy. Ordered protein aggregates with a strikingly uniform fibrillar morphology were observed (figure 10.3). The fibrils had a diameter of 25 nm and a length of several micrometres. At higher magnification, small spherical particles could be detected on the surface of the fibrils, presumably representing the GST tag (Smith & Johnson 1988).

Proteolytic cleavage of GST–HD30ΔP and GST–HD51ΔP fusion proteins

In our previous study (Scherzinger *et al.* 1997), we showed that proteolytic digestion of the GST–HD51 fusion protein with trypsin or factor Xa results in the formation of insoluble protein aggregates with a fibrillar or ribbon-like morphology. Such filaments were not produced by proteolysis of the GST–HD20 or GST–HD30 fusion proteins. To examine whether the deletion of the proline-rich region from the C-terminus of the HD exon 1 protein has any effect on protein aggregation and/or the morphology of the resulting fibrils, the purified fusion proteins GST–HD30ΔP and GST–HD51ΔP were digested with trypsin. Electron microscopy of the undigested GST–HD30ΔP and GST–HD51ΔP proteins revealed nearly spherical particles with diameters of 6–7 nm, consistent with an oligomeric form of the fusion proteins (figure 10.4*a*,*c*). In contrast, the preparations obtained by tryptic digestion of GST–HD51ΔP showed numerous clusters of high-molecular-mass fibrils with a ribbon-like morphology (figure 10.4*d*). These ribbon-like structures were very similar to the fibrils that have been previously observed after tryptic digestion of full-length GST–HD51 fusion protein (Scherzinger *et al.* 1997). In strong contrast with GST–HD51ΔP, tryptic digestion of GST–HD30ΔP, which contained only 30 glutamine residues, did not result in the

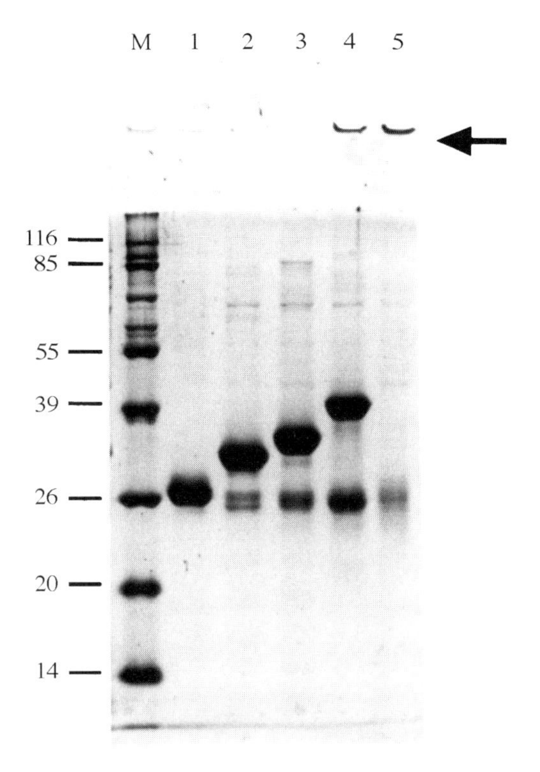

Figure 10.2 SDS–PAGE analysis of GST–HD fusion proteins. Aliquots (15 μl) of eluates from the glutathione–agarose column were subjected to SDS–PAGE (12.5% (w/v) gel) and analysed by staining with Coomassie blue R. Lanes 1–5 contained GST, GST–HD20ΔP, GST–HD30ΔP, GST–HD51ΔP and GST–HD83ΔP, respectively; lane M contained molecular mass standards (molecular masses are indicated in kDa at the left). The arrow indicates insoluble high-molecular-mass aggregates.

formation of any more highly ordered fibrillar structures (figure 10.4*b*), although clots of various sizes containing small particles were frequently detected in this protein preparation.

Discussion

In this study we analysed by SDS–PAGE and electron microscopy the aggregation of truncated GST–HD exon 1 fusion proteins, which lack the proline-rich region at the C-terminus of HD exon 1. We found that the truncated fusion protein GST–HD51ΔP readily self-assembled into fibrillar protein aggregates without prior proteolytic cleavage, whereas the fusion protein GST–HD51, containing the proline-rich region, showed little or no tendency to self-aggregate (Scherzinger *et al.* 1997). This indicates that the

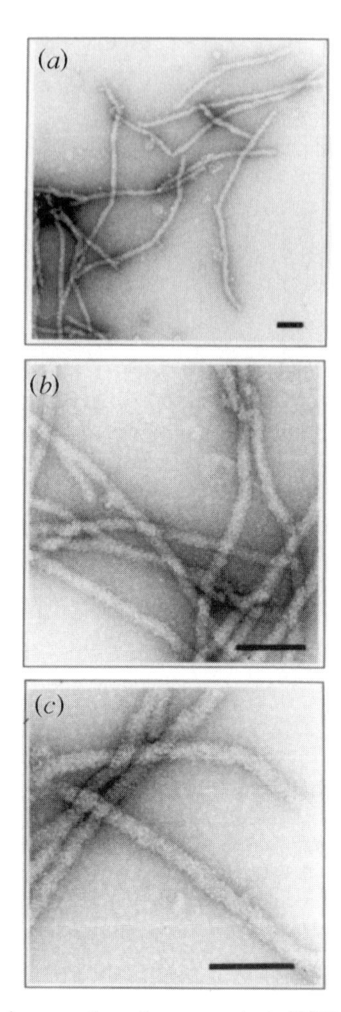

Figure 10.3 Electron micrographs of aggregated GST–HD51ΔP fusion protein. Aggregated GST–HD51ΔP fusion protein was separated from soluble protein by centrifugation, negatively stained with uranyl acetate, and viewed by electron microscopy. The scale bars in (*a*)–(*c*) are 100 nm.

presence of the proline-rich region located C-terminal to the polyQ tract increases the solubility of the fusion protein.

Electron micrographs of the GST–HD51ΔP protein aggregates revealed a fibrillar morphology that was different from that obtained after proteolytic cleavage of GST–HD51ΔP (figure 10.4*d*). All GST–HD51ΔP fibrils seemed to have a uniform structure and a diameter of 25 nm. No cross-links between single fibrils were detected. In contrast, the protein fractions obtained by proteolytic digestion of GST–HD51ΔP showed large clusters of high-molecular-mass fibrils with a typical ribbon-like morphology similar to that

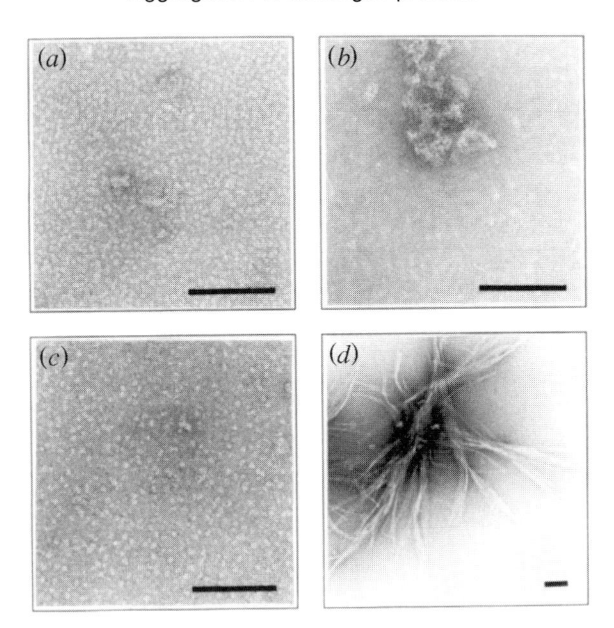

Figure 10.4 Electron micrographs of native GST–HD fusion proteins and their trypsin cleavage products. Purified GST fusion proteins were treated with protease, negatively stained with uranyl acetate and viewed by electron microscopy. The undigested GST–HD30ΔP (*a*) and GST–HD51ΔP (*c*) molecules appear as a homogeneous population of small, round particles. Digestion of GST–HD30ΔP with trypsin shows no evidence for the formation of ordered structures (*b*). In contrast, proteolytic cleavage of GST–HD51ΔP results in the formation of more highly ordered amyloid-like fibrils (*d*). The scale bars in (*a*)–(*d*) are 100 nm.

previously observed for trypsin-digested GST–HD51 (Scherzinger *et al.* 1997). This indicates that the GST tag present in the GST–HD51ΔP protein aggregates has a strong influence on the aggregation behaviour of the HD exon 1 protein. At higher magnification, we detected tightly packed spherical particles with a diameter of *ca.* 6–7 nm on the surface of the GST–HD51ΔP fibrils. This suggests that the soluble oligomeric GST–HD51ΔP particles (figure 10.4*c*) are assembled into ordered fibrillar structures owing to the formation of hydrogen-bonded hairpins between the polyQ chains within HD exon 1 (Perutz 1996). In good agreement with our previous results, the undigested or trypsin-digested GST–HD20ΔP and GST–HD30ΔP fusion proteins did not form any ordered high-molecular-mass protein aggregates. These results substantiate our previous findings that only polyQ tracts beyond a critical length might be capable of forming polar zippers (Perutz 1996), resulting in the formation of amyloid-like fibrils. Experiments to determine the critical length of the glutamine repeat necessary for the self-assembly of HD exon 1 proteins into more highly ordered structures are in progress.

The work described in this paper was supported by grants from the Deutsche Forschungsgemeinschaft (Wa1151/1), the European Union (Project: Eurohunt, BMH4-CT96-0244) and the HDSA.

References

Becher, M. W., Kotzuk, J. A., Sharp, A. H., Davies, S. W., Bates, G. P., Price, D. L. & Ross, C. A. 1998 Intranuclear neuronal inclusions in Huntington's disease and dentatorubral and pallidoluysian atrophy: correlation between the density of inclusions and *IT15* CAG triplet repeat length. *Neurobiol. Dis.* **4**, 387–397.

Davies, S. W., Trumaine, M., Cozens, B. A., DiFiglia, M., Sharp, A. H., Ross, C. A., Scherzinger, E., Wanker, E. E., Mangiarini, L. & Bates, G. P. 1997 Formation of neuronal intranuclear inclusions (NII) underlies the neurological dysfunction in mice transgenic for the HD mutation. *Cell* **90**, 537–548.

DiFiglia, M., Sapp, E., Chase, K. O., Davies, S. W., Bates, G. P., Vonsattel, J. P. & Aronin, N. 1997 Aggregation of huntingtin in neuronal intranuclear inclusions and dystrophic neurites in brain. *Science* **277**, 1990–1993.

Harper, P. S. 1996 *Huntington's disease*, 2nd edn. London: W. B. Saunders Co.

Holmberg, M. (and 10 others) 1998 Spinocerebellar ataxia type 7 (SCA7): a neurodegenerative disorder with neuronal intranuclear inclusions. *Hum. Mol. Genet.* **7**, 913–918.

Huntington's Disease Collaborative Research Group 1993 A novel gene containing a trinucleotide repeat that is unstable on Huntington's disease chromosomes. *Cell* **72**, 971–983.

Igarashi, S. (and 18 others) 1998 Suppression of aggregate formation and apoptosis by transglutaminase inhibitors in cells expressing truncated DRPLA protein with an expanded polyglutamine stretch. *Nature Genet.* **18**, 111–117.

Mangiarini, L. (and 10 others) 1996 Exon 1 of the Huntington's disease gene containing a highly expanded CAG repeat is sufficient to cause a progressive neurological phenotype in transgenic mice. *Cell* **87**, 493–506.

Matilla, A., Koshy, B. T., Cummings, C. J., Isobe, T., Orr, H. T. & Zoghbi, H. Y. 1997 The cerebellar leucine-rich acidic nuclear protein interacts with ataxin-1. *Nature* **389**, 974–978.

Paulson, H. L., Perez, M. K., Tottier, Y., Trojanowski, J. Q., Subramony, S. H., Das, S. S., Vig, P., Mandel, J.-L., Fischbeck, K. H. and Pittman, R. N. 1997 Intranuclear inclusions of expanded polyglutamine protein in spinocerebellar ataxia type 3. *Neuron* **19**, 333–344.

Perutz, M. 1996 Glutamine repeats and inherited neurodegenerative diseases: molecular aspects. *Curr. Opin. Struct. Biol.* **6**, 848–858.

Rubinsztein, D. C. (and 36 others) 1996 Phenotypic characterisation of individuals with 30–40 CAG repeats in the Huntington's disease (HD) gene reveals HD cases with 36 repeats and apparently normal elderly individuals with 36–39 repeats. *Am. J. Hum. Genet.* **59**, 16–22.

Sathasivam, K., Amaechi, I., Mangiarini, L. & Bates, G. P. 1997 Identification of an HD patient with a $(CAG)_{180}$ repeat expansion and the propagation of highly expanded CAG repeats in lambda phage. *Hum. Genet.* **99**, 692–695.

Scherzinger, E., Lurz, R., Trumaine, M., Mangiarini, L., Hollenbach, B., Hasenbank, R., Bates, G. P., Davies, S. W., Lehrach, H. & Wanker, E. E. 1997 Huntingtin-encoded polyglutamine expansions form amyloid-like protein aggregates *in vitro* and *in vivo*. *Cell* **90**, 549–558.

Skinner, P. J., Koshy, B. T., Cummings, C. L., Klement, I. A., Helin, K., Servadio, A., Zoghbi, H. Y. & Orr, H. T. 1997 Ataxin-1 with an expanded glutamine tract alters nuclear matrix-associated structures. *Nature* **389**, 971–974.

Smith, D. B. & Johnson, K. S. 1988 Single-step purification of peptides expressed in *Escherichia coli* as fusions with glutathione S-transferase. *Gene* **67**, 31–40.

Vonsattel, J.-P., Myers, R. H., Stevens, T. J., Ferrante, R. J., Bird, E. D. & Richardson, E. P. 1985 Neuropathological classification of Huntington's disease. *J. Neuropathol. Exp. Neurol.* **44**, 559–577.

Wanker, E. E., Scherzinger, E., Heiser, V., Sittler, A., Eickhof, H. & Lehrach, H. 1999 Membrane filter assay for the detection of amyloid-like protein aggregates. *Methods Enzymol.* **309**, 375–386

11

The localization and interactions of huntingtin

A. Lesley Jones

Introduction

Huntington's disease (HD) is an autosomal dominant, incurable, progressive neurodegeneration. The disease is associated with an expanded CAG repeat giving expanded polyglutamine in the protein product, huntingtin. Such expanded polyglutamine tracts give rise to a number of inherited neurodegenerations (LaSpada *et al.* 1991; Huntington's Disease Collaborative Research Group 1993; Orr *et al.* 1993; Koide *et al.* 1994; Nagafuchi *et al.* 1994; Kawaguchi *et al.* 1994; Imbert *et al.* 1996; David *et al.* 1997; Zhuchenko *et al.* 1997) distinguished by the presence of neuronal inclusions found in cells known to degenerate in six out of the eight such diseases characterized (Davies *et al.* 1997; DiFiglia *et al.* 1997; Paulson *et al.* 1997; Skinner *et al.* 1997; Igarashi *et al.* 1998; Li *et al.* 1998; Lunkes & Mandel, 1998) althought the relationship of the inclusions to the disease aetiology is unknown (Klement *et al.* 1998; Saudou *et al.* 1998). Interestingly, ataxin 6 recently has been shown to form cytoplasmic aggregates rather than intranuclear inclusions, in the Purkinje cells where it is most abundant (Ishikawa *et al.* 1999). Ataxin 6 is an ion channel protein, and the neurodegenerative phenotype is associated with a disease-causing CAG repeat which is notably shorter than for the rest of the polyglutamine diseases; it is possible that the constraints on protein structure caused by multimerization and insertion in the membrane allow the polyglutamine tract to become toxic at shorter lengths in this protein. Each disease shows a characteristically distinct, although overlapping, pattern of neuronal degeneration (Ross 1995); in HD, it is the medium spiny neurons of the basal ganglia which are most vulnerable (Hedreen & Folstein 1995). Possible reasons for such specificity include the protein context of the repeat, cell-specific post-translational modifications and interactions with other proteins.

Huntingtin has been immunolocalized in a series of studies which have shown a strong cytoplasmic immunoreactivity (DiFiglia *et al.* 1995; Gutekunst *et al.* 1995; Trottier *et al.* 1995; Sapp *et al.* 1997), although there have been reports of nuclear localization (de Rooij *et al.* 1996). The intranuclear inclusions, however, have only shown immunoreactivity associated with epitopes in the furthermost N-teminal regions of huntingtin, the region within which the polyglutamine sequence resides (DiFiglia *et al.* 1997; Becher *et al.* 1998). Antibodies detecting more C-terminal epitopes of huntingtin have not been localized to inclusions. Although in most other polyglutamine diseases, apart form SCA6, the nuclear intranuclear inclusions (NIIs) appear to be the only consistent pathological lesion, in HD dystrophic neurites have been observed, and it is possible that these play a role in degeneration, particularly early in the disease process (Maat-Schieman *et al.* 1999; Sapp *et al.* 1999)

To investigate whether proteins interacting with huntingtin are important in HD pathology and contribute to the specificty of the observed neurodegeneration, we have been looking for interactions of the polyglutamine-bearing N-terminal region of huntingtin in the yeast two-hybrid system. A number of such associations have been detected previously, either using the yeast two-hybrid or other affinity-based methods, and known interactors include huntingtin-associated protein 1 (HAP1) (Li *et al.* 1995), glyceraldehyde 3-phosphate dehydrogenase (GAPDH) (Burke *et al.* 1996), huntingtin-interacting protein 1 (HIP1) (Kalchman *et al.* 1997; Wanker *et al.* 1997), a ubiquitin-conjugating enzyme (Kalchman *et al.* 1996), a series of WW domain proteins (Faber *et al.* 1998), an SH3 domain-bearing protein, SH3GL3 (Sittler *et al.* 1998), and PQBP-1, a protein which binds to the polyglutamine region of several disease-associated polyglutamine repeat proteins (Waragai *et al.* 1999). We have detected two further interactions of huntingtin, with cystathionine β-synthase (Boutell *et al.* 1998) and the nuclear receptor co-repressor (N-CoR) (Boutell *et al.* 1999).

We have now immunolocalized both of these proteins in human control and HD brain along with a comprehensive localization of huntingtin itself, using an affinity-purified polyclonal serum to the N-terminal region of huntingtin and a series of monoclonal antibodies to three different, more C-terminal, regions.

The localization of huntingtin

The characteristics of the antibodies used to localize huntingtin are given in table 11.1, and the characteristics of the HD tissue used are shown in table 11.2. Figure 11.1 (plate 11) demonstrates that the more C-terminal monoclonal antibodies (mAbs), HDA3E10, HDB4E10 and HDC8A4, show a similar pattern of immunoreactivity, with strong cytoplasmic staining of neurones, sparing of nuclei and some staining of neuronal processes (figure 11.1*a*(i,ii) (plate 11)). This is similar to observations made previously (DiFiglia *et al.* 1995; Trottier *et al.* 1995; Gutekunst *et al.* 1996). The N-terminal polyclonal antiserum, N-675, shows a different staining pattern (figure 11.1*b* (plate 11)), with only faint cytoplasmic staining (figure 11.1*b*(i,ii) (plate 11)),

Table 11.1 *Anti-huntingtin antibody characteristics*

antibody	*type immunogen*	*huntingtin sequence detected*	*epitope*	*source*
N-675	pAb	residues 1–17	unknown	in house
HDA3E10	mAb	residues 997–1276	1173-7	MRIC
HDB4E10	mAb	residues 1841–2131	unknown	MRIC
HDC8A4	mAb	residues 2703–2911	unknown	MRIC
ubiquitin	mAb	–	–	Chemicon Intl
GFAP	pAb	–	–	Dr J. Newcombe

(pAb, polyclonal antibody.)

Table 11.2 *Details of HD brains used for immunohistochemistry*

no.	*Vonsattel grade*	*repeats (yr)*	*age-at death*
1	3	42/18	41
2	3	42/8	52
3	3	44/16	38
4	4	39/18	66
5	4	43/19	49
6	3	44/19	47

although this is similar to that observed with the more C-terminal mAbs. What this antibody does reveal, however, is the presence of strongly huntingtin immunoreactive NIIs (figure 11.1*b*(i) (plate 11)), a subset of which are also ubiquitin immunoreactive (figure 11.1*b*(ii) (plate 11)), as previously observed (DiFiglia *et al.* 1997; Becher *et al.* 1998). These immunoreactive NIIs are never observed in control brain tissue. The other notable difference between the immunoreactivity revealed by this antiserum and the more C-terminal antibodies is the immunostaining of neuronal processes (figure 11.1*b*(iii,iv) (plate 11)); the more C-terminal sera do not give strong immunoreactivity in the axon tracts of the subcortical white matter nor such intense staining of processes within the grey matter. This strong staining and blebbing in axons is similar to that observed by Sapp *et al.* (1999) and Maat-Schieman *et al.* (1999) in HD grade III brain. This immunoreactivity is less intense in control than HD cortex (compare figure 11.1*b*(ii) and (iv), (plate 11)).

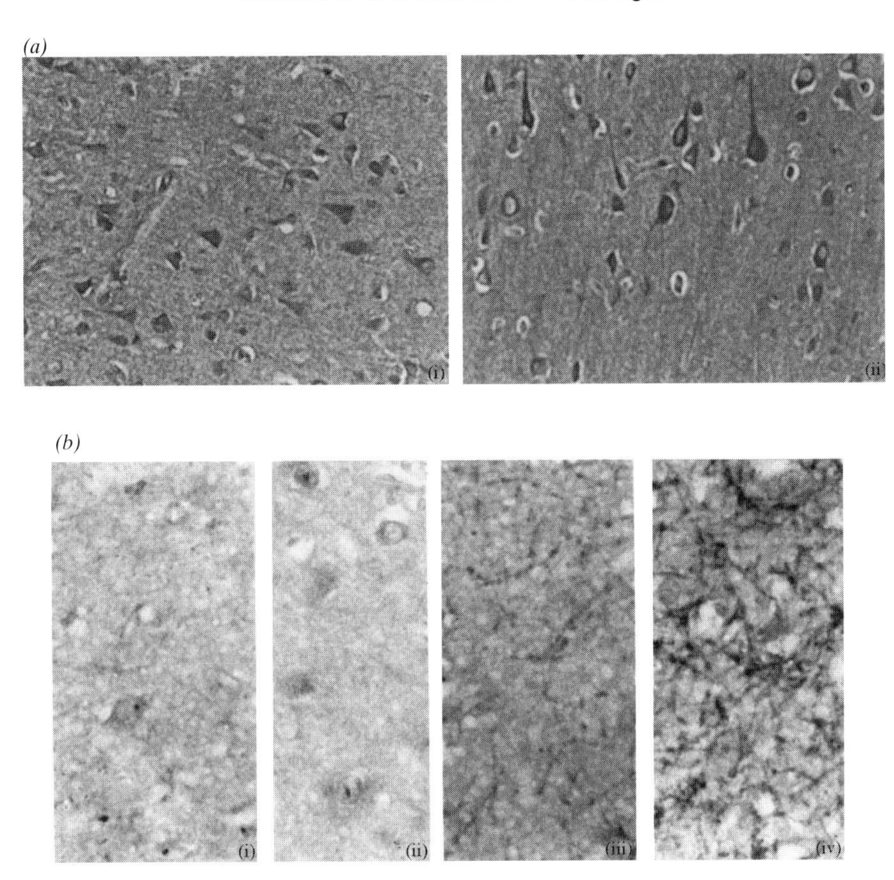

Figure 11.1 (*a*) Immunostaining of HD cortex with monoclonal antibodies (i) HDA3E10 and (ii) HDC8A4. Magnification x100. (*b*) Immunostaining of HD cortex with N-675 and ubiquitin: (i) N675 demonstrating the presence of inclusions; (ii) immunostained with ubiquitin (see table 11.1); (iii) control; and (iv) HD brain in the cortical layer VI immediately above the white matter, showing strong staining of processes and increased intensity of such staining in HD brain. (See also colour plate section.)

HD caudate, at this point in the disease (Vonsattel grade III or IV), has such an altered morphology compared with normal caudate that it is difficult to provide a meaningful comparison of these tissues. However, the most notable features are that the reactive astrocytes in HD caudate do show huntingtin immunoreactivity with all the more C-terminal antibodies (data not shown), confirming our initial observation of such immunoreactivity (Singhrao *et al.* 1998); the few neurons surviving show cytoplasmic immunoreactivity with the more C-terminal antibodies and the N-terminal antiserum demonstrates the presence of intranuclear inclusions (figure 1*b*(i) (plate 11). Inclusions in the caudate are less frequent than in the cortex, and it has been noted that NII numbers in the cortex correlate much better with disease stage than those in the caudate (Becher *et al.* 1998; Sieradzan *et al.* 1999). This could be a reflection of the earlier atrophy in the caudate than in the cortex (Hedreen & Folstein 1995), with affected inclusion-containing cells cleared from the tissue, or due to anteriograde or retrograde atrophy of the caudate neurons. The cortical neurons with inclusions are mainly in the lower cortical layers, particularly layer V, which have input into the caudate. Conversely, the death of the caudate neurons may be preventing retrograde messages to the neurons with cortical inputs into the caudate, which could lead to cortical atrophy and the formation of nuclear inclusions.

The localization of interacting proteins

Cystathionine β-synthase

We have detected an interaction of huntingtin with cystathionine β-synthase (CBS) (Boutell *et al.* 1998). CBS is a key enzyme in the generation of cysteine from methionine, catalysing the formation of cystathionine by the condensation of homocysteine and serine. The absence of CBS activity is associated with homocystinuria, a recessive disorder first recognized in the 1960s (Mudd *et al.* 1964); the gene was isolated in 1990 (Kraus 1990), and a number of mutations have been detected in homocystinureic patients. A detailed account of this disease can be found in Mudd *et al.* (1995). The initial metabolic consequence of CBS deficiency is the intracellular accumulation of the enzyme's substrate, homocysteine, followed by its export from the cell and an increase in the level of homocysteine and its derivatives in plasma, interstitial fluid and urine (Refsum *et al.* 1994). Mental retardation is often the first symptom that brings CBS deficiency to clinical attention through developmental delay in the early years; it is the most frequent abnormality of the central nervous system (CNS) (Mudd *et al.* 1985) and there is a high prevalence of psychiatric disorders including depression, behavioural abnormalities and personality disorders (Abbott *et al.* 1987). Pathology in the brain shows infarcts caused by cerebrovascular inclusions. The possibility that the neurological effects of CBS deficiency may be caused by excitotoxicity in the CNS has been suggested as the basis for the mental retardation and seizures seen in CBS-deficient patients (Schwarz & Zhou 1991). Two of the oxidation products of homocysteine, L-homocysteate and L-homocysteine sulphinate, are known to be potent agonists for *N*-methyl-D-aspartate (NMDA) receptors and thus to exert excitotoxic effects on neurons, which can be blocked by NMDA receptor antagonists (Schwarz *et al.* 1990). Both compounds can be detected in the urine of homocystinureic patients but not in normal controls (Ohmori *et al.* 1972).

One of the hypotheses for the selective neuronal death seen in HD is that it occurs through excitotoxic insult (DiFiglia 1990; Beal *et al.* 1991). There are a number of pieces of evidence that support this idea. HD initially affects the striatal area of the basal ganglia, which receives a major glutamatergic input from the cortex, thalamus and subthalamic nucleus (Hedreen & Folstein 1995). The neurochemical and neuropathological characteristics of HD can be mimicked in animals using glutamate receptor agonists such as kaininic and quinolinic acids, which are both excitotoxic amino acids. HD is characterized by a loss of striatal projection neurons and a sparing of striatal interneurons containing acetylcholine and somatostatin, and this pattern of selective vulnerability is also seen in rodents with quinolinic acid lesions of the striatum (Beal *et al.* 1991). As the oxidation products of homocysteine are known to be powerful excitotoxins, this could provide an explanation for some of the damage seen in HD brain. Huntingtin could, therefore, bind CBS and inhibit its activity, either directly or by preventing processing of the enzyme to its active form (Skovby *et al.* 1984).

Whilst the neuronal death in HD is initially specific, as the disease progresses virtually all neurons in the caudate and putamen are affected and other areas of the striatum and the cortex also atrophy; these are the areas with cells that contain huntingtin-imunoreactive aggregates. Although the focal inclusions in patients are nuclear, the huntingtin-positive dystrophic neurites also seen in HD brain and in transgenic mice show aggregation in the cytoplasm, and this aggregation may be be followed by translocation to the nucleus (Davies *et al.* 1997; DiFiglia *et al.* 1997; Maat-Schieman *et al.* 1999; Sapp *et al.* 1999). Although there is strong circumstantial evidence that the aggregation observed in HD and other polyglutamine repeat diseases (Paulson *et al.* 1997) is part of the pathological effect of the polyglutamine expansion, the actual primary pathological event remains unknown. Because of this evidence, we examined the localization of CBS in HD brain, in particular examining whether antibodies to CBS localized to inclusions. We used polyclonal antisera raised against recombinant human CBS from *Escherichia coli*, and a CBS antiserum which was a kind gift from Professor Jan Kraus. We found CBS in the brain to be ubiquitous, and no differences between the distribution in HD and control brain were observed. Neither antiserum localized to NIIs, nor was any strong staining of neuronal processes seen. As this is a metabolic enzyme, its ubiquitous localization is unsurprising. This indicates that it is unlikely that CBS is involved in HD pathology, although assays of its function in relation to CAG repeat length should be carried out to confirm this.

Nuclear receptor co-repressor

Both the yeast two-hybrid system and *in vitro* studies using His-tagged fusion proteins indicate that C-terminal N-CoR from rat or human brain binds specifically to rat and human huntingtin. This interaction occurred between rat and human N-CoR and rat and human huntingtin, as full-length huntingtin could be attached to recombinant rat or human C-terminal N-CoR (Boutell *et al.* 1999). N-CoR acts as a repressor of transcription in a complex which is known to be common to a number of sequence-specific DNA-binding transcriptional repressors, including the unliganded thyroid hormone–retinoic acid/retinoid X receptor dimers, Mad:Max dimers, RevErb and Dax1 orphan receptors, Pit1 and Ume6 (Alland *et al.* 1997; Hassig *et al.* 1997; Heinzel *et al.* 1997; Kadosh & Struhl 1997; Laherty *et al.* 1997; Nagy *et al.* 1997; Zhang *et al.* 1997; Crawford *et al.* 1998; Xu *et al.* 1998). The *Mus musculus* N-CoR was isolated by a yeast two-hybrid screen using thyroid hormone receptors by Horlein *et al.* (1995), and since then there has been an explosion of work on transcriptional regulation. It seems inevitable that more DNA-binding or transcriptional regulatory proteins will prove to be controlled by this repression system (Lavinsky *et al.* 1998). N-CoR and the homologous silencing mediator of retinoid and thyroid hormone receptor (SMRT) repress transcription by linking the sequence-specific DNA-binding moieties with proteins known to possess histone deacetylase activity (Pazin & Kadonga 1997; Crawford *et al.* 1998). This link is mediated by interaction between these co-repressors and the mSin3A and mSin3B proteins, mammalian homologues of the yeast Sin3 protein (Taunton *et al.* 1996; Yang *et al.* 1996) which interact with a series of proteins which include the histone deacetylases 1 and 2 (Alland *et al.* 1997; Hassig *et al.* 1997; Heinzel *et al.* 1997; Kadosh & Struhl 1997; Laherty *et al.* 1997; Nagy *et al.* 1997; Pazin & Kadonga 1997), although N-CoR also interacts directly with the basal transcription factors TFIIB, $TAF_{II}32$ and $TAF_{II}70$ (Muscat *et al.* 1998). Histone deacetylation is thought to repress transcription by condensation of chromatin, thus preventing access of the basal transcription factors to promotors.

The interaction of huntingtin and N-CoR is dependent on repeat length. Mapping of the domains within N-CoR has already revealed that the nuclear receptor interaction domain is located at the C-terminus of the protein (Horlein *et al.* 1995) (figure 11.2) and this is the domain that interacts with huntingtin. The two major repression domains responsible for the transcriptional silencing are localized to the N-terminal region of the protein, neither of which is included in the interacting constructs; these are the regions known to carry the Sin3/RPD3-interacting domains (figure 11.2) (Horlein *et al.* 1995). As huntingtin interacts with the C-terminal N-CoR domain, in this modular protein it is possible that the more N-terminal domains remain accessible to their normal binding partners. The binding of N-CoR to mutant huntingtin is likely to have altered dissociation kinetics; if N-CoR is sequestered to huntingtin then it may be excluded form its normal function, and the proteins with which it interacts similarly may be prevented from carrying out their normal role. This hypothesis is made more likely as the N-CoR-containing complexes are probably the rate limiting requirements in the action of specific nuclear receptors (Laherty *et al.* 1998)

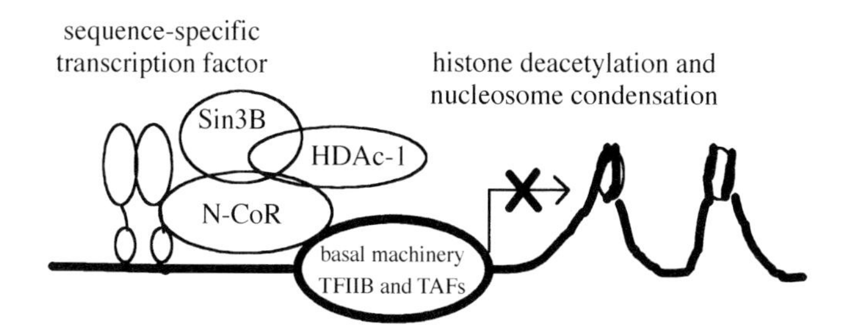

Figure 11.2 Simplified diagram showing the interactions of N-CoR and other proteins thought to be involved in the repression complex.

Because of this possibility, we examined the localization of N-CoR and a number of its associated complex proteins in immunohistochemcal studies in HD and control brain. The brains were those characterized in the study of huntingtin localization (table 11.2). We found that although commercial antibodies were available against a number of the complex proteins, few of these were suitable for immunohistochemistry in human brain. Of those that had no cross-reactions on Western blotting of human brain, we illustrate results from N-CoR and mSin3A (Santa Cruz Biotechnology) (figure 11.3 (plate 11)). Both of these antisera demonstrate localization in nuclei and cytoplasm of cortical neurons in control brain, but immunoreactivity is notably excluded from the nucleus in HD brain (figure 11.3*b*,*d* (plate 13)). We did not find N-CoR immunoreactivity localized to the NIIs in HD brain, but as the antiserum used was directed against the C-terminus of N-CoR and this is the region that interacts with huntingtin, the appropriate epitope may be masked. mSin3 immunoreactivity occasionally localized to structures which looked like NIIs, but with a much lower frequency than that with which NIIs themselves occurred (figure 11.3*d* (plate 13)). There also appeared to be increased immunostaining for both N-CoR and mSin 3A in neuronal processes (figure 11.3*b*,*f* (plate 13)), some of which appeared morphologically abnormal (figure 11.3*f* (plate 13)), in a similar way to HD brain immunostained with huntingtin N-terminal antibodies (Sapp *et al.* 1999). It must be noted that all the brain used was late-stage HD, but there appears to be a relocalization of transcriptional complex proteins.

It is also possible that the interaction with N-CoR plays a role in transport of aggregated or soluble mutated huntingtin into the nucleus. N-CoR immunoreactivity, like huntingtin immunoreactivity, is markedly increased in HD neuronal processes compared with control brain (figure 11.1*b*(iii,iv) (plate 11) and figure 11.3*b*,*f* (plate 13)). Huntingtin is thought to have a role in transport along the cytoskeleton (DiFiglia *et al.* 1995; Block-Galzara *et al.* 1997; Engelender *et al.* 1997; Kalchman *et al.* 1997), and thus might be blocked at the entrance to the nucleus. The only real way to answer this question is in cell culture or animal models.

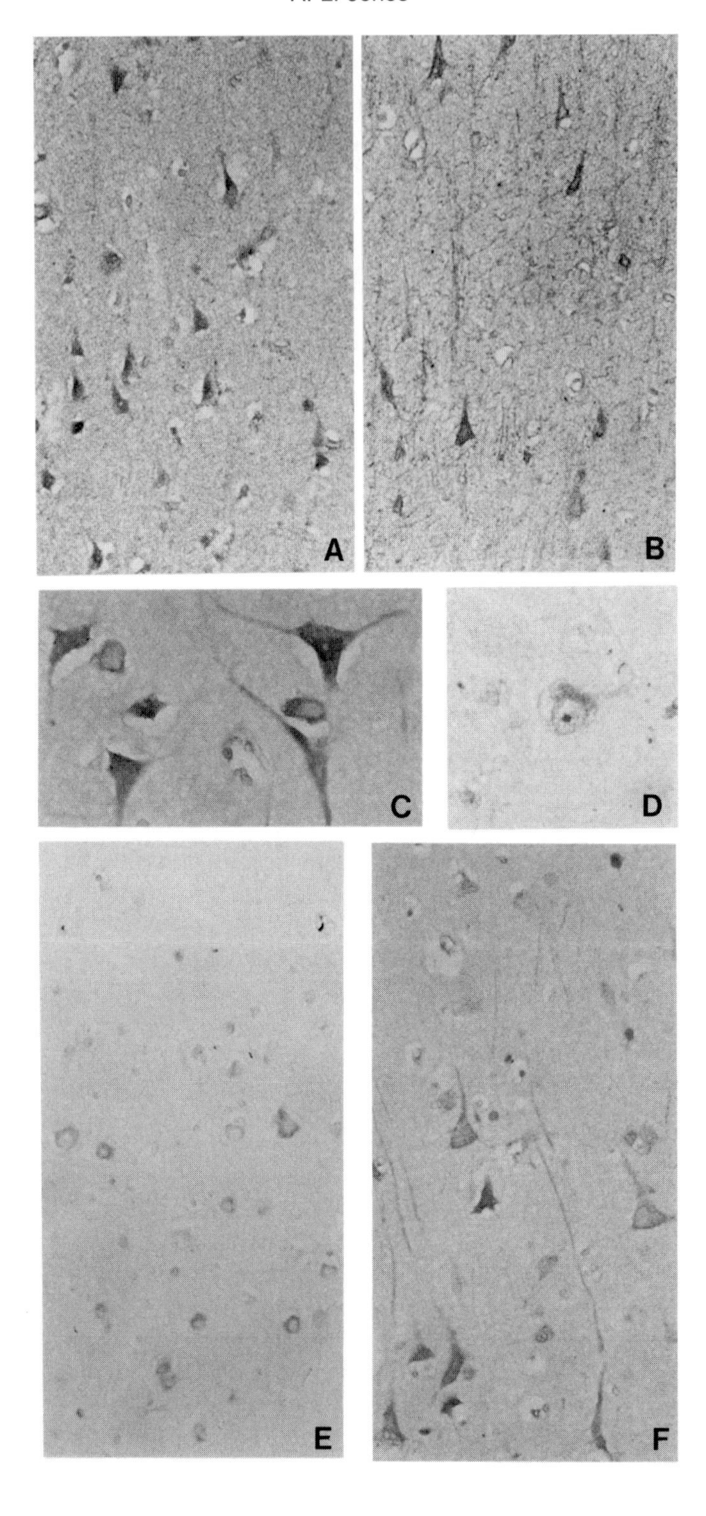
A
B
C
D
E
F

The role of huntingtin and the relationship of its interactions to disease pathology remain unclear. Further work on the nature of the interactions and their roles in the normal and pathological function of HD need to be carried out in animal and cellular models of the disease. This should provide clues as to which interactions are important in pathology, and which may be suitable targets for therapeutic intervantions.

This work was supported by the MRC (UK) and the Welsh Scheme for Health and Social Research. The mAbs HDA3E10, HDB4E10 and HDC8A4 were a kind gift from Professor G. E. Morris, Ms F. Wilkinson and N. thi Man, North East Wales Institute, Wrexham.

References

Abbott, M. H., Folstein, S. E., Abbey, H. & Pyeritz, R. E. 1987 Psychiatric manifestations of homocystinuria due to cystathionine β-synthase deficiency: prevalence, natural history, and relationship to neurologic impairment and vitamin B6-responsiveness. *Am. J. Med. Genet.* **26**, 959–969.

Alland, L., Muhle, R., Hou Jr, H., Potes, J., Chin, L., Schreiber-Agus, N. & DePinho, R. A. 1997 Role for N-CoR and histone deacetylase in Sin3-mediated transcriptional repression. *Nature* **387**, 49–55.

Beal, M. F., Ferrante, R. J., Swartz, K. J. & Kowall, N. W. 1991 Chronic quinolinic acid lesions in rats closely resemble Huntington's disease. *J. Neurosci.* **11**, 1649–1659.

Becher, M. W., Kotzuk, J. A., Sharp, A. H., Davies, S. W., Bates, G. P., Price, D. L. & Ross, C. A. 1998 Intranuclear neuronal inclusions in Huntington's disease and dentatorubral and pallidoluysian atrophy: correlation between the density of inclusions and *IT15* triplet repeat length. *Neurobiol. Dis.* **4**, 387–397.

Block-Galarza, J., Chase, K. O., Sapp, E., Vaughn, K. T., Vallee, R. B., DiFiglia, M. & Aronin, N. 1997 Fast transport and retrograde movement of huntingtin and HAP1 in axons. *NeuroReport* **8**, 2247–2251.

Boutell, J. M., Wood, J. D., Harper, P. S. & Jones, A, L. 1998 Huntingtin interacts with cystathionine beta-synthase. *Hum. Mol. Genet.* **7**, 371–378.

Boutell, J. M., Thomas, P., Neal, J. W., Weston, V. J., Duce, J., Harper, P. S. & Jones, A. L. 1999 Aberrant interactions of transcriptional repressor proteins with the Huntington's disease gene product, huntingtin. *Hum. Mol. Genet.* **8**, 1647–1655.

Burke, J. R., Enghild, J. J., Martin, M. E., Jou, Y.-S., Myers, R. M., Roses, A. D., Vance, J. M. & Strittmatter, W. J. 1996 Huntingtin and DRPLA proteins selectively interact with the enzyme GAPDH. *Nature Med.* **2**, 347–350.

Figure 11.3 (*a,b*) Control (*a*) and HD (*b*) cortex immunostained with N-CoR C20 (Santa Cruz Biotechnology), showing the lack of nuclear immunostaining and increased staining of processes in HD brain for N-CoR Magnification ×200. (*c–f*) Control (c) and HD (*d,e,f*) brain immunostained with mSin3A polyclonal serum (AK11) (Santa Cruz Biotechnology). Magnification ×400. Panel (*c*) shows nuclear staining of cortical neurons in control brain, whilst, generally, nuclear staining is less prevalent in HD cortex (*e*). Panel (*d*) demonstrates the mSin3A immunoreactivity of a nuclear inclusion (NII) and panel (*f*) shows some dystrophic neurites and abnormal neuronal processes in HD brain. Magnification ×400 (*c,d*); ×200 (*e,f*). (See also colour plate section.)

Crawford, P. A., Dorn, C., Sadovsky, Y. & Milbrandt, J. 1998 Nuclear receptor DAX-1 recruits nuclear receptor corepressor N-CoR to steroidogenic factor 1. *Mol. Cell. Biol.* **18**, 2949–2956.

David, G. (and 18 others) 1997 Cloning of the SCA7 gene reveals a highly unstable CAG repeat expansion. *Nature Genet.* **17**, 65–70.

Davies, S. W., Turmaine, M., Cozens, B. A., DiFiglia, M., Sharp, A. H., Ross, C. A., Scherzinger, E., Wanker, E. E., Mangiarini, L. & Bates, G. P. 1997 Formation of neuronal intranuclear inclusions underlies the neurological dysfunction in mice transgenic for the HD mutation. *Cell* **90**, 537–548.

de Rooij, K. E., Dorsman, J. C., Smoor, M. A., Den Dunnen, J. T. & Van Ommen, G. J. 1996 Subcellular localization of the Huntington's disease gene product in cell lines by immunofluorescence and biochemical subcellular fractionation. *Hum. Mol. Genet.* **5**, 1093–1099.

DiFiglia, M. 1990 Excitotoxic injury in the neostriatum: a model for Huntington's disease. *Trends Neurosci.* **13**, 286–289.

DiFiglia, M. (and 11 others) 1995 Huntingtin is a cytoplasmic protein associated with vesicles in human and rat brain neurons. *Neuron* **14**, 1075–1081.

DiFiglia, M., Sapp, E., Chase, K. O., Davies, S. W., Bates, G. P., Vonsattel, J. P. & Aronin, N. 1997 Aggregation of huntingtin in neuronal intranuclear inclusions and dystrophic neurites in brain. *Science* **277**, 1990–1993.

Engelender, S., Sharp, A. H., Colomer, V., Tokito, M. K., Lanahan, A., Worley, P., Holzbaur, E. L. F. & Ross, C. A. 1997 Huntingtin associated protein 1 (HAP1) interacts with dynactin $p150^{Glued}$ and other cytoskeletal related proteins. *Hum. Mol. Genet.* **6**, 2205–2212.

Faber, P. W., Barnes, G. T., Srinidhi, J., Chen, J., Gusella, J. F. & MacDonald, M. E. 1998 Huntingtin interacts with a family of WW proteins. *Hum. Mol. Genet.* **7**, 1463–1474.

Gutekunst, C. A., Levey, A. I., Heilman, C. J., Whaley, W. L., Yi, H., Nash, N. R., Rees, H. D., Madden, J. J. & Hersch, S. M. 1995 Identification and localization of huntingtin in brain and human lymphoblastoid cell-lines with anti-fusion protein antibodies. *Proc. Natl Acad. Sci. USA* **92**, 8710–8714.

Hassig, C. A., Fleischer, T. C., Billin, A. N., Schreiber, S. L. & Ayer, D. E. 1997 Histone deacetylase activity is required for full transcriptional repression by mSin3A. *Cell* **89**, 341–347.

Hedreen, J. C. & Folstein, S. E. 1995 Early loss of neostriatal striosome neurons in Huntington's disease. *J. Neuropathol. Exp. Neurol.* **54**, 105–120.

Heinzel, T. (and 14 others) 1997 A complex containing N-CoR, mSin3 and histone deacetylase mediates transcriptional repression. *Nature* **387**, 43–48.

Horlein, A. J. (and 10 others) 1995 Ligand-ndependent repression by the thyroid-hormone receptor mediated by a nuclear receptor co-repressor. *Nature* **377**, 397–404.

Huntington's Disease Collaborative Research Group 1993 A novel gene containing a trinucleotide repeat that is expanded and unstable on Huntington's disease chromosomes. *Cell* **72**, 971–983.

Igarashi, S. (and 18 others) 1998 Suppression of aggregate formation and apoptosis by transglutaminase inhibitors in cells expressing truncated DRPLA protein with an expanded polyglutamine stretch. *Nature Genet.* **18**, 111–117.

Imbert, G. (and 14 others) 1996 Cloning of the gene for spinocerebellar ataxia 2 reveals a locus with high sensitivity to expanded CAG/glutamine repeats. *Nature Genet.* **13**, 285–291.

Ishikawa, Takahashi, K., Nakazawa, M., Watanabe, Y., Konagaya, A. 1999 Automated processing of 2-D gel electrophoretograms of genomic DNA for hunting pathogenic DNA molecular changes. *Genome Inform. Ser. Workshop Genome Inform.* **10**, 121–132.

Kadosh, D. & Struhl, K. 1997 Repression by Ume6 involves recruitment of a complex containing Sin3 corepressor and Rpd3 histone deacetylase to target promoters. *Cell* **89**, 365–371.

Kalchman, M. A., Graham, R. K., Xia, G., Koide, H. B., Hodgson, J. G., Graham, K. C., Goldberg, Y. K., Gietz, R. D., Pickart, C. M. & Hayden, M. R. 1996 Huntingtin is ubiquitinated and interacts with a specific ubiquitin conjugated enzyme. *J. Biol. Chem.* **271**, 19385–19394.

Kalchman, M. A. (and 13 others) 1997 *HIP1*, a human homologue of *S. cerevisiae Sla2p*, interacts with membane-associated huntingtin in the brain. *Nature Genet.* **16**, 44–53.

Kawaguchi, Y. (and 12 others) 1994 CAG expansions in a novel gene for Machado–Joseph disease at chromosome 14q32.1. *Nature Genet.* **8**, 221–228.

Klement, I. A., Skinner, P. J., Kaytor, M. D., Yi, H., Hersch, S. M., Clark, H. B., Zoghbi, H. Y. & Orr, H. T. 1998 Ataxin-1 nuclear localization and aggregation: role in polyglutamine-induced disease in *SCA1* transgenic mice. *Cell* **95**, 41–53.

Koide, R. (and 15 others) 1994 Unstable expansion of CAG repeat in hereditary dentatorubral–pallidoluysian atrophy (DRPLA). *Nature Genet.* **6**, 9–13

Kraus, J. P. 1990 Molecular analysis of cystathionine β-synthase—a gene on chromosome 21. *Prog. Clin. Biol. Res.* **360**, 201.

Laherty, C. D., Yang, W. M., Sun, J. M., Davie, J. R., Seto, E. & Eisenman, R. N. 1997 Histone deacetylases associated with the mSin3 corepressor mediate mad transcriptional repression. *Cell* **89**, 349–356.

La Spada, A. R., Wilson, E. M., Lubahn, D. B., Harding, A. E. & Fischbeck, K. H. 1991 Androgen receptor gene mutations in X-linked spinal andbulbar muscular atrophy. *Nature* **352**, 77–79.

Lavinsky, R. M. (and 14 others) 1998 Diverse signalling pathways modulate nuclear receptor recruitment of N-CoR and SMRT complexes. *Proc. Natl Acad. Sci. USA* 95, 2920–2925.

Li, M., Miwa, S., Kobayashi, Y., Merry, D. E., Yamamoto, M., Tanaka, F., Doyu, M., Hashizume, Y., Fischbeck, K. H. & Sobue, G. 1998 Nuclear inclusions of the androgen receptor protein in spinal and bulbar muscular atrophy. *Ann. Neurol.* **44**, 249–254.

Li, X.-J., Li, S.-H., Sharp, A. H., Nucifora J,, F. C. Schilling, G., Lanahan, A., Worley, P., Snyder, S. H. & Ross, C. A. 1995 A huntingtin-associated protein enriched in brain with implications for pathology. *Nature* **378**, 398–402.

Lunkes, A. & Mandel, J.-L. 1998 A cellular model that recapitulates major pathogenic steps of Huntington's disease. *Hum. Mol. Genet.* **7**, 1355–1361.

Maat-Schieman, M. L., Dorsman, J. C., Smoor, M. A.,Siesling, S., Van Duinen, S. G., Verschuuren, J. J., den Dunnen, J. T., Van Ommen, G. J. & Roos, R. A. 1999 Distribution of inclusions in neuronal nuclei and dystrophic neurites in Huntington disease brain. *J. Neuropathol. Exp. Neurol.* **58**, 129–137.

Mudd, S. H., Finkelstein, J. D., Irreverre, F. & Laster, L. 1964 Homocystinuria: an enzymatic defect. *Science* **143**, 1443.

Mudd, S. H. (and 13 others) 1985 The natural history of homocystinuria due to cystathionine β-synthase deficiency. *Am. J. Hum. Genet.* **37**, 1–31.

Mudd, S. H., Levy, H. L. & Skovby, F. 1995 Disorders of transsulfuration. In *The metabolic and molecular bases of inherited disease*, 7th edn (ed. C. R. Scriver, A. L. Beaudet, W. S. Sly & D. Valle), pp. 1279–1327. New York: McGraw-Hill.

Muscat, G. E., Burke, L. J. & Downes, M. 1998 The corepressor N-CoR and its variants RIP13a and RIP13Delta1 directly interact with the basal transcription factors TFIIB, TAFII32 and TAFII70. *Nucleic Acids Res.* **26**, 2899–2907.

Nagafuchi, S. (and 22 others) 1994 Dentatorubral and pallidoluysian atrophy expansion of an unstable CAG trinucleotide on chromosome-12p. *Nature Genet.* **6**, 14–18.

Nagy, L., Kao, H. Y., Chakravarti, D., Lin, R. J., Hassig, C. A., Ayer, D. E., Schreiber, S. L. & Evans, R. M. 1997 Nuclear receptor repression mediated by a complex containing SMRT, mSin3A, and histone deacetylase. *Cell* **89**, 373–380.

Ohmori, S., Kodama, H., Ikegami, T., Mizuhara, S., Oura, T., Isshiki, G. & Uemera, I. 1972 Unusual sulfur-containing amino acids in the urine of homocystinuric patients: III. Homocysteic acid, homocysteine sulfinic acid, *S*-(carboxymethylthio) homocysteine, and *S*-(3-hydroxy-3-carboxy-*n*-propyl)homocysteine. *Physiol. Chem. Phys.* **4**, 286.

Orr, H. T., Chung, M. Y., Banfi, S., Kwiatkowski, T. J., Servadio, A., Beaudet, A. L., Mccall, A. E., Duvick, L. A., Ranum, L. P. W. and Zoghbi, H. Y. 1993 Expansion of an unstable trinucleotide CAG repeat in spinocerebellar ataxia type-1. *Nature Genet.* **4**, 221–226

Paulson, H. L., Perez, M. K., Trottier, Y., Trojanowski, J. Q., Subramony, S. H., Das, S. S., Vig, P., Mandel, J.-L., Fischbeck, K. H. & Pittman, R. N. 1997 Intranuclear inclusions of expanded polyglutamine protein in spinocerebellar ataxia type 3. *Neuron* **19**, 333–344.

Pazin, M. J. & Kadonga, J. T. 1997 What's up and down in histone deacetylation and transcription? *Cell* **89**, 325–328.

Refsum, H., Helland, S. & Ueland, P. M. 1985 Radioenzymic determination of homocysteine in plasma and urine. *Clin. Chem.* **31**, 624.

Ross, C. A. 1995 When more is less: pathogenesis of glutamine repeat neurodegenerative diseases. *Neuron* **15**, 493–496.

Sapp, E., Schwarz, C., Chase, K., Bide, P. G., Young, A. B., Penney, J., Vonsattel, J. P., Aronin, N. & DiFiglia, M. 1997 Huntingtin localization in brains of normal and Huntington's disease patients. *Ann. Neurol.* **42**, 604–612.

Sapp, E., Penney, J., Young, A., Vonsattel, J. P. & DiFiglia, M. 1999 Axonal transport of N-terminal huntingtin suggests early pathology of corticostriatal projections in Huntington's disease. *J. Neuropathol. Exp. Neurol.* **58**, 165–173.

Saudou, F., Finkbeiner, S., Devys, D. & Greenberg, M. E. 1998 Huntingtin acts in the nucleus to induce apoptosis but death does not correlate with the formation of intranuclear inclusions. *Cell* **95**, 55–66.

Schwarz, S. & Zhou, G. Z. 1991 *N*-Methyl-D-aspartate receptors and CNS symptoms of homocystinuria. *Lancet* **337**, 1226–1227.

Schwarz, S., Zhou, G. Z., Katki, A. G. & Rodbard, D. 1990 L-Homocysteate stimulates [^{3}H]MK-801 binding to the phencyclidine recognition site and thus is an agonist for the *N*-methyl-D-aspartate operated cation channel. *Neuroscience* **37**, 193–200.

Sieradzan, K. A., Mechan, A. O., Jones, L., Wanker, E. E., Nukina, N., Mann. D. M. 1999 Huntington's disease intranuclear inclusions contain truncated, ubiquitinated huntingtin protein. *Exp. Neurol.* **156**, 92–9.

Singhrao, S. K., Thomas, P., Wood, J. D., MacMillan, J. C., Neal, J. W., Harper, P. S. & Jones, A. L. 1998 Huntingtin protein colocalizes with lesions of neurodegenerative disease: an investigation in Huntington's, Alzheimer's and Pick's diseases. *Exp. Neurol.* **150**, 213–222.

Sittler, A., Walter, S., Wedemeyer, N., Hasenbank, R., Scherzinger, E., Eickhoff, H., Bates, G. P., Lehrach, H. & Wanker, E. E. 1998 SH3GL3 associates with the huntingtin exon 1 protein and promotes the formation of polygln-containing protein aggregates. *Mol. Cell* 2, 427–436.

Skinner, P. J., Koshy, B. T., Cummings, C. J., Klement, I. A., Helin, K., Servadio, A., Zoghbi, H. Y. & Orr, H. T. 1997 Ataxin-1 with an expanded glutamine tract alters nuclear matrix-associated structures. *Nature* **389**, 971–978.

Skovby, F., Kraus, J. P. & Rosenberg, L. E. 1984 Biosynthesis and proteolytic activation of cystathionine β-synthase in rat liver. *J. Biol. Chem.* **259**, 588–593.

Taunton, J., Hassig, C. A. & Schreiber, S. L. 1996 A mammalian histone deacetylase related to the yeast transcriptional regulator Rpd3p. *Science* **272**, 408–411.

Trottier, Y., Devys, D., Imbert, G., Saudou, F., An, I., Lutz, Y., Weber, C., Agid, Y., Hirsch, E. C. & Mandel, J.L. 1995 Cellular-localization of the Huntingtons disease protein and discrimination of the normal and mutated form. *Nature Genet.* **10**, 104–110.

Wanker, E. E., Rovira, C., Scherzinger, E., Hasenbank, R., Wälter, S., Tait, D., Colicelli, J. & Lehrach, H. 1997 HIP-1: a huntingtin interacting protein isolated by the yeast two-hybrid system. *Hum. Mol. Genet.* **6**, 487–495.
Waragai, M., Lammers, C. H., Takeuchi, S., Imafuku, I., Udagawa, Y., Kanazawa, I., Kawabata, M., Mouradian, M. M. & Okazawa, H. 1999 PQBP-1, a novel polyglutamine tract-binding protein, inhibits transcription activation by Brn-2 and affects cell survival. *Hum. Mol. Genet.* **8**, 977–987.
Xu, L. (and 13 others) 1998 Signal-specific co-activator domain requirements for Pit-1 activation. *Nature* **395**, 301–306.
Yang, W. M., Inouye, C., Zeng, Y., Bearss, D. & Seto, E. 1996 Transcriptional repression by YY1 is mediated by interaction with a mammalian homolog of the yeast global regulator RPD3. *Proc. Natl Acad. Sci. USA* **93**, 12845–12850.
Zhang, Y., Iratni, R., Erdjument-Bromage, H., Tempst, P. & Reinberg, D. 1997 Histone deacetylases and SAP18, a novel polypeptide, are components of a human Sin3 complex. *Cell* **89**, 357–364.
Zhuchenko, O., Bailey, J., Bonnen, P., Ashizawa, T., Stockton, D.W., Amos, C., Dobyns, W. B., Subramony, S. H., Zoghbi, H. Y. & Lee, C. C. 1997 Autosomal dominant cerebellar ataxia (SCA6) associated with small polyglutamine expansions in the α_{1A}-voltage-dependent calcium channel. *Nature Genet.* **15**, 62–69.

12

Analysis of the subcellular localization of huntingtin with a set of rabbit polyclonal antibodies in cultured mammalian cells of neuronal origin: comparison with the distribution of huntingtin in Huntington's disease autopsy brain

J. C. Dorsman, M. A. Smoor, M. L. C. Maat-Schieman, M. Bout, S. Siesling, S. G. van Duinen, J. J. G. M. Verschuuren, J. T. den Dunnen, R. A. C. Roos and G. J. B. van Ommen

Introduction

Huntington's disease (HD) is a neurodegenerative disorder with a midlife onset. The neuropathology involves a loss of neurons in specific parts of the brain, with the striatum being most severely affected. The biochemical basis of the disease is not understood and so far no cure is available.

The gene involved in Huntington's disease encodes a 348 kDa protein called huntingtin. The mutations found in patients imply that HD belongs to the so-called triplet repeat expansion diseases; a $(CAG)_n$ repeat coding for a polyglutamine stretch is expanded in HD patients. The normal range of the repeat sizes is between 11 and 36 units, whereas patients have 37–121 copies (de Rooij *et al.* 1993; Huntington's Disease Collaborative Research Group 1993; Read 1993). In patients, both alleles (normal and mutated) are expressed (Jou & Myers 1995), suggesting a gain of function of the mutant protein. In this model, the expanded polyglutamine part of the protein confers a new property that disturbs the normal activity of huntingtin or causes an interaction with novel partners. In line with the latter proposal, a protein, designated Hap1, has been identified that interacts preferentially with huntingtin with an extended repeat (Li *et al.* 1995). However, another protein has been identified, Hip1, that binds preferentially to normal-sized huntingtin (Kalchman *et al.* 1997; Wanker *et al.* 1997).

The suggestion that the observed alteration of the CAG-encoded polyglutamine stretch of huntingtin in HD patients has a crucial role in the molecular pathogenesis of this neurodegenerative disease is strengthened by the fact that transgenic mice that express only the first exon of the HD gene with an extended repeat of around 140 units display a severe neurological disorder with lethal outcome (Mangiarini *et al.* 1996). In the brain of these mice, abnormal protein aggregates have been detected in the nuclei of neurons called neuronal intranuclear inclusions (NIIs) (Davies *et al.* 1997). The sensitivity of neuronal cells to this type of mutation is also suggested by the fact that at least seven other neurodegenerative disorders are associated with an expansion of polyglutamine-encoding CAG repeats. Furthermore, similar NIIs have been found in various brain regions of HD patients (DiFiglia *et al.* 1997) and of patients with spinocerebellar ataxia type 3, another polyglutamine repeat disorder (Paulson *et al.* 1997).

Nevertheless, these neurodegenerative diseases affect different parts of the brain, which suggests that other parts of the mutated proteins also contribute to the observed specific pathologies. Furthermore, in the transgenic mice that express only the first exon of the HD gene with the extended repeat, the specific cell death of neuronal cells in the striatum that is characteristic of HD is not observed. Accordingly, both the extended glutamine repeat and other domains of huntingtin might contribute to HD. Therefore, studies of the function of the glutamine-repeat region and an analysis of the function of the other segments of huntingtin are both required for a full understanding of HD pathology. Furthermore, detailed studies on the subcellular localization of this protein under various conditions might provide useful information. Essential tools for such studies include the availability of a well-characterized set of antibodies raised against various parts of huntingtin.

Here we describe the characterization of a panel of rabbit polyclonal antibodies raised against different regions of huntingtin. Purified polyclonal antibodies were used in immunofluorescence and in biochemical subcellular fractionation experiments with cells of neuronal origin. In addition, we studied the subcellular localization of huntingtin during neuronal differentiation *in vitro* and in HD autopsy brain.

Materials and methods

Generation of fusion proteins and antibodies

Different huntingtin–glutathione *S*-transferase (huntingtin–GST) fusion genes were inserted into derivatives (pRP261, 265 and 269) of the pGEX vectors (Smith & Johnson 1988). GST fusion proteins were purified by using standard procedures and were subsequently used for the immunization of rabbits. Huntingtin–GST fusion proteins encompassing amino acid residues 596–1030, 852–1193 and 1929–2421 gave rise to serum 97, serum 98 and serum 93, respectively. Synthetic peptides used for immunization were coupled to carrier proteins (bovine serum albumin (BSA), lysozyme or thyroglobulin) by using standard procedures (Harlow & Lane (1988), pp. 313–315). The distinct peptides used for this study contained residues 1–19 (sera 7, 10, 1859 and 1862), residues 11–19 (sera 2 and 5), residues 701–744 (sera 1359 and 1495) and residues 3114–3141 (sera 1356 and 1358). Rabbits were injected and bled monthly. Sera were tested with ELISA and Western blotting experiments.

Affinity purification of antibodies raised against peptides

Antibodies no. 7 and no. 10 directed against the first 19 residues of huntingtin were affinity purified with an affinity column in accordance with the procedure supplied by the manufacturer (Pierce). In brief, a total of 2 mg of peptide was coupled to a column matrix and subsequently the sera were applied. After washing the column, bound antibodies were eluted with 0.1 M glycine, pH 2.5, and the collected fractions were immediately neutralized with 1 M Tris–HCl, pH 9.5. The purified antibodies were tested in Western blotting experiments.

Differentiation of neuroblastoma cells in vitro

All cell lines were propagated as described (de Rooij *et al.* 1996). Mouse neuroblastoma cells (N1E-115) were grown under standard conditions in minimal essential medium supplemented with 2% (v/v) foetal calf serum (Life Technologies). Differentiation was induced by adding dimethyl sulphoxide to a final concentration of 1% (v/v). The subcellular localization of huntingtin was analysed in immunofluorescence experiments (days 0, 3 and 6 after the induction of differentiation).

Immunofluorescence, subcellular fractionation and Western blotting experiments

Immunofluorescence, subcellular fractionation and Western blotting experiments were performed as described (de Rooij *et al.* 1996). Gels were scanned and analysed with the use of Gelworks (Ultra Violet Products, NonLinear Dynamics). Staining with 4',6-diamidino-2-phenylindole (DAPI) was performed by the addition of this compound to the mounting medium (1 $\mu g\ ml^{-1}$). VectaShield (Vector) was used as an anti-fading agent and mounting medium.

Immunohistochemistry

Immunolabelling was performed on sections 5 μm thick cut from tissue blocks fixed in 10% (v/v) formalin and embedded in paraffin wax. Before pretreatment by heating in citrate buffer (pH 6.0) for 20 min, all sections were preincubated in 0.3% (v/v) H_2O_2 to block endogenous peroxidase. The primary antiserum no. 7 was incubated on the section overnight at room temperature. Next the section was incubated with biotinylated pig anti-rabbit immunoglobulin (Dako) for 30 min, followed by peroxidase-conjugated streptavidin (Dako) for 30 min. Peroxidase activity was detected with 3,3'-diaminobenzidine as the chromagen. Controls were obtained by omission of the primary antisera.

Results

Generation and characterization of antibodies

Rabbit polyclonal antibodies (pAbs) were generated against various regions of huntingtin by using either synthetic peptides or GST–huntingtin fusion proteins spread over the total huntingtin sequence (figure 12.1*a*). A total of 11 pAbs were obtained, raised against seven different antigens. Huntingtin can be cleaved by the pro-apoptotic protease apopain, generating an N-

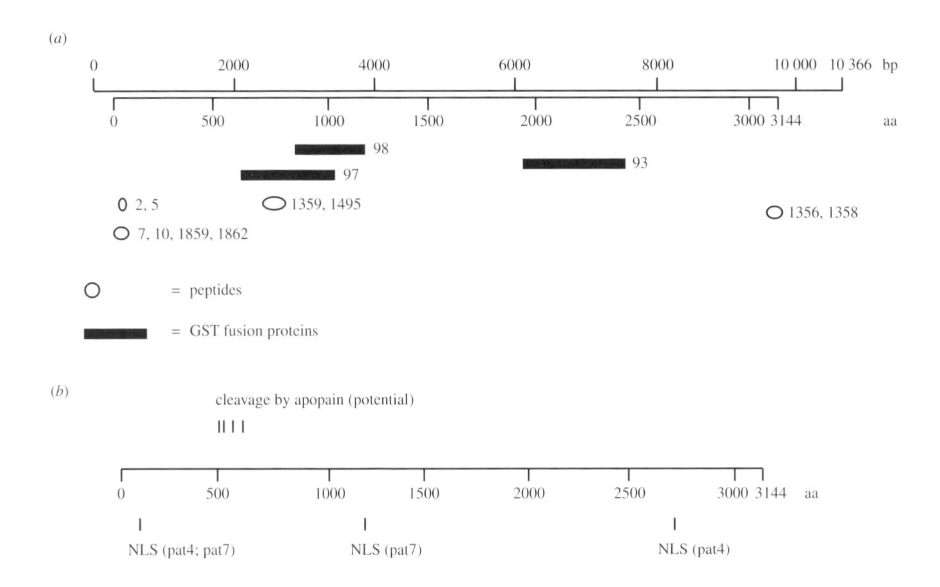

Figure 12.1 The localization of the antigens used for the generation of pAbs in relation to motifs present in huntingtin. (*a*) Antigens. The localization of the different peptides (ovals) and GST fusion proteins (bold lines) used for the immunization of rabbits are indicated. The numbers signify the names of the different sera. (*b*) Motifs. The putative nuclear localization signals for human huntingtin have been indicated (PSORT II; http://psort.nibb.ac.jp:8800/). Pat4 motifs are present at residue positions 89 (RPKK) and 2662 (RKHR); pat7 motifs are present at positions 86 (PLHRPKK) and 1182 (PIRRKGK). In addition, potential cleavage sites for apopain have been indicated (SWISS-PROT: P42858; http://expasy.hcuge.ch/). Potential cleavage sites are present at residue positions 513–514, 530–531, 552–553 and 589–590. N-terminal cleavage products of huntingtin with an expanded repeat might accumulate preferentially in HD brain (Davies *et al.* 1997; DiFiglia *et al.* 1997; Maat-Schieman *et al.* 1999).

terminal fragment (Goldberg *et al.* 1996; see also figure 12.1*b*). In principle, only the pAbs raised against peptides present in the N-terminus can detect N-terminal cleavage products (i.e. nos 2, 5, 7, 10, 1859 and 1862; see also figure 12.1*a*). In figure 12.1*b*, the position of amino acid sequences conforming to nuclear localization signals are also indicated.

All generated antisera were tested for reaction against huntingtin by using either ELISA (anti-peptide antibodies) or Western blots (anti-GST fusion proteins). All sera showed high reactivity against the antigens used for immunization. Subsequently we tested the capacity of the pAbs to recognize huntingtin in cell extracts with the use of Western analysis. A representative experiment is shown in figure 12.2. Antisera no. 7 and no. 10 were raised against a peptide encompassing the first 19 residues of huntingtin (see figure 12.1*a*). Both antibodies could detect one major band in the 200–400 kDa region of the blot (figure 12.2*a*). The same antisera were also purified by affinity chromatography. Figure 12.2*b* shows the results of the analysis of various fractions containing purified pAb no. 7. The results for the purification of pAb no. 10 were similar to those for pAb no. 7 (results not shown). The specificity of the observed interaction was underscored by the detection of two bands, i.e. huntingtin with a normal repeat (15) and an extended repeat (50)

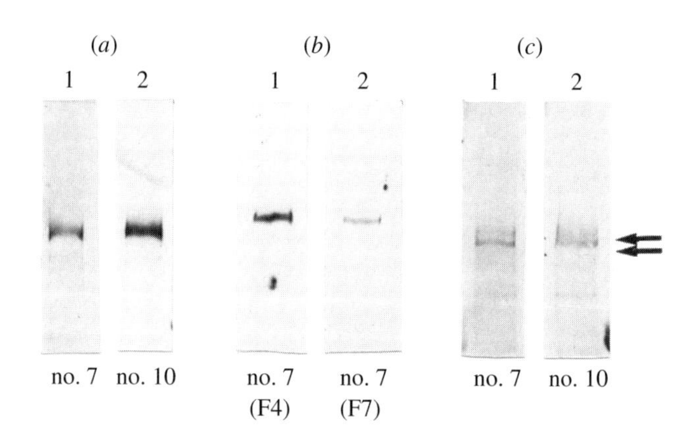

Figure 12.2 Western analysis with the use of pAbs no. 7 and no. 10 raised against a peptide encompassing the first 19 residues of huntingtin. (*a*) Analysis of crude sera. The results of Western experiments with enhanced chemiluminescence detection by using extracts prepared from human VH10 cells are shown for pAbs no. 7 and no. 10. (*b*) Affinity purification of pAb no. 10. Antibodies present in the serum were bound to a peptide column; eluted fractions were tested in Western blotting experiments with enhanced chemiluminescence detection with the use of extracts prepared from VH10 cells. F4 and F7 indicate fractions obtained after elution. (*c*) Detection of huntingtin with an expanded repeat. The results of Western blotting experiments with enhanced chemiluminescence detection with the use of extracts prepared from an HD patient (15–50-glutamine repeat) are shown for pAbs no. 7 and no. 10. Arrows indicate the position of full-length normal and mutated huntingtins.

in extracts prepared from patient-derived lymphoblasts (figure 12.2*c*). In addition, pAbs no. 7 and no. 10 could also detect a clear expression in *Caenorhabditis elegans* transgenic for exon 1 or exons 1–7 of the HD gene with $(CAG)_{17}$ and $(CAG)_{73}$ (results not shown). These results suggest that both pAbs efficiently recognize the N-terminus of the HD protein in the presence of relatively long repeat sizes and that these pAbs therefore constitute suitable reagents for the analysis of HD gene expression in human autopsy brain. All other antibodies were also capable of recognizing a protein in the expected size range, although with different affinities (results not shown).

Studies on the subcellular localization of huntingtin

The pAbs were used to extend our studies on the subcellular localization of huntingtin. All (affinity-purified) pAbs showed a comparable huntingtin signal in immunofluorescence experiments with human and mouse cells, including cells of neuronal origin. A representative immunofluorescence experiment is shown in figure 12.3. For this study, the mouse neuroblastoma cell line N1E-115 was used. These cells constitute a suitable model system to study neuronal differentiation *in vitro* (see, for example, Kimhi *et al.* 1976; De Laat & Van der Saag 1982; Kranenburg *et al.* 1995). Cells were stimulated to differentiate by the addition of 1% (v/v) dimethyl sulphoxide, after which the cells flattened and neurites began to form from day 2 onwards. At day 6 the cells were fully differentiated. Immunofluorescence experiments with pAb no. 10 showed both a nuclear and cytoplasmic signal in each distinct stage of differentiation (figure 12.3*a–c*). On differentiation we observed no drastic changes in amount, appearance or subcellular localization of the staining.

To confirm the immunofluorescence results, a biochemical subcellular fractionation experiment was performed on undifferentiated N1E-115 cells. A procedure was followed that resulted in a cytoplasmic protein fraction, a nuclear protein fraction and nuclear remainders. The protein extracts were separated by SDS–PAGE, and Western blots were analysed with different huntingtin antibodies. The efficiency of the fractionation procedure was verified by analysing the distribution of the nuclear protein p53 and the cytoplasmic proteins Raf and tubulin. Films were scanned and the distribution of the proteins in the various fractions was calculated. As expected, p53 was detected exclusively in the nuclear fraction and Raf in the cytoplasm. Most tubulin was also detected in the cytoplasmic fraction, but a small amount was found in the nuclear fraction, indicating some carry-over of cytoplasmic proteins into the nuclear fraction during fractionation. The results obtained for tubulin were used to correct the scans of the other proteins, i.e. all calculated values for the nuclear fraction were lowered with a percentage equal to the percentage of tubulin found in the nuclear fraction (figure 12.4). With the use of pAb no. 10, after this correction *ca.* 85% of huntingtin is found in the cytoplasm and 10–15% is detected in the nucleus.

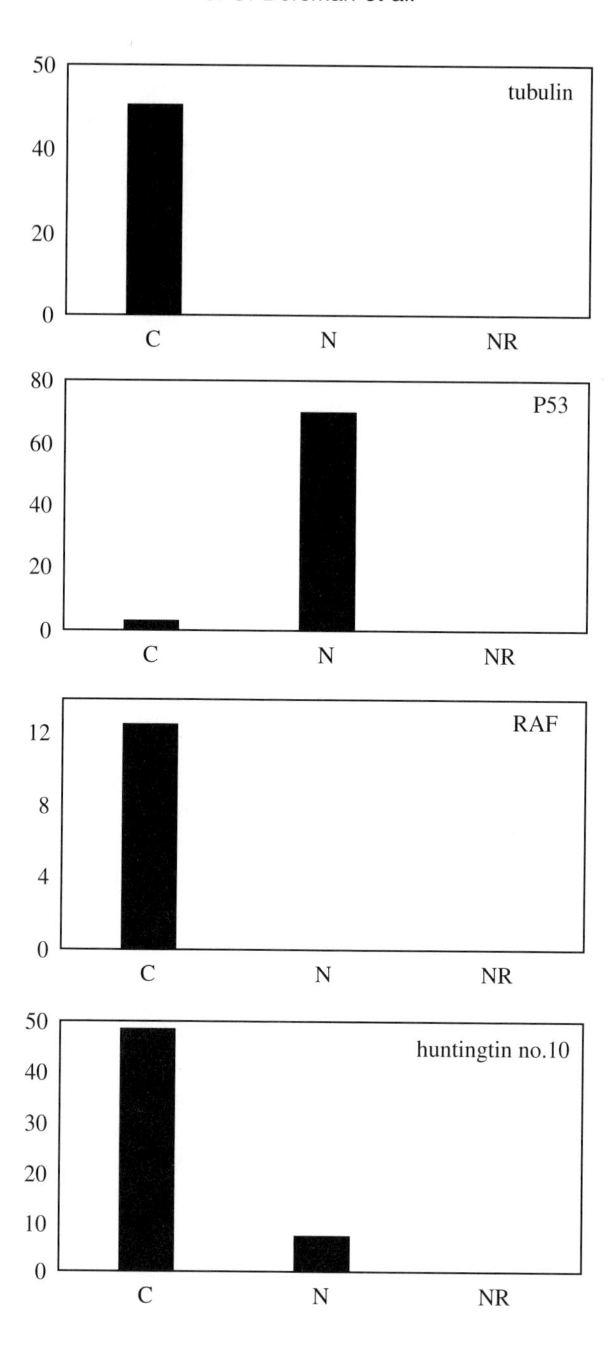

50
40
20
0
tubulin
C
N
NR
80
60
40
20
0
P53
C
N
NR
12
8
4
0
RAF
C
N
NR
50
40
30
20
10
0
huntingtin no.10
C
N
NR

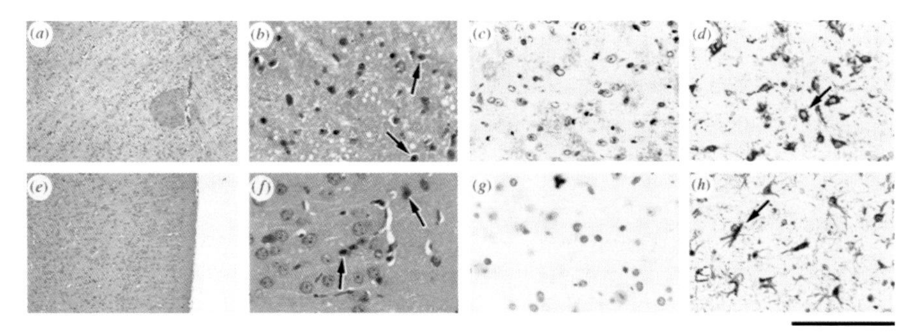

Plate 1 Chapter 4 Neuronal loss and gliosis in HD transgenic striatum and cortex. See p.50 for detailed caption.

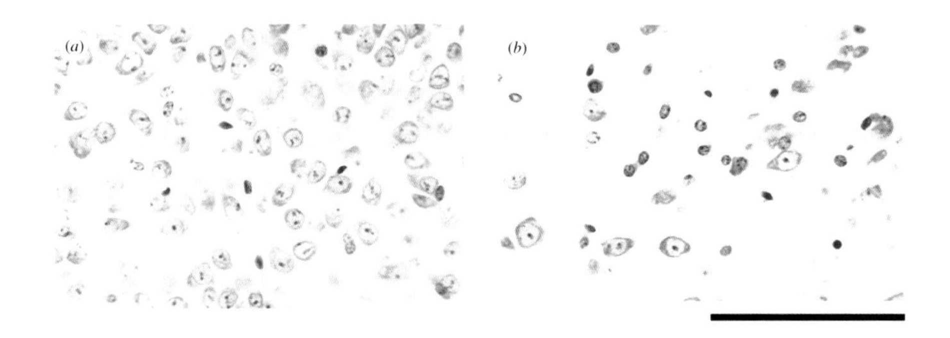

Plate 2 Chapter 4 Neuronal loss in HD transgenic mice. See p.52 for detailed caption.

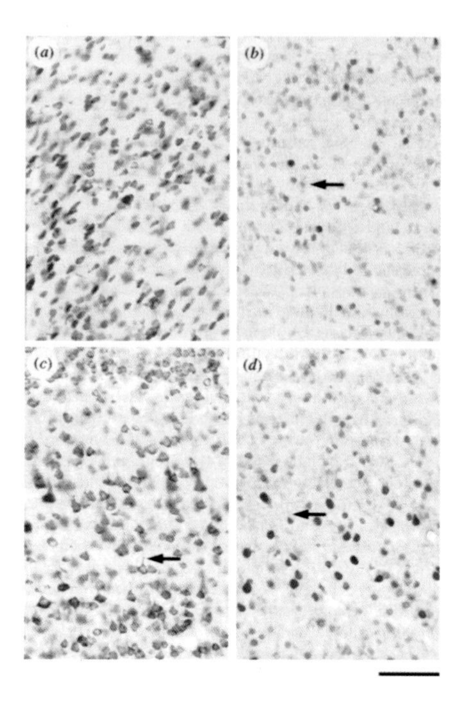

Plate 3 Chapter 4 Decreased immunoreactivity for Neu-N antibody in HD transgenic mice. See p.53 for detailed caption.

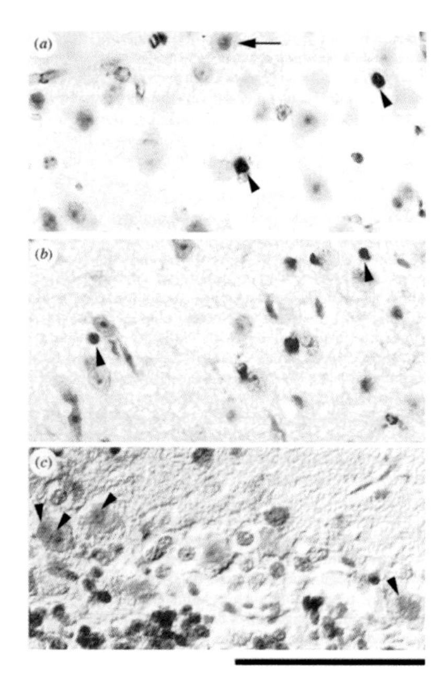

Plate 4 Chapter 4 Immunoreactivity with anti-huntingtin and anti-ubiquitin. See p.54 for detailed caption.

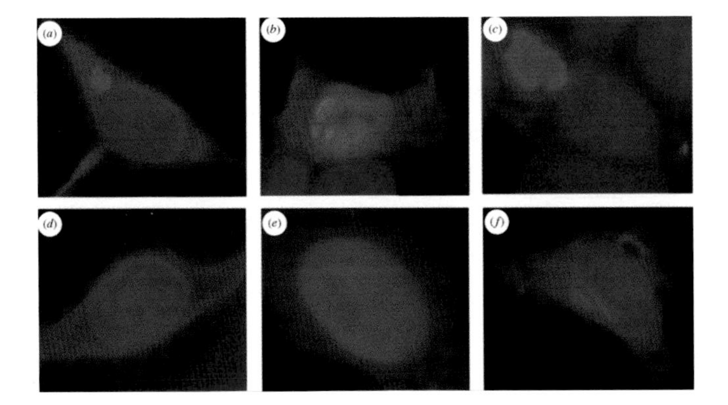

Plate 5 Chapter 5 The localization of huntingtin is altered by the addition of an active NLS sequence. Huntingtin, detected by mAb 2166, appears as red stain, the nucleus is counter-stained in blue. Nuclear huntingtin stain is pink when the red stain is overlapped with blue. The huntingtin aggregates appear as large clumped masses, easily differentiated from normal diffuse stain. See p.68 for detailed caption.

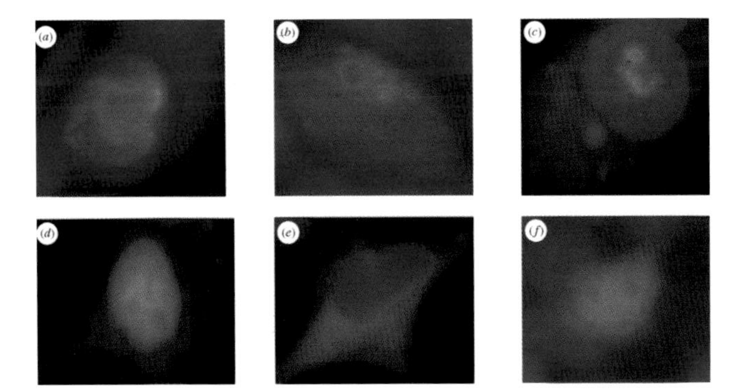

Plate 6 Chapter 5 The localization of huntingtin is altered by the addition of an active NES sequence. Huntingtin is shown in red, the nucleus is counter-stained in blue, and nuclear huntingtin stain is pink when the red stain is overlapped with blue. See p.70 for detailed caption.

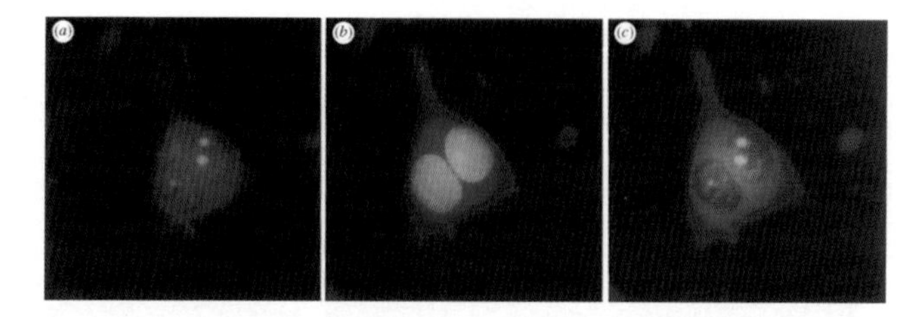

Plate 7 Chapter 8 Co-localization of N-terminal huntingtin and SUG1, a subunit of the PA700 complex of the proteasome, in inclusions observed in NG108-15 cells expressing FL-hd116. See p.121 for detailed caption.

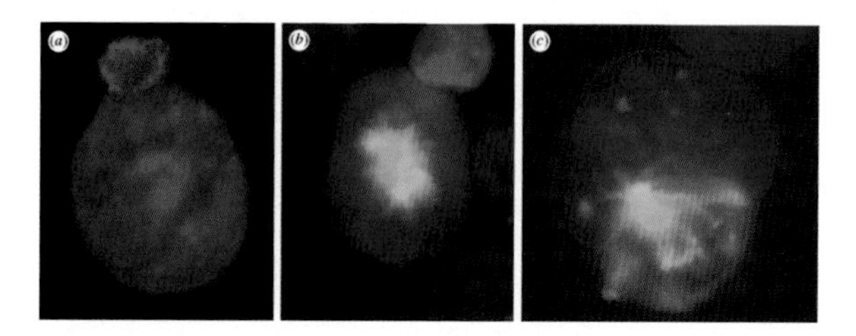

Plate 8 Chapter 9 Extended polyQ constructs expressed by transient transfection in Cos-1 cells, 48 h after transfection. See p.132 for detailed caption.

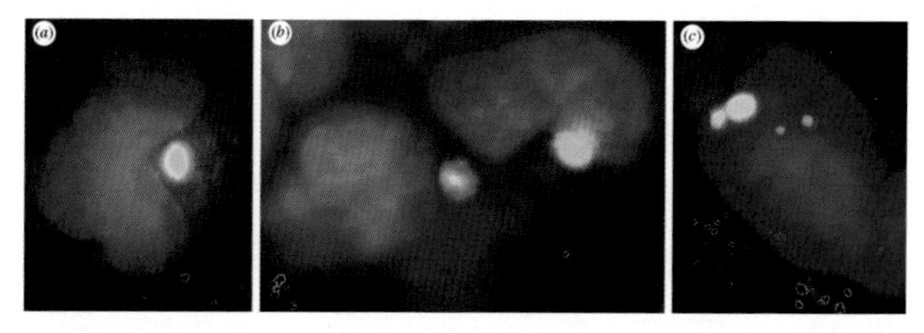

Plate 9 Chapter 9 Co-expression and co-aggregation of normal length and extended polyQ tracts. See p.135 for detailed caption.

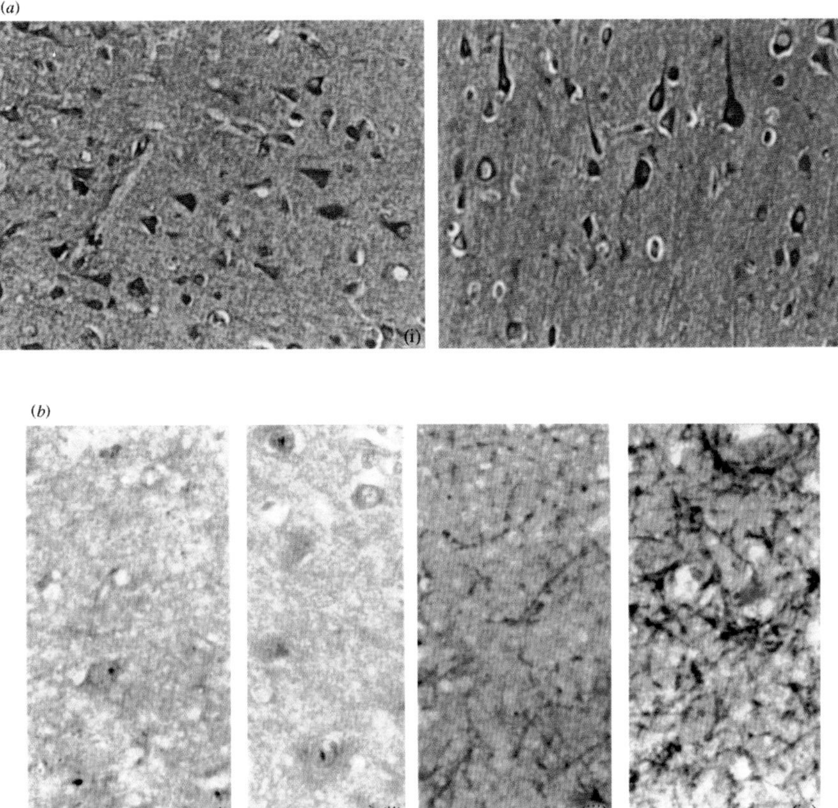

Plate 10 Chapter 11 (*a*) Immunostaining HD cortex with monoclonal antibodies (i) HDA3E10 and (ii) HDC8A4. Magnification ×100. (*b*) Immunostaining of HD cortex with N-675 and ubiquitin: (i) N675 demonstrating the presence of inclusions; (ii) immunostained with ubiquitin (see table 11.1); (iii) control; and (iv) HD brain in the cortical layer VI immediately above the white matter, showing strong staining of processes and increased intensity of such staining in HD brain.

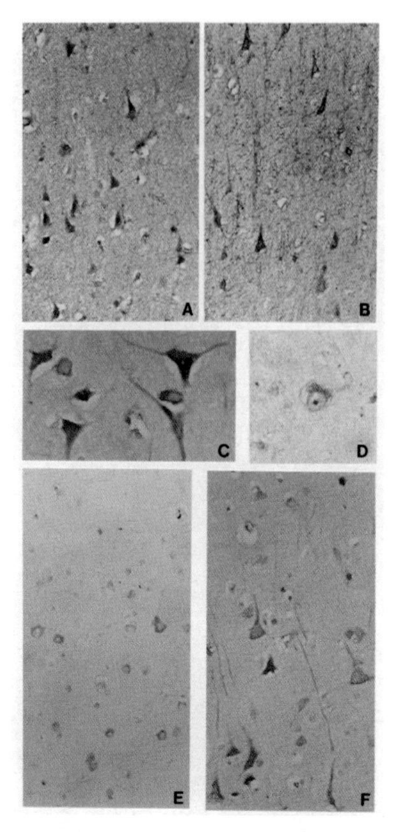

Plate 11 Chapter 11 Control (*a*) and HD (*b*) (Santa Cruz Biotechnology), Control (c) and HD (*d,e,f*) (Santa Cruz Biotechnology). See p.161 for detailed caption.

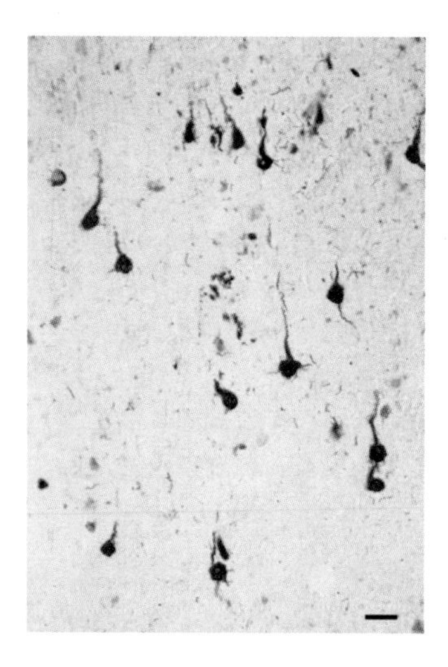

Plate 12 Chapter 21 Neurofibrillary lesions in cerebral cortex from an Alzheimer's disease patient revealed with a phosphorylation-dependent anti-tau antibody.

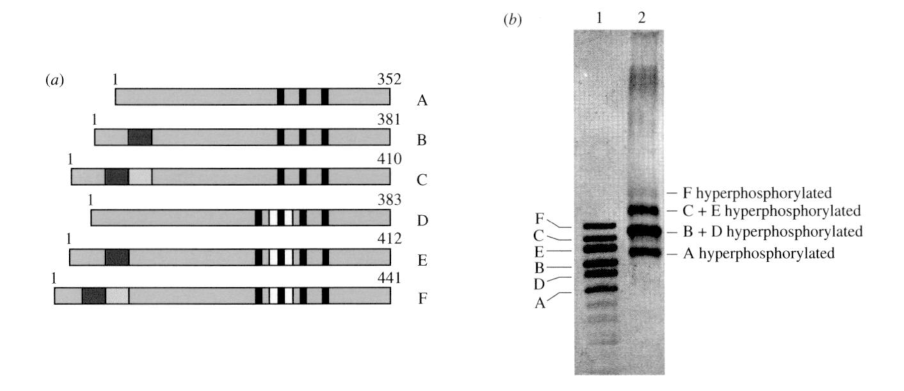

Plate 13 Chapter 21 Isoforms of human tau. See p. 278 for detailed caption.

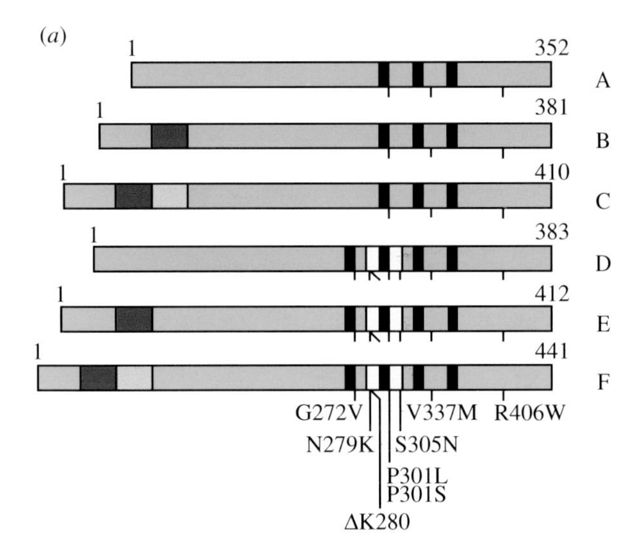

Plate 14 Chapter 21 Mutations in the tau gene in frontotemporal dementia and Parkinsonism linked to chromosome 17 (FTDP-17). (*a*) Schematic diagram of the six tau isoforms (A-F) that are expressed in adult human brain. See p. 284 for detailed caption.

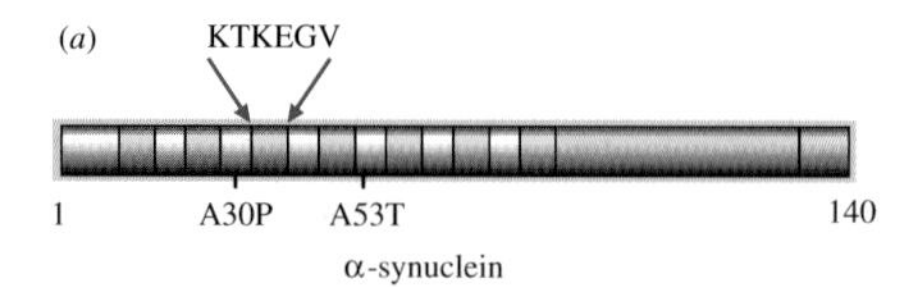

Plate 15 Chapter 21 Mutations in the α-synuclein gene in familial Parkinson's disease. (*a*) Schematic diagram of human α-synuclein. (Reproduced from Goedert & Spillantini (1998).) See p. 290 for detailed caption.

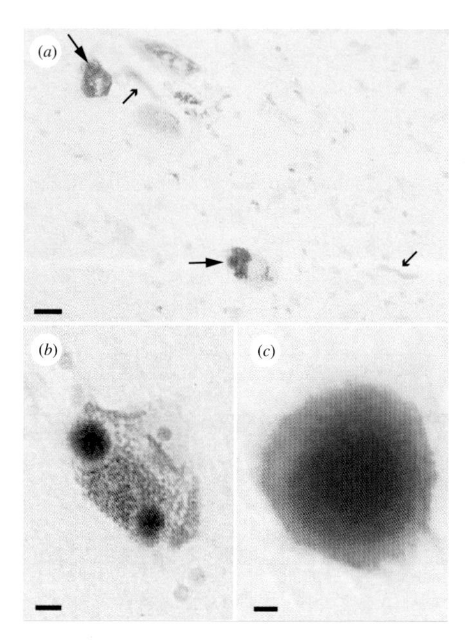

Plate 16 Chapter 21 Substantia nigra from patients with Parkinson's disease immunostained for α-synuclein. (Reproduced from Spillantini *et al.* (1997*a*).) See p. 292 for detailed caption.

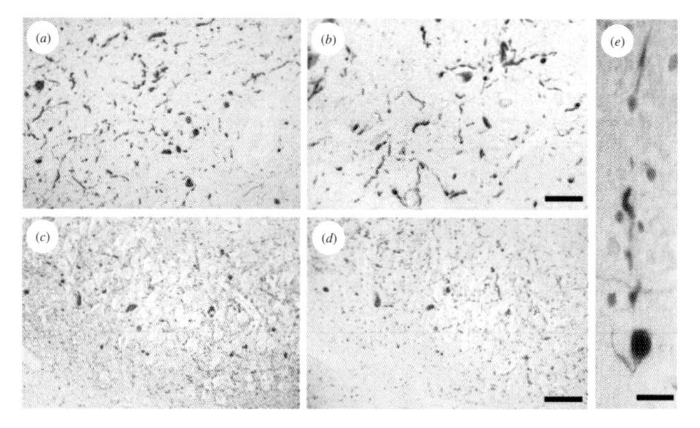

Plate 17 Chapter 21 Brain tissue from patients with dementia with Lewy bodies immunostained for α-synuclein. (Reproduced from Spillantini *et al.* (1998*a*).) See p. 293 for detailed caption.

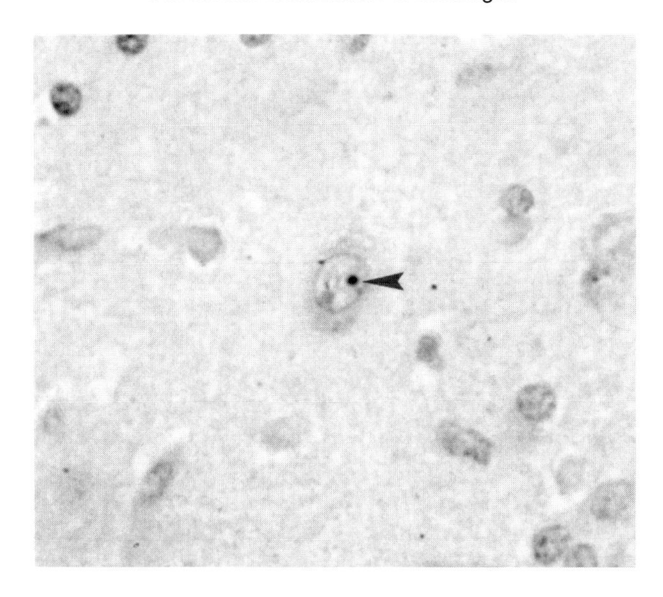

Figure 12.5 Immunohistochemical analysis of HD autopsy brain: a strip of HD neocortex showing neurons immunostained with pAb no. 7. Bound peroxidase-conjugated antibodies were revealed in a peroxidase reaction with 3,3'-diaminobenzidine as the chromagen. Controls were obtained by omission of the primary antiserum.

NIIs in the neostriatum and neocortex, might contribute to the observed neurological defects in HD (Davies *et al.* 1997; DiFiglia *et al.* 1997). In an extension of these studies, NIIs were also detected in the HD-affected allocortex, but not in the pallidum, cerebellum and substantia nigra (Maat-Schieman *et al.* 1999). In all brain regions tested so far, only the N-terminal fragment of huntingtin seems to be present in the NIIs. The accumulation of full-length huntingtin with an expanded repeat, probably in other regions of the nucleus, has nevertheless also been reported (DiFiglia *et al.* 1997; Lunkes & Mandel 1998).

Figure 12.4 Subcellular fractionation studies. Undifferentiated mouse neuroblastoma N1E-115 cells were fractionated into cytoplasmic (C), nuclear (N) and nuclear remnant (NR) fractions, which were analysed in Western blotting experiments with pAb no. 10. The validity of the fractionation procedure followed was tested with antibodies against two different cytoplasmic proteins (tubulin and the kinase Raf) and against one nuclear protein (p53). After enhanced chemiluminescence detection of the signals, the films were scanned. A small amount of tubulin could be detected in the nuclear fractions, whereas the p53 protein and Raf were present exclusively in the nucleus and the cytoplasm, respectively. The scanning data have been corrected for trapping of a small amount of tubulin in the nuclear fraction. In agreement with the data of the immunofluorescence experiments, pAb no. 10 detects huntingtin both in the cytoplasm and the nucleus.

Various features of the HD protein might influence its subcellular localization and intranuclear distribution. Full-length huntingtin contains four putative nuclear localization signals (see figure 12.1*b*) and many phosphorylation sites (Hoogeveen *et al.* 1993; J. C. Dorsman and M. A. Smoor, unpublished data). It has been reported that phosphorylation in the vicinity of nuclear localization signals might alter the efficiency of import of proteins into the nucleus (see, for example, Hubner *et al.* 1997; Xiao *et al.* 1997). In particular, the cellular stress caused by the expression of huntingtin with an expanded repeat might affect the activities of many kinases and phosphatases, possibly including enzymes capable of modifying huntingtin. It is quite plausible that in cells expressing huntingtin with an expanded repeat, it is differently modified, leading to alterations in its subcellular distribution.

In addition, cleavage by the pro-apoptotic protease apopain seems to increase when huntingtin contains an expanded repeat (Goldberg *et al.* 1996) (see also figure 12.1*b*). For such N-terminal fragments the probability of a nuclear localization is higher than for the full-length protein, as calculated by the PSORTII program (http://psort.nibb.ac.jp:8800/). An increased nuclear import of huntingtin with an expanded glutamine repeat or N-terminal fragments thereof, coupled with the increased propensity for aggregation of the latter (see also Scherzinger *et al.* 1997), might well result in the formation of NIIs.

Whether the observed increase in accumulation of the N-terminal fragments of huntingtin with an expanded repeat in the NIIs and the concomitant functional consequences have a pivotal role in the aetiology of HD remains to be elucidated. It still remains possible that alterations in cytoplasmic and nuclear functions of (full-length) huntingtin contribute to the observed pathology, whereas its accumulation of NIIs is a parallel corollary of the alterations.

We thank Professor R. Plasterk for the kind gift of pRP vectors. We also thank Professor R. Plasterk and Dr H. van Luenen for collaboration on the *C. elegans* project. This work was supported by the Netherlands Organization for Scientific Research (NWO) through the Foundation for Medical Sciences in the Netherlands (MW) and by the European Community (PL960244).

References

Davies, S. W., Turmaine, M., Cozens, B. A., DiFiglia, M., Sharp, A. H., Ross, C. A., Scherzinger, E., Wanker, E. E., Mangiarini, L. & Bates, G. P. 1997 Formation of neuronal intranuclear inclusions underlies the neurological dysfunction in mice transgenic for the HD mutation. *Cell* **90**, 537–548.

De Laat, S. W. & Van der Saag, P. T. 1982 The plasma membrane as a regulatory site in growth and differentiation of neuroblastoma cells. *Int. Rev. Cytol.* **74**, 1–54.

dc Rooij, K. E., Dc Koning Gans, P. A., Skraastad, M. I., Belfroid, R. D., Vegter-van der Vlis, M., Roos, R. A. C, Bakker, E., Van Ommen, G. J. B, Den Dunnen,

J. T. & Losekoot, M. 1993 Dynamic mutation in Dutch Huntington's disease patients: increased paternal repeat instability extending to within the normal size range. *J. Med. Genet.* **30**, 996–1002.

de Rooij, K. E., Dorsman, J. C., Smoor, M. A., Den Dunnen, J. T. & Van Ommen, G. J. 1996 Subcellular localization of the Huntington's disease gene product in cell lines by immunofluorescence and biochemical subcellular fractionation. *Hum. Mol. Genet.* **5**, 1093–1099.

DiFiglia, M., Sapp, E., Chase, K. O., Davies, S. W., Bates, G. P., Vonsattel, J. P. & Aronin, N. 1997 Aggregation of huntingtin in neuronal intranuclear inclusions and dystrophic neurites in brain. *Science* **277**, 1990–1993.

Goldberg, Y. P. (and 10 others) 1996 Cleavage of huntingtin by apopain, a proapoptotic cysteine protease, is modulated by the polyglutamine tract. *Nature Genet.* **13**, 442–449.

Harlow, E. & Lane, D. 1988 *Antibodies: a laboratory manual.* Cold Spring Harbor, NY: Cold Spring Harbor Laboratory Press.

Huntington's Disease Collaborative Research Group 1993 A novel gene containing a trinucleotide repeat that is expanded and unstable on Huntington's disease chromosomes. *Cell* **72**, 971–983.

Hoogeveen, A. T., Willemsen, R., Meyer, N., de Rooij, K. E., Roos, R. A., Van Ommen, G. J. & Galjaard, H. 1993 Characterization and localization of the Huntington disease gene product. *Hum. Mol. Genet.* **2**, 2069–2073.

Hubner, S., Xiao, C. Y. & Jans, D. A. 1997 The protein kinase, CK2 site (Ser111/112) enhances recognition of the simian virus 40 large T-antigen nuclear localization sequence by importin. *J. Biol. Chem.* **272**, 17191–17195.

Jou, Y.-S. & Myers, R. M. 1995 Evidence from antibody studies that the CAG repeat in the Huntington disease gene is expressed in the protein. *Hum. Mol. Genet.* **4**, 465–469.

Kalchman, M. A. (and 13 others) 1997 HIP1, a human homologue of *S. cerevisiae* Sla2p interacts with membrane-associated huntingtin in the brain. *Nature Genet.* **16**, 44–53.

Kimhi, Y., Palfrey, C., Spector, I., Barak, Y. & Littauer, U. Z. 1976 Maturation of neuroblastoma cells in the presence of dimethylsulfoxide. *Proc. Natl Acad. Sci. USA* **73**, 462–466.

Kranenburg, O., Scharnhorst, V., van der Eb, A. J. & Zantema, A. 1995 Inhibition of cyclin-dependent kinase activity triggers neuronal differentiation of mouse neuroblastoma cells. *J. Cell Biol.* **131**, 227–234.

Li, X. J., Li, S. H., Sharp, A. H., Nucifora, F. C. Jr, Schilling, G., Lanahan, A., Worley, P., Snyder, S. H. & Ross, C. A. 1995 A huntingtin-associated protein enriched in brain with implications for pathology. *Nature* **378**, 398–402.

Lunkes, A. & Mandel, J. L. 1998 A cellular model that recapitulates major pathogenic steps of Huntington's disease. *Hum. Mol. Genet.* **7**, 1355–1361.

Maat-Schieman, M. L. C., Dorsman, J. C., Smoor, M. A., Siesling, S., Van Duinen, S. G., Verschuuren, J. J. G. M., Den Dunnen, J. T., Nan Ommen, G. J. B. & Roos, R. A. C. 1999 Distibution of inclusions in neuronal nuclei and dystrophic neurites in Huntington disease brain. *J. Neuropathol. Exp. Neurol.* **58**, 129–137.

Mangiarini, L. (and 10 others) 1996 Exon 1 of the HD gene with an expanded CAG repeat is sufficient to cause a progressive neurological phenotype in transgenic mice. *Cell* **87**, 493–506.

Paulson, H. L., Perez, M. K., Trottier, Y., Trojanowski, J. Q., Subramony, S. H., Das, S. S., Vig, P., Mandel, J. L., Fischbeck, K. H. & Pittman, R. N. 1997 Intranuclear inclusions of expanded poly-glutamine protein in spinocerebellar ataxia type 3. *Neuron* **19**, 333–344.

Read, A. P. 1993 Huntington's disease: testing the test. *Nature Genet.* **4**, 329–330.

demonstrates an inherent toxicity of extended polyglutamine sequences. However, the neuropathology was widespread, whereas in each of the trinucleotide-repeat diseases specific brain regions are predominantly affected. Therefore we believe that the polyglutamine sequences confer on the parent protein altered properties culminating in cell-selective disease.

Thesis

Our current knowledge of the pathogenesis of Huntington's disease is based on its neuropathology in the juvenile- and adult-onset forms as well as insights from transgenic models of Huntington's disease and cultured striatal neurons transfected with mutant huntingtin. Changes in huntingtin localization occur in the Huntington's disease brain and most frequently involve a diffuse nuclear and focal perinuclear accumulation of the protein (Sapp *et al.* 1997). Additionally, in the largest group of Huntington's disease patients, those with adult onset (90–95% of patients) (Snell *et al.* 1993), aggregation of the N-terminal region of mutant huntingtin in neurites predominates. In the remaining 5–10% of Huntington's disease patients, those with juvenile onset, nuclear inclusions are abundant in the cortex and striatum. The transgenic models of Huntington's disease reveal that expanded polyglutamine repeats serve to target the mutant huntingtin to the nucleus and that small mutant huntingtin fragments are delivered more effectively than large fragments or full-length huntingtin (Davies *et al.* 1997; Laforet *et al.* 1998; Reddy *et al.* 1998). These findings demonstrate, in the human, that cleavage of mutant huntingtin precedes its aggregation and neuropathological consequences. However, even abundant nuclear inclusions do not inevitably cause striatal neurodegeneration (Davies *et al.* 1997). Studies *in vitro* confirm this, with the important caveat that such experiments test short-term, harmful effects of overexpressed mutant huntingtin (Saudou *et al.* 1998; Kim *et al.* 1999*a*). We propose that mutant huntingtin contributes to a number of nuclear and cytoplasmic events that lead to neuronal cell death; the handling of mutant huntingtin in juvenile-onset Huntington's disease probably differs from that in adult-onset, resulting in two presentations of the disease.

Neuronal changes in Huntington's disease

Localization of huntingtin and its implication for function in mammalian brain

We organized our research plan to determine the localization of wild-type huntingtin and the expression of the mutant protein. The protein coded by the Huntington's disease gene, huntingtin (*ca.* 350 kDa) appears early in the development of the mammalian brain, by day 15 of gestation (Bhide *et al.* 1996). The amount of huntingtin increases in the rodent newborn to achieve adult levels by three weeks. In the adult rodent, wild-type huntingtin resides in the cytoplasm (DiFiglia *et al.* 1995; Sharp *et al.* 1995; Trottier *et al.* 1995). Ultrastructural and biochemical studies in brain point to an association of huntingtin with tubular cisternae and vesicles of the Golgi complex (figure 13.1*a*), synaptic vesicles (figure 13.1*b*) (see also DiFiglia *et al.* 1995) and

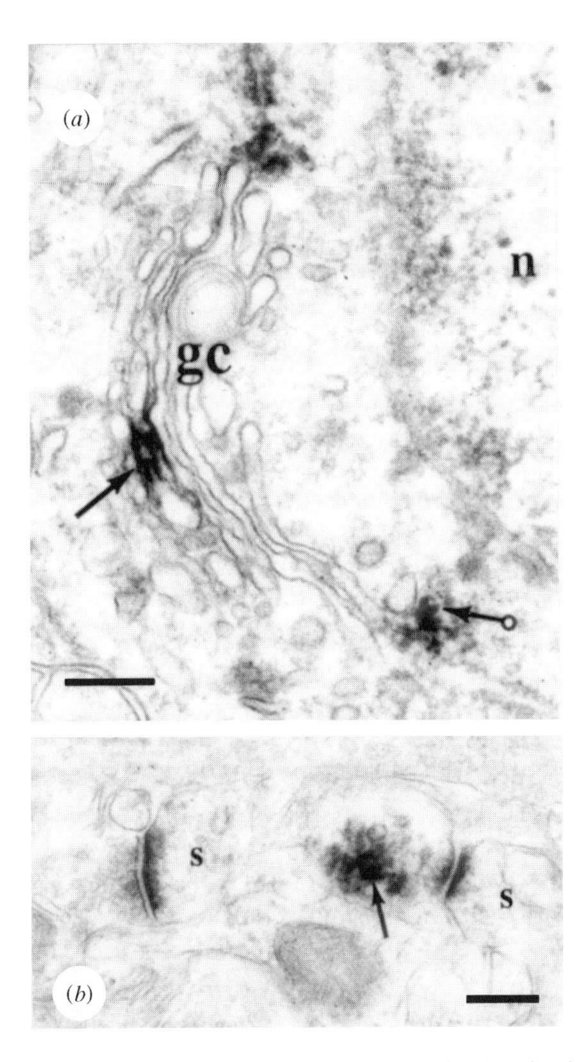

Figure 13.1 Ultrastructural localization of immunoreactive huntingtin in rat cortex. (*a*) Labelling occurs on membranes of tubular cisternae (arrow) and vesicles (ringed arrow) of the Golgi complex (gc). n, Nucleus. (*b*) Immunoreactivity in the axon terminal on the right is associated with vesicles (arrow). The labelled terminal forms an asymmetric synapse with a dendritic spine (s). The terminal on the left has no immunoreactivity and also contacts a spine (s). Ab1 antiserum against huntingtin was used. Scale bars, 0.25 μm. (Taken from DiFiglia *et al.* (1995).)

microtubules (Gutenkunst *et al.* 1995; Tukamoto *et al.* 1997), in addition to huntingtin in soluble fractions (DiFiglia *et al.* 1995; Sharp *et al.* 1995). Huntingtin co-distributes with clathrin in differentiating neurons (figure 13.2*d*) (Kim *et al.* 1999*b*) and decorates clathrin-coated pits and mature clathrin-coated vesicles (figure 13.2*a*,*b*) (Velier *et al.* 1998) and uncoated vesicles (figure 13.2*c*). Co-immunoisolation of huntingtin with synaptophysin in synaptosomes is consistent with the transport of huntingtin to presynaptic sites (DiFiglia *et al.* 1995). In axoplasmic flow studies in peripheral nerve, huntingtin is anterogradely and retrogradely transported (Block-Galarza *et al.* 1996). On the basis of these results, we speculate that wild-type huntingtin participates in vesicle transport.

Identifying partners for huntingtin has been used to elucidate functions of wild-type huntingtin, as well as possible changes in function of the mutated huntingtin. Huntingtin interacts biochemically with numerous, predominantly cytoplasmic proteins, including huntingtin associated protein 1 (HAP-1) (X.-J. Li *et al.* 1995), huntingtin interacting protein (HIP-1) (Kalchman *et al.* 1997; Wanker *et al.* 1997), a ubiquitin-conjugating enzyme (Kalchman *et al.* 1996), calmodulin (Bao *et al.* 1996), apopain or caspase-3 (Goldberg *et al.* 1996), α-adaptin (Faber *et al.* 1998; Gusella & MacDonald 1998), glyceraldehyde-3-phosphate dehydrogenase (GAPDH) (Burke *et al.* 1996) and cystathionine β-synthase (Boutell *et al.* 1998). HAP-1 attaches to dynactin (Engelender *et al.* 1997; S. H. Li *et al.* 1998), which contributes to the movement of particles along microtubules; HIP-1 is structurally similar to Sla2P, also implicated in cytoskeletal function in yeast; and adaptin binds to clathrin-coated vesicles. There is also evidence that huntingtin interacts with WW and SH3 protein domains (Faber *et al.* 1998; Sittler *et al.* 1998). Collectively, these studies form the framework that huntingtin might participate, directly or indirectly, in vesicle transport.

Expression of mutant huntingtin in Huntington's disease

Three general mechanisms have been used to explain the pathogenesis of diseases with autosomal dominant inheritance: (i) the mutated gene might not be transcribed, resulting in a decrease in active protein to levels insufficient to perform important cellular functions; (ii) a protein might be

Figure 13.2 (*a*–*c*) Electron microscopic localization of immunoreactive huntingtin in dendrites of mouse cortical neurons. Immunoperoxidase labelling is associated with a coated pit (*a*), a coated vesicle (*b*) and with a cluster of vesicles ((*c*), arrows) near the plasma membrane. Staining was performed with anti-huntingtin antiserum Ab585. Scale bars, 0.1 μm. (Taken from Velier *et al.* (1998)). (*d*) Co-localization of huntingtin (with antiserum Ab1) and clathrin in immortalized striatal cells *in vitro*. Both proteins distribute to the same regions of cell bodies (top, arrows) and neurites (bottom, arrows) of differentiated striatal cells. Confocal immunofluorescence microscopy. (Taken from Kim *et al.* (1999*b*).)

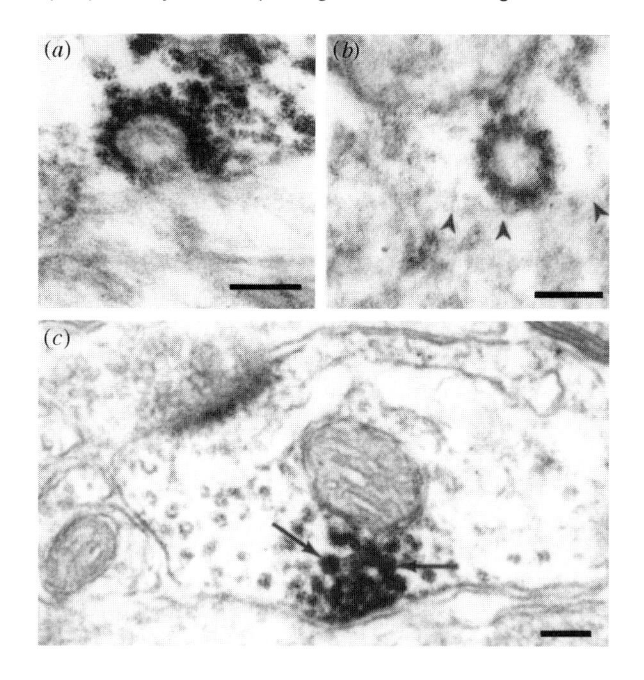
(a)
(b)
(c)

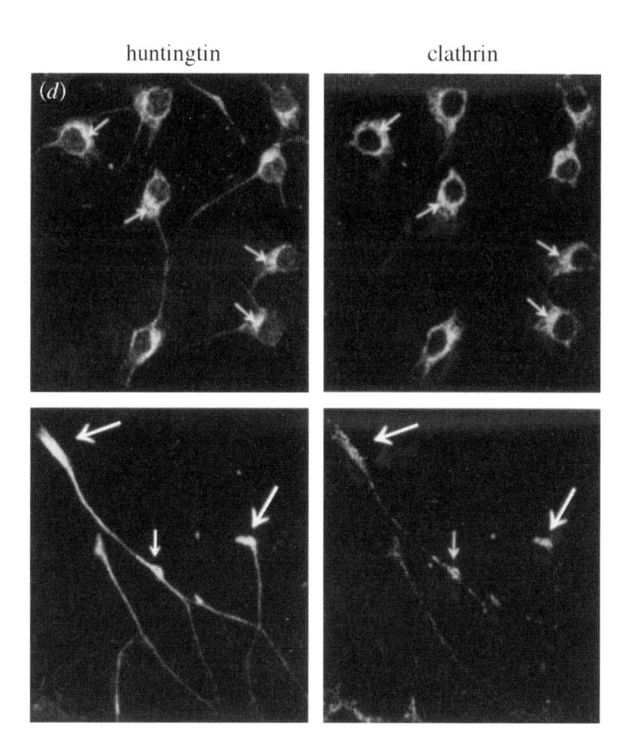
huntingtin
clathrin
(d)

translated, but be unable to perform its activities and block the activities of the wild-type protein; (iii) the protein might change its function or non-specifically render the cell less hardy.

Current information provides evidence that the mutant huntingtin is expressed and maintains its normal activity (which is yet to be defined). We found that mutant huntingtin is expressed throughout the brain in patients with Huntington's disease (Aronin *et al.* 1995). A mechanism for cell death in Huntington's disease needs to consider that unaffected neurons also express the mutant protein. The absence of normal protein does not produce a disease phenotype. Crucially, patients homozygous for Huntington's disease have a phenotype similar to the hemizygous patients (Wexler *et al.* 1987). Collectively, these findings support the idea that the vulnerable neurons should handle the mutant protein differently from protected cells.

Nuclear inclusions and dystrophic neurites

We examined the morphology and ultrastructure of neurons in the striatum and cortex in Huntington's disease. Our idea was that study of the human condition would best guide our future experiments. Our findings revealed a complex of subcellular changes, involving cytoplasmic and nuclear structures (DiFiglia *et al.* 1997; Sapp *et al.* 1997), possibly portending a series of molecular aberrations leading to apoptotic neuronal death in Huntington's disease (Portera-Cailliau *et al.* 1995; Butterworth *et al.* 1998). Nuclear changes were readily apparent, but we shall first describe findings in the cytoplasm.

Changes in the cytoplasm pervade the neuropathology in Huntington's disease. With an antiserum directed against an internal epitope of huntingtin (DiFiglia *et al.* 1995), we identified large, immunoreactive granules in the somatodendritic cytoplasm; under electron microscopy, these granules resembled multivesicular bodies, organelles dually involved in retrograde transport and protein degradation (LaVail & Lavail 1974; Hollenbeck 1993; Sapp *et al.* 1997) (figure 13.3). Huntingtin heavily concentrated in the perinuclear cytoplasm accumulated in part over endosomal–lysosomal organelles and tubulovesicular membranes. Some of the full-length protein or large fragments of huntingtin recognized by the antibody might have been destined for the nucleus but was unable to penetrate its borders because it was irreversibly bound to subcompartments of the endoplasmic reticulum. In additional studies, mutant huntingtin tranfected into clonal striatal cells was incorporated into vacuoles, which had ultrastructural features of autophagosomes (Kegel *et al.* 2000). Thus, huntingtin is degraded in the endosomal-lysosomal system. We propose that autophagy contributes to neuronal death in Huntington's disease.

Another aspect of huntingtin's cytoplasmic localization in the brain with Huntington's disease was revealed by an N-terminally directed antiserum; huntingtin immunoreactivity resided in dystrophic neurites. The dense accumulations of huntingtin overlapped with ubiquitin (DiFiglia *et al.* 1997; Sapp *et al.* 1999) and were embedded in neurofilament-labelled processes (figure 13.4*a*,*b*). Jackson *et al.* (1995) reported dystrophic neurites containing ubiquitin in Huntington's disease. Consisting largely of degenerating axons, the dystrophic neurites marked the striatal and cortical landscape in the adult brain in patients with Huntington's disease and were much more prevalent in

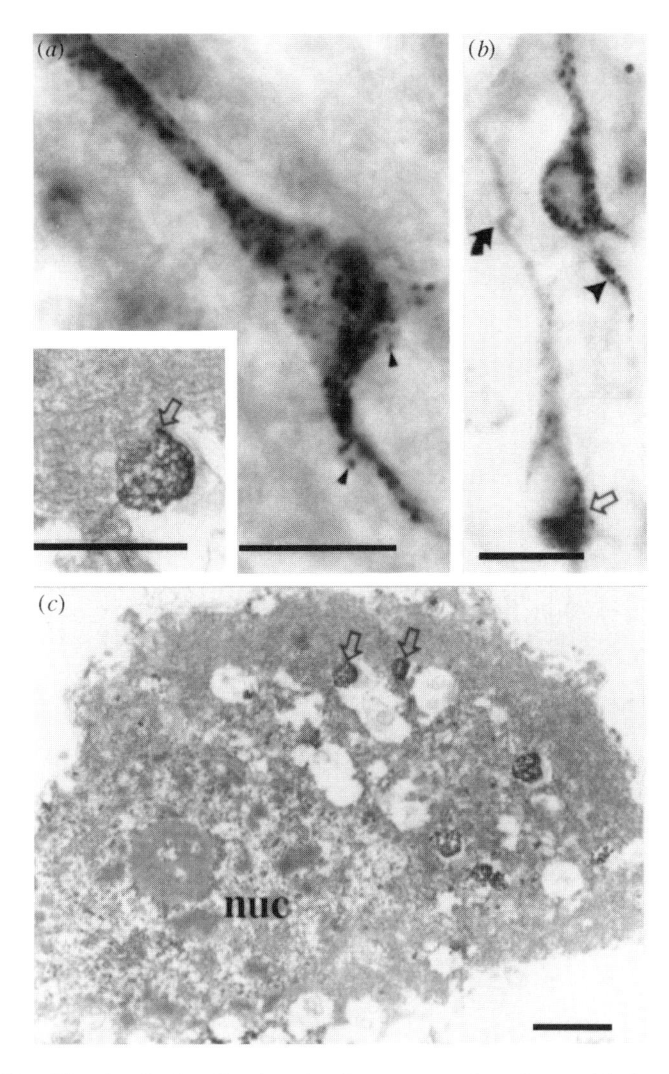

Figure 13.3 Huntingtin localization in the cortex of patient with juvenile-onset Huntington's disease. (*a*) Immunoreactive granules fill the somatodendritic cytoplasm (arrowheads). (*b*). Huntingtin staining is concentrated in the perinuclear cytoplasm of the soma on the left (open arrow) and in punctate granules in the cell on the right. The curved arrow shows the retraction of an apical dendrite, and the arrowhead identifies a possible axon initial segment. Staining was performed with anti-huntingtin antiserum Ab585. (*c*) Electron micrograph of a pyramidal neuron; huntingtin-labelled granules (arrows) appear throughout the cytoplasm. nuc, Nucleus. The inset in (*a*) shows a higher magnification of a granule that resembles a multivesicular body. The scale bars in (*a*–*c*) are 25 μm and that in the inset is 5 μm. (Taken from Sapp *et al.* (1995).)

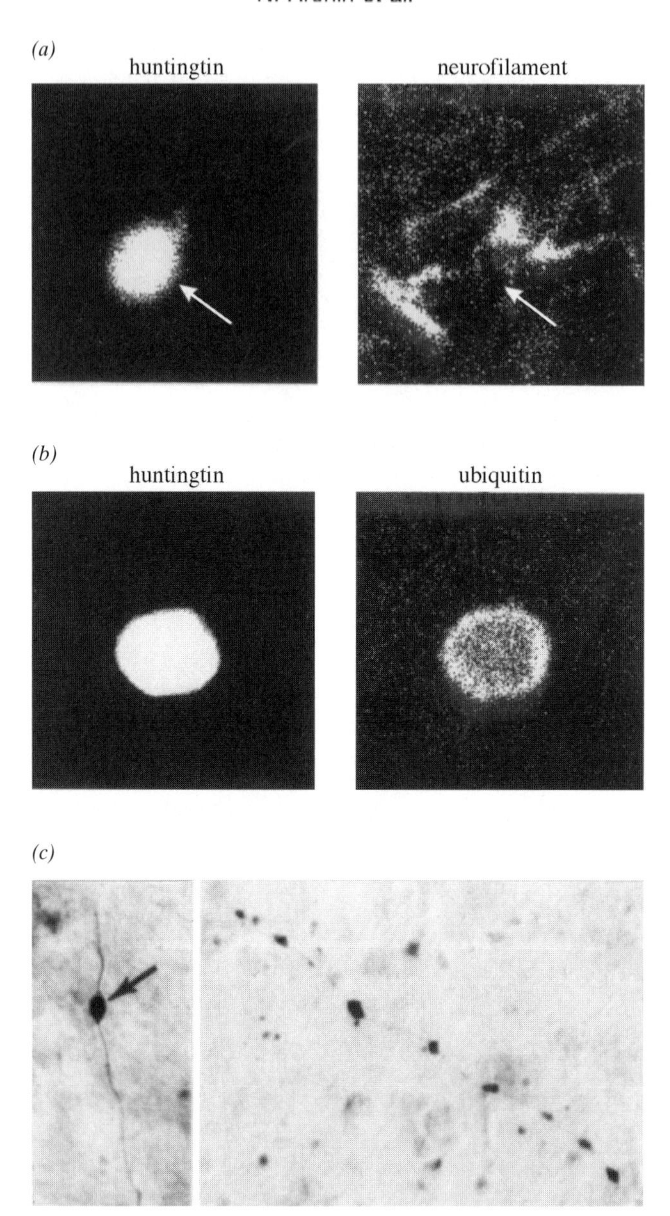

Figure 13.4 N-terminal mutant huntingtin in dystrophic axons in the Huntington's disease brain. (*a*) Double-label immunofluorescence shows a huntingtin immunoreactive dystrophic neurite (arrow) positioned within a neurofilament-labelled axonal process in the cortex. (*b*) Co-localization of huntingtin with ubiquitin in a cortical dystrophic neurite. (*c*) Nascent dystrophic axons traverse the striatal neuropil of an adult-onset patient with low-grade striatal pathology. Arrow identifies abnormal swelling. ((*a*,*b*) Taken from DiFiglia *et al.* (1997); (*c*) from Sapp *et al.* (1999).)

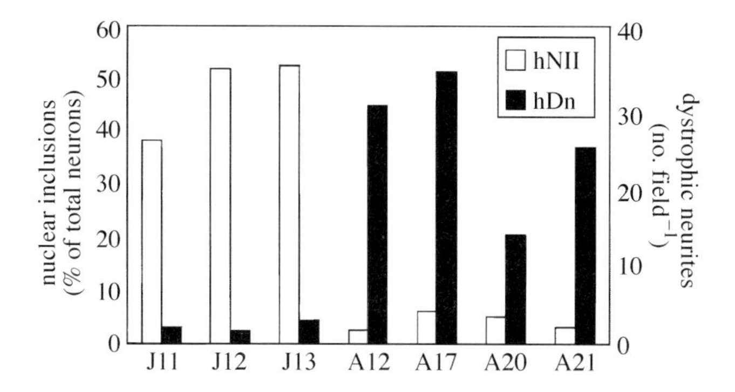

Figure 13.5 Frequency of nuclear inclusions and dystrophic neurites in patients with Huntington's disease with juvenile onset (J11, J12, J13) and adult onset (A12, A17, A20, A21). Nuclear inclusions are more frequent in juveniles, and dystrophic neurites are more abundant in adults. (Taken from DiFiglia *et al.* (1997).)

adult-onset Huntington's disease than in the juvenile-onset disease (figure 13.5). A substantial portion of the dystrophic neurites comprised cortical–striatal projections that degenerate early in the disease (Sapp *et al.* 1999) (figure 13.4*c*).

Wild-type huntingtin is absent from, or rare in, the nucleus in neurons in control brains, on the basis of most studies (DiFiglia *et al.* 1995; Sharp *et al.* 1995; Trottier *et al.* 1995), although there is a report of nuclear localization with appropriate biochemical correlates (de Rooij *et al.* 1996). In post-mortem human striatum in Huntington's disease, we found moderately intense, diffuse nuclear localization of huntingtin detected with the use of antiserum against an internal epitope of the protein (Sapp *et al.* 1997). The N-terminal antiserum unmasked huntingtin in dense intranuclear inclusions in brains with juvenile-onset Huntington's disease (figure 13.6), and to a lesser extent in adult-onset Huntington's disease (DiFiglia *et al.* 1997). The nuclear inclusions were dispersed in neurons of cortex and striatum (figure 13.6), regions of most active disease, but not in globus pallidus or cerebellum (figure 13.6*e*,*f*), which are largely unaffected. Like the dystrophic neurites, the nuclear inclusions contained ubiquitin (figure 13.6*a*). Ultrastructural review revealed granular and filamentous configurations (figure 13.6*g*). Contemporaneously, nuclear inclusions were identified in other CAG trinucleotide repeat diseases (Paulson *et al.* 1997; Becher *et al.* 1998; Holmberg *et al.* 1998; M. Li *et al.* 1998; Merry *et al.* 1998).

The formation of nuclear inclusions was polyglutamine-dependent and was more frequent with larger expansions; 30–50% of neurons in the cortex in juvenile-onset Huntington's disease contained nuclear inclusions, compared with *ca.* 3–6% of cortical neurons in the adult-onset disease (DiFiglia *et al.* 1997; Vonsattel & DiFiglia 1998) (figure 13.5). This finding was not related to the neuropathological grade of striatal degeneration (DiFiglia *et al.* 1997).

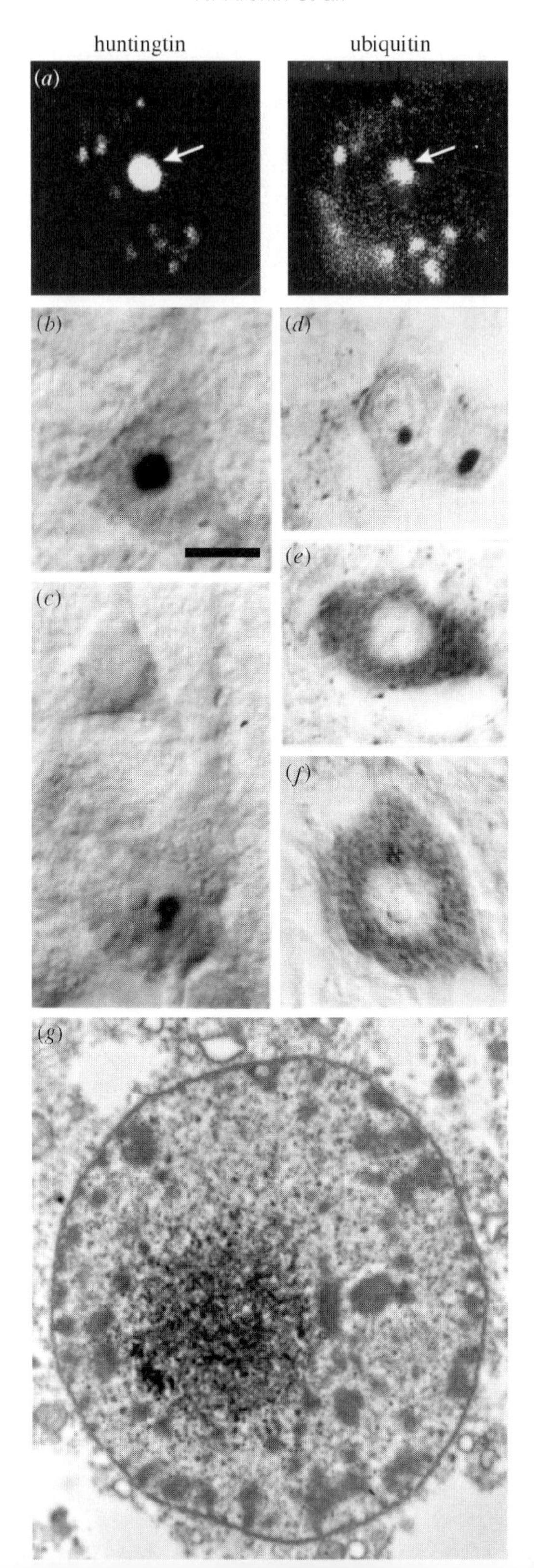
huntingtin
ubiquitin
(a)
(b)
(c)
(d)
(e)
(f)
(g)

In the adult-onset patients with low-grade neuropathology (grades 1 and 2), nuclear inclusions were absent from, or rare in, the striatum when dystrophic neurites had formed (Sapp *et al.* 1999).

How does huntingtin enter the nucleus? The large size of huntingtin would predict its nuclear entry either by active transport or by its being processed to form smaller fragments. Measurement of huntingtin immunoreactivity in isolated nuclei demonstrated that much of the huntingtin in the nuclei consisted of N-terminal fragments of lower molecular mass, in the 20–40 kDa range (DiFiglia *et al.* 1997) (it should be noted that approximating the molecular mass of proteins containing lengthy polyglutamine sequences is risky). Proteins of this approximate size could enter the nucleus passively (Alpert *et al.* 1994). Smaller, cleaved products of mutant huntingtin were shown to enter the nucleus preferentially (Hackam *et al.* 1998; Martindale *et al.* 1998). Detloff and co-workers showed that highly extended polyglutamine sequences can promote the nuclear entry of an otherwise cytoplasmic protein (Ordway *et al.* 1997). We found full-length huntingtin in nuclear extracts of brain with Huntington's disease. Larger proteins have access to the nucleus, but enter passively at a low rate, or enter by facilitated transport or active import. We contend that over a time frame of years to decades, with the lengthened polyglutamine sequences as facilitator, small and large huntingtin fragments and full-length protein have an opportunity to gain access to the nucleus.

Is nuclear aggregation of huntingtin necessary or sufficient to trigger cell death in striatal neurons?

Lessons from transgenic models of Huntington's disease

Several laboratories have generated transgenic mice expressing either an N-terminal fragment or a full-length mutant huntingtin (Mangiarini *et al.* 1996, Chapter 5; Reddy *et al.* 1998, Chapter 4; Davies *et al.*, Chapter 3). Differences between the transgenic mice include the length of the polyglutamine expansion, the size of the huntingtin, the promoter and the background strain. The initial transgenic mice with Huntington's disease (as previously described) used a human Huntington's disease promoter to express mutant huntingtin with an expanded polyglutamine repeat (115–150 repeats) and a

Figure 13.6 Nuclear inclusions formed by N-terminal mutant huntingtin in the brain with Huntington's disease. (*a*) Immunofluorescence shows co-localization of huntingtin with ubiquitin (arrows) in the same inclusion of a cortical pyramidal cell. (*b,c*) Immunoperoxidase labelling shows a nuclear inclusion distinct from the adjacent unlabelled nucleolus in cortical pyramidal neurons. (*d*) Two medium-sized striatal neurons have nuclear inclusions. (*e,f*) Neurons in the globus pallidus (*e*) and cerebellum (*f*) are devoid of nuclear inclusions and show labelling in the cytoplasm. (*g*) Electron micrograph showing immunoperoxidase labelling associated with granules and filaments within an inclusion. The scale bar in (*b*) is 10 μm. (Taken from DiFiglia *et al.* (1997).)

small huntingtin N-terminal fragment (exon 1, 69 residues) in a CBL/6 mouse strain. This transgenic strain was very instructive. Despite extensive nuclear inclusions, striatal cell loss was absent or minimal. The animals succumbed within three months to general inanition, although recent evidence implicates diabetes mellitus (with hypoinsulinaemia) as a cause (Hurlbert *et al.* 1998). Detloff and co-workers created transgenic mice expressing HPRT with a stretch of 146 glutamine residues. HPRT is ordinarily localized in the cytoplasm, but these animals showed abundant nuclear inclusions throughout the brain. No cell death was observed. The animals were short-lived and exhibited abundant seizures. These studies indicate that small proteins with expanded polyglutamine sequences are targeted to neuronal nuclei and accumulate in sufficient abundance to form large aggregates, detected as inclusions. Transgenic animals harbouring expanded polyglutamine stretches in the context of small proteins manifest phenotypic changes consistent with generalized brain dysfunction. However, the presence of inclusions does not predict cell-specific or region-specific neurodegeneration under these conditions.

Tagle and co-workers (Reddy *et al.* 1998, Chapter 4) and our own laboratory (Laforet *et al.* 1998) made transgenic models of Huntington's disease expressing a substantially larger mutant huntingtin than in the initial transgenic mice with Huntington's disease. Our transgenic mice expressed a 46 or 100 CAG repeat (one base change at position 70) in the first 3221 bases of huntingtin complementary DNA (cDNA). The host strain was SJL/B6 F1 hybrid. The animals exhibited motor deficits, lost *ca.* 20% of striatal cells and lived beyond one year (Laforet *et al.* 1999; G. A. Laforet, M. DiFiglia and N. Aronin, unpublished data). Nuclear inclusions were apparent but not abundant in the striatum of normal, affected mice in lineages with 46 or 100 glutamine repeats. Tagle and co-workers generated transgenic mice with 16, 48 and 89 glutamine repeats in a full-length huntingtin in the FVB/N murine strain. Mutant huntingtin expression varied, from nearly equivalent to native protein to about fivefold greater. Nuclear inclusions were detected, but not abundantly, and the animals had a predictable 20% cell loss in the striatum and cortex (Reddy *et al.* 1998, Chapter 4).

These studies *in vivo* reveal that the size of the huntingtin protein substantially influences the longevity of the transgenic mice with Huntington's disease and also the presence of striatal cell loss. Neither the background strain of the mice nor the abundance of nuclear inclusions accounts for neurodegeneration.

Revelations in the short-term: studies in neurons and other cells in culture

Experiments in cell culture have provided important clues to the role of inclusions of huntingtin in the pathogenesis of cell death. We studied effects of mutant huntingtin expression in cultured hybrid striatal, medium-sized neurons, generated from clonal mouse fetal striatal–neuroblastoma fusions (Kim *et al.* 1999*a*). These cells maintain many fundamental characteristics of striatal spiny neurons, which are the initial targets of neurodegeneration in Huntington's disease (Wainwright *et al.* 1995). Constructs containing 18, 46 or 100 CAG repeats in either a full-length huntingtin cDNA or a large fragment of huntingtin cDNA (1–3321 bases) were transfected into the clonal striatal neurons. The cDNAs were intact or modified as N-terminal FLAG–huntingtin fusion proteins or huntingtin–green fluorescent protein (GFP) fusion proteins; huntingtin and FLAG were detected by immunohistochemistry and GFP was viewed by fluorescence (Kim *et al.* 1999*a*). Full-length and 3 kb mutant huntingtin cDNAs (irrespective of molecular tags) produced nuclear inclusions. Transfection with the 100 CAG repeat huntingtin cDNA resulted in a greater number of inclusions than with the 46 repeat cDNA; the 18 repeat produced none. Therefore the clonal striatal neurons recapitulated the polyglutamine-dependent formation of inclusions seen in the patient brain in Huntington's disease.

We have found that transfection of partial or full-length mutant huntingtin cDNAs into clonal striatal cells generated N-terminal fragments; the sizes varied between 70 and 100 kDa (Kim *et al.* 1999*a*). Because huntingtin can be a target of caspase cleavage (Goldberg *et al.* 1996), we introduced caspase inhibitors into the cultures of transfected cells. With the caveats that caspase inhibitors lack true specificity and probably affect separate components of apoptotic cascades, we considered the inhibitors primarily as tools to block huntingtin degradation.

The caspase inhibitor Z-DEVD-FMK (Enzyme Systems) prevented inclusion formation after transfection with mutant huntingtin cDNAs. No abatement in cell death was observed (Kim *et al.* 1999*a*). An implication of this result is that huntingtin nuclear inclusions are not necessary for mutant huntingtin to induce striatal cell survival. In contrast, we found that the general caspase inhibitor Z-VAD-FMK (Enzyme Systems) did not alter mutant huntingtin nuclear inclusions, but (surprisingly) did improve the survival of the transfected clonal striatal cells. In a separate model, Saudou *et al.* (1998) reported that entry of an N-terminal fragment of mutant huntingtin (171 residues, 68 glutamine repeats) into the nucleus reduced survival of the transfected primary rat striatal neurons. By placing a nuclear export signal in the mutant huntingtin, accumulation of the transfected mutant huntingtin was abrogated and cell survival unaffected. They found that inclusion formation was not requisite for cell death, but that delivery of a mutant huntingtin fragment to the nucleus was needed to induce apoptosis. Similarly, in neuroblastoma cell lines stably expressing mutant huntingtin, Lunkes & Mandell (1998; Lunkes *et al.*, Chapter 9) found that apoptosis was increased by the expression of mutant huntingtin but was not tightly correlated with the formation of nuclear and cytoplasmic inclusions.

The above studies *in vitro* suggest that nuclear inclusions are not necessary or sufficient to predict striatal cell death, which is reminiscent of the neuropathology seen in adult-onset Huntington's disease. Other compartments that accumulate huntingtin might be more critical for cell death. In striatal hybrid neurons, we found that cell death (marked by significant cell shrinkage) occurred when N-terminal huntingtin fragments of wild-type or mutant proteins accumulated into perinuclear vacuoles (Kim *et al.* 1999*a*), not unlike the perinuclear huntingtin accumulation associated with degenerating neurons in the brain with Huntington's disease (Sapp *et al.* 1997). That N-terminal fragments of mutant huntingtin can influence cellular toxicity has support from studies in the laboratories of Hayden (Hackam *et al.* 1998) and Ross (Cooper *et al.* 1998), who showed that introducing N-terminal fragments of mutant huntingtin into neuroblastoma cell sensitized the cells to apoptosis. We found that a blockade of several N-terminal huntingtin products by caspase inhibition was correlated with increased cell survival (Kim *et al.* 1999*a*). N-terminal products of either wild-type or mutant huntingtin are toxic at high levels in non-striatal and striatal cells (Hackam *et al.* 1998; Kim *et al.* 1999*a*). How, then, can N-terminal mutant huntingtin fragments be selectively disruptive to cellular function? Insight comes from crucial studies in which full-length huntingtin was overexpressed in cultured cells (Lunkes & Mandell 1998; Kim *et al.* 1999*a*). Under these conditions, N-terminal fragments accumulated only when the protein contained an expanded polyglutamine tract.

Multiple pathogenic mechanisms in Huntington's disease

We present evidence that mutant huntingtin accumulates in multiple subcellular compartments in selected neurons in Huntington's disease and is handled differently in the juvenile-onset patient compared with the adult-onset patient. Available information points to two general pathways leading to cell death, with an emphasis on nuclear disruption in juvenile-onset Huntington's disease and the axon as a major site of pathology in adult-onset Huntington's disease. In general, the larger is the CAG repeat, the more likely it is that mutant huntingtin will be targeted to the nucleus in addition to its accumulation in the cytoplasm. Neuronal death is exacerbated by the accumulation of N-terminal fragments of huntingtin in either the nucleus or cytoplasm but does not depend on the presence of nuclear or cytoplasmic inclusions. The N-terminal region of huntingtin is cleaved from the full-length protein and robustly amasses in cells when the polyglutamine tract is expanded. Mutant huntingtin's N-terminal domain interacts abnormally with cytoplasmic proteins involved in vesicle trafficking (HIP1, HAP1, GAPDH and SH3GL3). The early appearance of dystrophic axons enriched in N-terminal mutant huntingtin in adult-onset patients conceivably coincides with disturbances of vesicle membrane transport in these axons. Huntingtin also interacts at its N-terminus with resident nuclear proteins (Faber *et al.* 1998). Alterations in nuclear function caused by targeting of the cleaved or fully intact protein to the nucleus, with the tendency of huntingtin to form filaments (Perutz *et al.* 1994; Scherzinger *et al.* 1997), are especially pronounced in patients with juvenile-onset Huntingtons' disease and even pre-empt changes in the cytoplasm. We believe that a single pathogenic event causing Huntington's disease might prove elusive.

This work was supported by grants from National Institutes of Health NS 31579 to N.A. and M.D. and NS 16367 and NS 35711 to M.D., the Hereditary Disease Foundation to N.A. and M.D., and the Huntington's Disease Society of America to M.D. G.L. is a fellow of the Howard Hughes Medical Institute.

References

Alpert, B., Bray, D., Lewis, J., Raff, M., Roberts, K. & Watson, J. D. 1994 *Molecular biology of the cell*, 3rd edn. New York: Garland Publishing.

Aronin, N., Cooper, P. E., Lorenz, L. J., Sagar, S. M., Bird, E. D., Leeman, S. E. & Martin, J. B. 1983 Somatostatin is increased in the basal ganglia in Huntington's disease. *Ann. Neurol.* **13**, 519–526.

Aronin, N. (and 17 others) 1995 CAG expansion affects the expression of mutant huntingtin in the Huntington's disease brain. *Neuron* **15**, 1193–1201.

Bao, J., Sharp, A. H., Wagster, M. V., Becher, M., Schilling, G., Ross, C. A., Dawson, V. L. & Dawson, T. M. 1996 Expansion of polyglutamine repeat in huntingtin leads to abnormal protein interactions involving calmodulin. *Proc. Natl Acad. Sci. USA* **93**, 5037–5042.

Bates, G. & Lehrach, N. 1994 Trinucleotide repeat expansions and human genetic disease. *Bioessays* **16**, 277–284.

Becher, M. W., Kotzuk, J. A., Sharp, A. H., Davies, S. W., Bates, G. P., Price, D. L. & Ross, C. A. 1998 Intranuclear neuronal inclusions in Huntington's disease and dentatorubral and pallidoluysian atrophy: correlation between the density of inclusions and IT15 CAG triplet repeat length. *Neurobiol. Dis.* **4**, 387–397.

Bhide, P. G. (and 10 others) 1996 Expression of normal and mutant huntingtin in the developing brain. *J. Neurosci.* **16**, 5523–5535.

Block-Galarza, J., Chase, K. O., Sapp, E., Vaughn, K. T., Vallee, R. B., DiFiglia, M. & Aronin, N. 1997 Fast transport and retrograde movement of huntingtin and HAP 1 in axons. *NeuroReport* **8**, 2247–2251.

Boutell, J. M., Wood, J. D., Harper, P. S. & Jones, A. L. 1998 Huntingtin interacts with cystathionine beta-synthase. *Hum. Mol. Genet.* **7**, 371–378.

Burke, J. R., Enghild, J. J., Martin, M. E., Jou, Y. S., Myers, R. M., Roses, A. D., Vance, J. M. & Strittmatter, W. J. 1996 Huntingtin and DRPLA proteins selectively interact with the enzyme GAPDH. *Nature Med.* **2**, 347–350.

Butterworth, N. J., Williams, L., Bullock, J. Y., Love, D. R., Faull, R. L. & Dragunow, M. 1998 Trinucleotide (CAG) repeat length is positively correlated with the degree of DNA fragmentation in Huntington's disease striatum. *Neuroscience* **87**, 49–53.

Cooper, J. K.. (and 12 others) 1998 Truncated N-terminal fragments of huntingtin with expanded glutamine repeats form nuclear and cytoplasmic aggregates in cell culture. *Hum. Mol. Genet.* **7**, 783–790.

Davies, S. W., Turmaine, M., Cozens, B. A., DiFiglia, M., Sharp, A. H., Ross, C. A., Scherzinger, E., Wanker, E. E., Mangiari, L. & Bates, G. P. 1997 Formation of neuronal intranuclear inclusion underlies the neurological dysfunction in mice transgenic for the HD mutation. *Cell* **90**, 537–548.

de Rooij, K. E., Dorsman, J. C., Smoor, M. A., Den Dunnen, J. T. & Van Ommen, G. J. 1996 Subcellular localization of the Huntington's disease gene product in cell lines by immunofluorescence and biochemical subcellular fractionation. *Hum. Mol. Genet.* **5**, 1093–1099.

DiFiglia, M. (and 11 others) 1995 Huntingtin is a cytoplasmic protein associated with vesicles in human and rat brain neurons. *Neuron* **14**, 1075–1081.

DiFiglia, M., Sapp, E., Chase, K. O., Davies, S. W., Bates, G. P., Vonsatell, J.-P. & Aronin, N. 1997 Aggregation of huntingtin in neuronal intranuclear inclusions and dystrophic neurites in brain. *Science* **277**, 1990–1993.

Engelender, S., Sharp, A. H., Colomer, V., Tokito, M. K., Lanahan, A., Worley, P., Holzbaur, E. L. & Ross, C. A. 1997 Huntingtin-associated protein 1 (HAP1) interacts with the p150Glued subunit of dynactin. *Hum. Mol. Genet.* **6**, 2205–2212.

Faber, P. W., Barnes, G. T., Srinidhi, J., Chen, J., Gusella, J. F. & MacDonald, M. E. 1998 Huntingtin interacts with a family of WW domain proteins. *Hum. Mol. Genet.* **7**, 1463–1474.

Goldberg, Y. P. (and 10 others) 1996 Cleavage of huntingtin by apopain, a proapoptotic cysteine protease, is modulated by the polyglutamine tract. *Nature Genet.* **13**, 442–449.

Graveland, G. A., Williams, R. S. & DiFiglia, M. 1985 Evidence for degenerative and regenerative changes in neostriatal spiny neurons in Huntington's disease. *Science* **227**, 770–773.

Gusella, J. F. & MacDonald, M. E. 1998 Huntingtin: a single bait hooks many species. *Curr. Opin. Neurobiol.* **8**, 425–430.

Gutenkunst, C. A., Levey, A. I., Heilman, C. J., Whaley, W. L., Yi, H., Nash, N. R., Rees, H. D., Madden, J. J. & Hersch, S. M. 1995 Identification and localization of huntingtin in brain and human lymphoblastoid cell lines with anti-fusion protein antibodies. *Proc. Natl Acad. Sci. USA* **92**, 8710–8714.

Hackam, A. S., Singaraja, R., Wellington, C. L., Metzler, M., McCutcheion, K., Zhang, T., Kalchman, M. & Hayden, M. R. 1998 The influence of huntingtin protein size on nuclear localization and cellular toxicity. *J. Cell Biol.* **141**, 1097–1105.

Holmberg, M.. (and 10 others) 1998 Spinocerebellar ataxia type 7 (SCA 7): a neurodegenerative disorder with neuronal intranuclear inclusions. *Hum. Molec. Genet.* **7**, 913–918.

Huntington, G. 1872 On chorea. *Med. Surg. Rep.* **26**, 317–321.

Huntington's Disease Collaborative Research Group 1993 A novel gene containing a trinucleotide repeat that is expanded and unstable on Huntington's disease chromosomes. *Cell* **72**, 971–983.

Hurlbert, M. S., Kaddis, F. G., Zhou, W., Bell, K. P., Hutt, C. J. & Freed, C. R. 1998 Neural transplantation in a transgenic mouse model of Huntington's disease. *Soc. Neurosci. Abstr.* **28**, abstract 380.9.

Jackson, M., Gentleman, G., Ward, L., Gray, T., Randall, K., Morell, K. & Lowe, J. 1995 The cortical neuritic pathology of Huntington's disease. *Neuropathol. Appl. Neurobiol.* **21**, 18–26.

Kalchman, M. A., Graham, R. K., Xia, G., Koide, H. B., Hodgson, J. G., Graham, K. C., Goldberg, Y. P., Gietz, R. D., Pickart, C. M. & Hayden, M. R. 1996 Huntingtin is ubiquinated and interacts with a specific ubiquitin-conjugating enzyme. *J. Biol. Chem.* **271**, 19385–19394.

Kalchman, M. A. (and 13 others) 1997 HIP 1, a human homologue of *S. cerevisiae* Sla2P, interacts with membrane-associated huntingtin in the brain. *Nature Genet.* **16**, 44–53.

Kegel, K. B., Kim, M., Sapp, E., McIntyre, C., Castano, J. G., Aronin, N. & DiFiglia, M., 2000 Huntingtin expression stimulates endosomal-lysosomal activity, endosome tubulation, and autophagy. *J. Neurosci.* **20**, 7268–7278

Kim, M. (and 14 others) 1999*a* Mutant huntingtin expression in immortalized striatal cells: dissociation of inclusion formation and neuronal survival by caspase inhibition. *J. Neurosci.* **19**, 964–973.

Kim, M., Velier, J., Chase, K., Laforet, G. Kalchman, M. A., Hayden, M. R., Won, L., Heller A., Aronin, N. & DiFiglia, M. 1999*b* Forskolin and dopamine D_1 receptor activation increase huntingtin's association with endosomes in immortalized neuronal cells of striatal origin. *Neuroscience* **89**, 1159–1167.

Laforet, G. A. (and 14 others) 1998 Development and characterization of a novel transgenic model of Huntington disease which recapitulates features of the human illness. *Soc. Neurosci. Abstr.* **24**, 972.

LaVail, J. H. & LaVail, M. M. 1974 The retrograde intraaxonal transport of horseradish peroxidase in the chick visual system: a light and electron microscopic study. *J. Comp. Neurol.* **157**, 303–358.

Li, M., Miwa, S., Kobayashi, Y., Merry, D. E., Yamamoto, M., Tanaka, F., Doyu, M., Hashizume, Y., Fischbeck, K. H. & Sobue, G. 1998 Nuclear inclusions of the androgen receptor protein in spinal and bulbar muscular atrophy. *Ann. Neurol.* **44**, 249–254.

Li, S. H., Gutekunst, C. A., Hersch, S. M. & Li, X. J. 1998 Interaction of huntingtin-associated protein with dynactin p150Glued. *J. Neurosci.* **18**, 1261–1269.

Li, X.-J., Li, S.-H., Sharp, A. H., Nucifora Jr, F. C., Schilling, G., Lanahan, A., Worley, P., Snyder, S. H. & Ross, C. A. 1995 A huntingtin-associated protein enriched in brain with implications for pathology. *Nature* **378**, 398–402.

Lunkes, A. & Mandel, J. L. 1998 A cellular model that recapitulates major pathogenic steps of Huntington's disease. *Hum. Mol. Genet.* **7**, 1355–1361.

Mangiarini, L. (and 10 other) 1996 Exon 1 of the HD gene with an expanded CAG repeat is sufficient to cause a progressive neurological phenotype in transgenic mice. *Cell* **87**, 493–506.

Martindale, D. (and 12 others) 1998 Length of huntingtin and its polyglutamine tract influences localization and frequency of intracellular aggregates. *Nature Genet.* **18**, 150–154.

Merry, D. E., Kobayashi, Y., Bailey, C. K., Taye, A. A. & Fischbeck, K. H. 1998 Cleavage, aggregation and toxicity of the expanded androgen receptor in spinal and bulbar muscular atrophy. *Hum. Mol. Genet.* **7**, 693–701.

Ordway, J. M. (and 11 others) 1997 Ectopically expressed CAG repeats cause intranuclear inclusions and a progressive late onset neurological phenotype in the mouse. *Cell* **91**, 753–763.

Paulson, H. L., Perez, M. K., Trottier, Y., Trojanowski, J. Q., Subramony, S. H., Das, S. S., Vig, P., Mandel, J. L., Fischbeck, K. H. & Pittman, R. N. 1997 Intranuclear inclusions of expanded polyglutamine protein in spinocerebellar ataxia type 3. *Neuron* **19**, 333–344.

Perutz, M. F., Johnson, T., Suzuki, M. & Finch, J. T. 1994 Glutamine repeats as polar zippers: their possible role in inherited neurodegenerative diseases. *Proc. Natl Acad. Sci. USA* **91**, 5355–5358.

Portera-Cailliau, C., Hedreen, J. C., Price, D. L. & Koliatsos, V. E. 1995 Evidence for apoptotic cell death in Huntington disease and excitotoxic animal models. *J. Neurosci.* **15**, 3775–3787.

Reddy, P. H., Williams, M., Charles, V., Garrett, L., Pike-Buchanan, L., Whetsell Jr, W. O. Miller, G. & Tagle, D. A. 1998 Behavioral abnormalities and selective neuronal loss in HD transgenic mice expressing mutated full-length HD cDNA. *Nature Genet.* **20**, 198–202.

Sapp, E., Schwarz, C., Chase, K., Bhide, P. G., Young, A. B., Penney, J., Vonsattel, J.-P., Aronin, N. & DiFiglia, M. 1997 Localization of huntingtin in multivesicular bodies in Huntington's disease. *Ann. Neurol.* **43**, 604–610.

Sapp, E., Penney, J., Young, A., Aronin, N., Vonsattel, J.-P. & DiFiglia, M. (1999) Axonal transport of N-terminal huntingtin suggests early pathology of corticostriatal projections in Huntington's disease. *J. Neuorpathol. Exp. Neurol.* **58**, 165–173.

Saudou, F., Finkbeiner, S., Devys, D. & Greenberg, M. E. 1998 Huntingtin acts in the nucleus to induce apoptosis but death does not correlate with the formation of intranuclear inclusions. *Cell* **95**, 55–66.

Scherzinger, E., Lurz, R., Turmaine, M., Mangiarini, L, Hollenbach, B., Hasenbank, R., Bates, G. P., Davies, S. W., Lehrach, H. & Wanker, E. E. 1997 Huntingtin-encoded polyglutamine expansions form amyloid-like protein aggregates *in vitro* and *in vivo*. *Cell* **90**, 549–558.

Sharp, A. H. (and 15 others) 1995 Widespread expression of Huntington's disease gene (IT 15) protein product. *Neuron* **14**, 1065–1074.

Sittler, A., Walter, S., Wedemeyer, N., Hasenbank, R., Scherzinger, E., Eickhoff, H., Bates, G. P., Lehrach, H. & Wanker, E. E. 1998 SH3GL3 associates with the huntingtin exon 1 protein and promotes the formation of polyGln-containing protein aggregates. *Mol. Cell* **2**, 427–436.

Snell, R. G., MacMillan, J. C., Cheadle, J. P., Fenton, I., Lazarou, L. P., Davies, P., MacDonald, M., Gusella, J. F., Harper, P. S. & Shaw, D. J. 1993 Relationship between trinucleotide repeat expansions and phenotypic variation in Huntington's disease. *Nature Genet.* **4**, 393–397.

Trottier, Y., Devys, D., Imbert, G., Saudou, F., An, I., Luta, Y., Weber, C., Agid, Y., Hirsch, E. C. & Mandel, J.-L. 1995 Cellular localization of the Huntington's disease protein and discrimination of the normal and mutated form. *Nature Genet.* **10**, 104–110.

Tukamoto, T., Nukina, N., Ide, K. & Kanazawa, I. 1997 Huntington's disease gene product, huntingtin, associates with microtubules *in vitro*. *Mol. Brain Res.* **51**, 8–14.

Velier, J., Kim, M., Schwarz, C., Kim, T. W., Sapp, E., Chase, K., Aronin, N. & DiFiglia, M. 1998 Wild-type and mutant huntingtins function in vesicle trafficking in the secretory and endocytic pathways. *Exp. Neurol.* **152**, 34–40.

Vonsattel, J.-P. G. & DiFiglia, M. 1998 Huntington disease. *J. Neuropathol. Exp. Neurol.* **57**, 369–384.

Wainwright, M. S., Perry, B. D., Won, L. A., O'Malley, K. L., Wang, W. Y., Ehrlich, M. E. & Heller, A. 1995 Immortalized murine striatal neuronal cell lines expressing dopamine receptors and cholinergic properties. *J. Neurosci.* **15**, 676–688.

Wanker, E. E., Rovira, C., Scherzinger, E., Hasenbank, R., Walter, S., Tait, D., Colicelli, J. & Lehrach, H. 1997 HIP-1; a huntingtin interacting protein isolated by the yeast two-hybrid system. *Hum. Mol. Genet.* **6**, 487–495.

Wexler, N. S. (and 18 others) 1987 Homozygotes for Huntington's disease. *Nature* **326**, 194–197.

Genomic influences on CAG instability

14

CAG repeat instability, cryptic sequence variation and pathogenicity: evidence from different loci

M. Frontali, A. Novelletto, G. Annesi and C. Jodice

Introduction

It is well known that several autosomal dominant disorders (Huntington's disease (HD), spinocerebellar ataxia (SCA)1, SCA2, SCA3, SCA6, SCA7, dentatorubral–pallioluysian atropy (DRPLA) and spinal and bulbar muscular atrophy (SBMA)) are associated with unstable expansions of CAG repeat sequences in the coding region of the corresponding genes. Three aspects of these mutations are explored here: (i) the process leading wild-type alleles to expand above the threshold of instability; (ii) the phenotypic effect of expanded alleles with an unusual type of interruption of the polyglutamine repeat; and (iii) the possibility that one of the mutations, namely SCA6, acts through a pathogenetic mechanism different from that hypothesized for the other disorders owing to $(CAG)_n$ expansions.

Length change and variability of normal HD alleles

In all expanded-polyglutamine disorders, new mutant alleles arise from wild-type alleles that show a length polymorphism of variable structure and degree (Jodice *et al.* 1997*a*). Expanded alleles are prone to further length variation (instability) during both meiotic and mitotic cell divisions (see, for example, Chong *et al.* 1995). At present, little is known about the mechanism leading to the elongation of alleles from the normal range into the unstable range. The presence of variant trinucleotides, interrupting the $(CAG)_n$ sequence, has been shown to have a stabilizing effect. For example, in most SCA1 and SCA2 normal alleles, the $(CAG)_n$ sequence is interrupted by CAT or CAA trinucleotides. In these genes, pure $(CAG)_n$ stretches longer than 21 units have never been observed (Chung *et al.* 1993; Imbert *et al.* 1996), whereas the unstable expanded alleles have stretches of 35 or more pure CAG units. It is the loss of the variant trinucleotide that presumably triggers the expansion process in SCA1 and SCA2. Pearson *et al.* (1998) have proposed a model in which CAG interruptions exert their stabilizing effect by inhibiting DNA strand slippage.

Other genes, such as HD, DRPLA or SCA3, have normal alleles with no interruptions, except for the presence of a variant trinucleotide at the 3' or 5' end of the repeat tract in both the normal and expanded alleles (McNeil *et al.* 1997). In SCA3 and DRPLA, the $(CAG)_n$ length distributions of normal and expanded alleles are widely separated, and events producing new mutant alleles must be assumed to involve jumps of several units. For HD, which shows continuity between the range of normal and expanded alleles, Rubinsztein *et al.* (1994) have proposed a model that favours single unit changes leading progressively towards the instability threshold. To investigate this issue further, the HD CAG repeat was analysed in a model system consisting of normal and tumour tissues from sporadic patients affected by colon cancer (Di Rienzo *et al.* 1998). It is known that microsatellites in colon cancer cells are particularly prone to instability, owing partly to defects in the mismatch repair system (Liu *et al.* 1995). Overall, ten tumour DNAs with variant $(CAG)_n$ sizes on a total of 115 pairs of matched normal-tumour DNAs were observed. The results, reported in figure 14.1, show that increments of a single unit account for only 20% of all changes and that the overall frequency of single unit variations is *ca.* 30%, under the conservative assumption that every variant allele was derived from the normal allele more similar to it. The frequencies were still lower when the distribution of

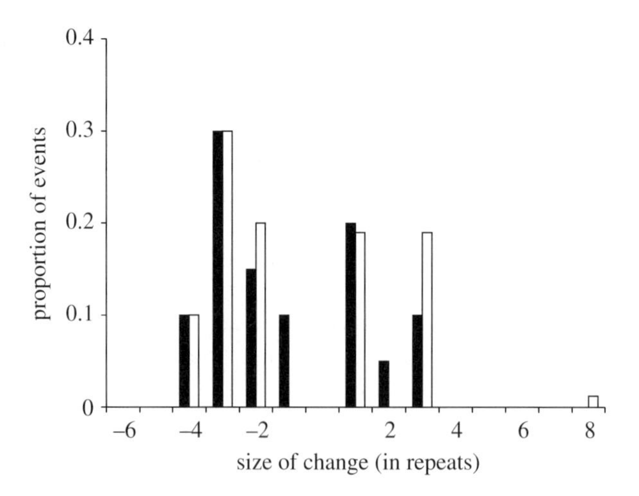

Figure 14.1 Histogram of HD $(CAG)_n$ size changes in colorectal tumour DNAs compared with non-tumour DNA. New sizes were scored as additional bands in the tumour DNA, compared with the patient genotype. Two methods were used to estimate the length change leading to the variant allele from either of the patient normal alleles: (i) each variant allele was assumed to derive from the normal allele more similar to it (dark bars); (ii) size change was estimated by the EM algorithm (Di Rienzo *et al.* 1998) (light bars).

estimated changes was obtained through the EM algorithm (Di Rienzo *et al.* 1998), which optimizes the relative probability of the variant size's being generated by one or the other of the parental alleles. In addition, a slight excess of shortenings was observed, mainly contributed by changes of –3. The present and the previous data on HD normal polymorphism in different world populations (Jodice *et al.* 1997*a*) fit the theory by Di Rienzo *et al.* (1998), which predicts that the square of length change observed in this model system is linearly related to the variance of the $(CAG)_n$ size in the population. The experimental verification (Di Rienzo *et al.* 1998) of this expectation strengthens the idea that the events scored in this model are similar to those occurring in meioses *in vivo*. It should be noted that large meiotic series were analysed by Brinkmann *et al.* (1998), showing that most changes in microsatellites involve a single unit. However, this report did not include the HD gene, leaving open the possibility of a locus-specific mutational pattern of changes larger than one unit. In this context, it should also be remembered that the frequency of expanded HD alleles in the population can be maintained not only through the length mutation of wild-type alleles but also by means of the increased fitness in carriers of alleles in the medium–low expansion range, as proposed by Frontali *et al.* (1996).

Phenotypic effect of interrupted expanded polyglutamine tracts

The role of interruptions of the $(CAG)_n$ sequences has frequently been analysed in connection with length instability, whereas little is known about the role of purity in the corresponding protein products. It has recently been shown, both *in vivo* and *in vitro*, that expanded polyglutamine stretches lead to the formation of intranuclear inclusions containing insoluble ubiquitinated aggregates of the protein (see, for example, Davis *et al.* 1997; Di Figlia *et al.* 1997; Paulson *et al.* 1997; Scherzinger *et al.* 1997). Several lines of evidence suggest that the intranuclear inclusions have a role in the pathogenesis of these disorders, rather than simply being a by-product: (i) the inclusions have been found post-mortem in specifically affected tissues, but not in those unaffected by the disorder (Di Figlia *et al.* 1997; Paulson *et al.* 1997); (ii) the inclusions have been found in transgenic animals before the onset of symptoms (Davis *et al.* 1997); (iii) inclusions have a greater frequency in juvenile cases characterized by large numbers of glutamine units and a more severe phenotype (Di Figlia *et al.* 1997); and (iv) transgenic mice, which express a hypoxanthine–guanine phosphoribosyltransferase (normally not containing polyglutamine) engineered with the insertion of a large polyglutamine tract, show intranuclear inclusions and have a progressive neurological deficit (Ordway *et al.* 1997). Little is known about the process underlying the formation of intranuclear inclusions. A first step is likely to be the proteolytic cleavage of the expanded polyglutamine tract. The cleaved sequence seems to have the role of recruiting the full-length protein, independently of its number of glutamine units, into insoluble aggregates (Paulson *et al.* 1997).

The question is whether a 'non-pure' expanded polyglutamine sequence can have the same pathogenetic role. The serendipitous finding of two members of an Italian family (figure 14.2) carrying an unusual SCA1 allele with 45 repeat units, i.e. well within the range of expansions, seems to provide a first clue in addressing this question. The family was ascertained not on the basis of an ataxic phenotype, but during a random population analysis. Subject II-1 (figure 14.2), 24 years old and in good health, carried a 45-repeat allele whose sequence showed two CAT–CAG–CAT interruptions (figure 14.2). At the protein level, this implies a similar arrangement of histidine residues within the polyglutamine stretch. The analysis of the subject's parents showed that the father, 66 years old and in good health, carried the same allele. Published data (table 14.1) on 315 SCA1 patients, molecularly tested, show that the age at onset is rarely more than 60 years, even when the number of CAG repeats is smaller than 45, and is never above 65. Had the interrupted allele had the same effect of pure $(CAG)_n$ stretches, then the oldest carrier of the unusual pattern would most probably have been already affected at his age. This would suggest that interrupted alleles have, at least, a delaying effect on age at onset, if not a non-pathogenetic role altogether. A similar pattern of interruptions

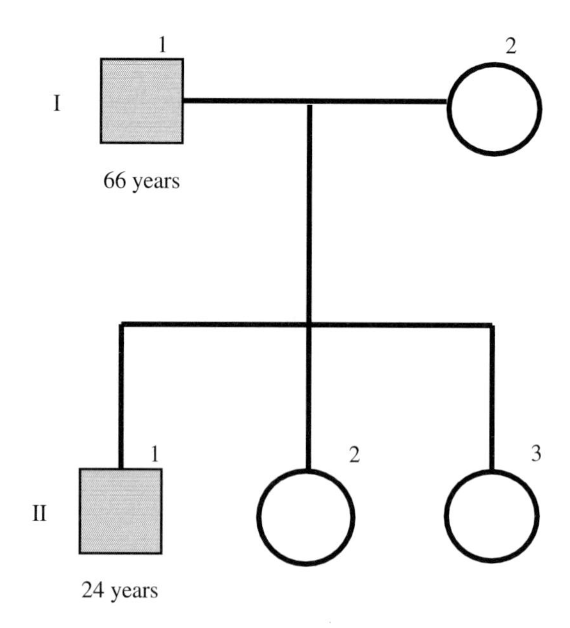

Figure 14.2 Family with an expanded SCA1 allele (45 repeats) containing unusual CAT interruptions. Black squares indicate heterozygotes for an allele $(CAG)_{15}$-CAT-CAG-CAT-$(CAG)_{12}$-CAT-CAG-CAT-$(CAG)_{12}$, showing no signs of disease. The ages of these subjects are also reported.

Table 14.1 *Published data on age at onset in SCA1 patients*

source	*no. of patients*	*range of CAG units*	*range of age at onset*	*no. with onset at 60–65 years*
Sasaki *et al.* (1996)	35	42–63	15–63	2
Dubourg *et al.* (1995)	42	42–67	21–52	0
Kameya *et al.* (1995)	20	42–58	19–55	0
Genis *et al.* (1995)	22	41–59	26–52	0
Ranum *et al.* (1994)	113	42–81	4–65	2
Jodice *et al.* (1994)	55	47–66	15–51	0
Orr *et al.* (1993)	27	43–81	3–65	1

in an expanded SCA1 allele, with 44 repeats, has been reported by Quan *et al.* (1995) in a family ascertained through a young member affected with an early-onset (at age two years) cerebellar ataxia. Both the child and her healthy 33-year-old father carried the same allele. The young age of both subjects and the presence of an early-onset ataxic phenotype (although incongruous with the relatively low number of repeats) could well be compatible with a pathogenetic role for the unusual allele. The present family, instead, provides stronger support for the hypothesis that expanded polyglutamine stretches interrupted by histidine residues have a low or null pathogenetic potential. The presence of histidine residues might prevent or decrease the formation of intranuclear inclusions by altering the structure of the polyglutamine stretches, which, according to Perutz *et al.* (1994), form β-strands, tending to aggregate by linking to one another through hydrogen bonds between their main chain and the side-chain amides. Alternatively, histidine residues could decrease the probability of polyglutamine being cleaved from the protein, or they might lower the affinity of polyglutamine for transglutaminase. This enzyme, according to Green (1993) and Khalem *et al.* (1996), favours aggregation by linking glutamine residues to the ε-amino group of lysine residues of other proteins, by means of isopeptide bonds. A closer study of the role of purity in expanded polyglutamines would be extremely important not only for a better evaluation of the predictive and diagnostic value of tests based on $(CAG)_n$ expansion, but also for the analysis of the processes involved in the pathogenesis of neurodegenerative disorders due to expanded polyglutamine tracts.

Role of the small $(CAG)_n$ expansions at the *CACNA1A* gene

In expanded polyglutamine disorders, the mutated $(CAG)_n$ stretches typically have a number of units ranging from 35 to over 100. Several lines of evidence indicate that the mutation confers on proteins a gain of function

(Quigley *et al.* 1992; Zeitlin *et al.* 1995; White *et al.* 1997). So far, the only exception seems to be SCA6. This disorder is caused by small expansions (20–30 units) of the CAG repeat sequence (4–20 units) at the *CACNA1A* gene, which codes for the α_{1A} subunit of the voltage-gated calcium channel type P/Q (figure 14.3). A complete sequence analysis of the coding region of the gene in a patient with 23 CAG repeats showed that the expansion was the only detectable mutation (Jodice *et al.* 1997*b*). In terms of the type of mutation and associated phenotype, SCA6 was thought (Zhuchenko *et al.* 1997) to differ from the other two disorders owing to point mutations at the same gene, i.e. familial hemiplegic migraine (FHM) and episodic ataxia type 2 (EA2) (Ophoff *et al.* 1996). Zhuchenko *et al.* (1997) described the SCA6 phenotype as a permanent and progressive ataxia differing from EA2, which is instead characterized by episodes of ataxia and/or vertigo and mild interictal cerebellar signs. Both disorders differ in turn from FHM, whose landmark is migraine with hemiplegic aura, even if its phenotype can also include mild cerebellar signs (Joutel *et al.* 1994).

Recent evidence, however, suggests that EA2 and SCA6 should be considered as expressions of the same disorder. This conclusion is supported by the observation of an unstable allele containing 20 or 25 CAG repeats in two branches of the same family (Jodice *et al.* 1997*b*). Patients with 25 repeats had a severe progressive ataxia similar to SCA6, whereas patients with 20 repeats had the typical features of EA2. Moreover, families with EA2 caused by point mutations were reported as also including patients with a permanent progressive ataxia (Yue *et al.* 1997; Trettel *et al.* 1998). EA2 and SCA6 therefore seem to have an identical, although highly variable, phenotype (table 14.2), which ranges from short episodes with mild interictal signs to severe progressive ataxia with cerebellar atrophy. Both extremes of this spectrum have been found in association with either point mutations or small $(CAG)_n$ expansions.

Several mutations are now known to be associated with the SCA6/EA2 phenotype. These also include mutations producing truncated proteins that are probably unable to exert any function at all. Should a common pathogenetic mechanism for all the EA2/SCA6 mutations be postulated, this would imply that the small $(CAG)_n$ expansions differ from all the other poly-

Table 14.2 *Main clinical features in the three disorders due to* CACNA1A *mutations*

symptoms	*FHM*	*EA2*	*SCA6*
migraine with hemiplegic aura	+	–	–
episodes of ataxia/vertigo	+/–	+	+/–
permanent nystagmus	+/–	+	+
permanent progressive ataxia	+/–	+/–	+/–
cerebellar atrophy	+/–	+/–	+/–

glutamine disorders by inducing a loss rather than a gain of function of the corresponding protein. In fact, the EA2/SCA6 mutations reported so far (table 14.3, figure 14.3) include: (i) the deletion of a highly conserved nucleotide in codon 1266 (exon 22), which alters the reading frame for a

Table 14.3 *Mutations at the* CACNA1A *gene causing the EA2/SCA6 phenotype*

mutation	*effect on protein*	*domain/segment*	*source*
1. del C (exon 22)	premature stop–truncated protein	III S1	Ophoff *et al.* (1996)
2. G→A (5' splice junction intron 24)	aberrant splicing–truncated protein	III S3	Ophoff *et al.* (1996)
3. C→T (exon 23)	Arg→Stop	III (S1/S2)	Yue *et al.* (1998)
4. T→C (exon 28)	Phe→Ser	III S6	Trettel *et al.* (1998)
5. G→A (exon 6)	Gly→Arg	I P(S5/S6)	Yue *et al.* (1997)
6. $(CAG)_n$ expansion	$(Gln)_{\geqslant 20} \rightarrow (Gln)_{\leqslant 30}$	C-terminal tail	Zhuchenko *et al.* (1997) Jodice *et al.* (1997b)

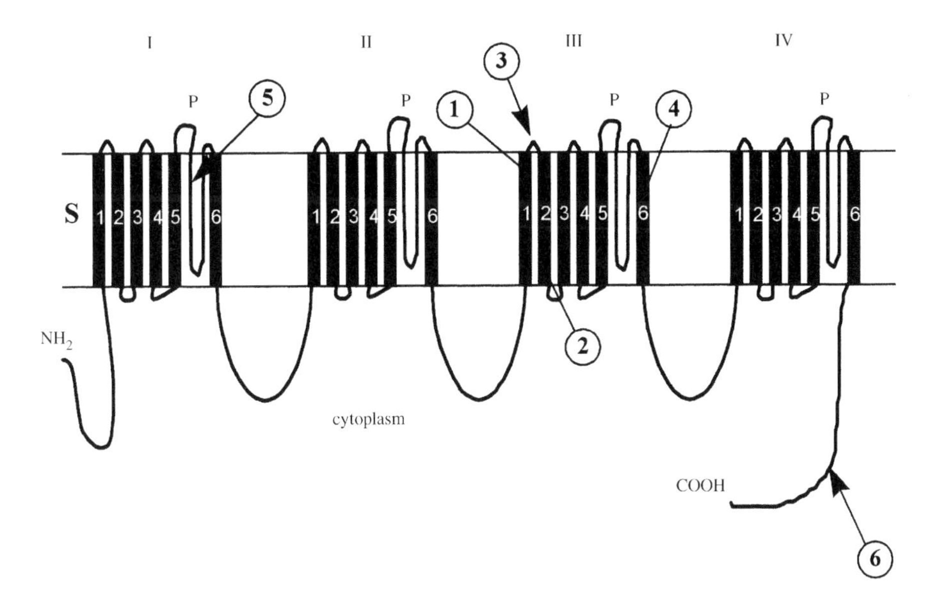

Figure 14.3 Structure model of the α_{1A} subunit of Ca^{2+} channel type P/Q. The protein possesses four domains (I–IV) containing six hydrophobic α-helical transmembrane regions (S1–S6) connected by hydrophilic links. The P segments connecting S5 and S6 of each domain form the pore of the channel. The locations of the mutations responsible for the EA2/SCA6 phenotype described in table 14.3 are reported.

substantial portion of the coding region (2400 codons) and leads to a premature stop signal at codon 1294; (ii) a single nucleotide substitution that abolishes the 5' splice site of intron 24, causing an aberrant splicing; and (iii) a nonsense mutation creating a premature stop signal at codon 1279 (exon 23). All these defects are probably producing truncated proteins that should severely impair the formation or the functioning of the channels, either through a mechanism of haploinsufficiency or by interfering with channel assembly in the cell membrane. Defective channel activity can also be postulated for the two substitutions of a highly conserved amino acid residue located in the P and S6 segments. These regions are, in fact, responsible for ion-binding (Yue *et al.* 1997) and channel-selectivity (Hockermann *et al.* 1977), respectively. The change from a non-polar amino acid residue (Phe) to a polar one (Ser) in S6 and from a non-charged residue (Gly) to a charged one (Arg) in P might well interfere with the role of these segments. As far as the small CAG repeat expansion is concerned (whose length is well within the normal ranges of other CAG repeat tracts), it should be noted that (i) the sequence is translated into a polyglutamine stretch only in some of the isoforms (Zhuchenko *et al.* 1997); (ii) the α_{1A} subunit is expressed in brain and kidney (Ophoff *et al.* 1996), but little is known about the ratio of the different isoforms in different cell types; and (iii) the polyglutamine tract, when present, is located in the intracytoplasmic C-terminal part of the protein, which is known to be involved in the tonic inhibition of the channel opening probability (Wei *et al.* 1994). On the basis of this evidence, a loss of function for the expanded alleles could be explained, at a pre-translational level, by their altering the RNA stability or by their interfering with the translation process. In both cases, the end result will be a decrease in protein synthesis acting through a mechanism of haploinsufficiency. Alternatively, at a post-translational level, expanded polyglutamine stretches could cause a defective functioning of the channel by interfering with the role of the C-terminal tail of the protein.

This work was supported by Telethon Italia grant E355 to M.F.

References

Brinkmann, B., Klintschar, M., Neuhuber, F., Huhne, J. & Rolf, B. 1998 Mutation rate in human microsatellites: influence of the structure and length of the tandem repeat. *Am. J. Hum. Genet.* **62**, 1408–1415.

Chong, S. S., McCall, A. E., Cota, J., Subramony, S. H., Orr, H. T., Hughes, M. R. & Zoghbi, H. Y. 1995 Gametic and somatic tissue-specific heterogeneity of the expanded SCA1 CAG repeat in spinocerebellar ataxia type 1. *Nature Genet.* **10**, 344–350.

Chung, M., Ranum, L. P. W., Duvick, L. A., Servadio, A., Zoghbi, H. Y. & Orr, H.T. 1993 Evidence for a mechanism predisposing to intergenerational CAG repeat instability in spinocerebellar ataxia type 1. *Nature Genet.* **5**, 254–258.

Davies, S. W., Turmaine, M., Cozens, B. A., DiFiglia, M., Sharp, A. H., Ross, C. A.,

Scherzinger, E., Wanker, E. E., Mangiarini, L. & Bates, G. P. 1997 Formation of neuronal intranuclear inclusions underlies the neurological dysfunction in mice transgenic for the HD mutation. *Cell* **90**, 537–548.

Di Figlia, M., Sapp, K. O., Davies, S. W., Bates, G. P., Vonsattel, J. P. & Aronin, N. 1997 Aggregation of huntingtin in neuronal intranuclear inclusions and dystrophic neurites in brain. *Science* **227**, 1990–1993.

Di Rienzo, A., Donnelly, P., Toomajian, C., Sisk, B., Hill, A., Petzl-Erler, M. L., Haines, G. K. & Barch, D. H. 1998 Heterogeneity of microsatellite mutations within and between loci, and implications for human demographic histories. *Genetics* **148**, 1269–1284.

Dubourg, O., Durr, A., Cancel, G., Stevanin, G., Chneiweiss, H., Penet, C., Agid, Y. & Brice, A. 1995 Analysis of the SCA1 CAG repeat in a large number of families with dominant ataxia: clinical and molecular correlations. *Ann. Neurol.* **37**, 176–180.

Frontali, M. *et al.* 1996 Genetic fitness in Huntington's disease and spinocerebellar ataxia 1: a population genetics model for CAG repeat expansions. *Ann. Hum. Genet.* **60**, 423–435.

Genis, D., Matilla, T., Volpini, V., Rosell, J., Davalos, A., Ferrer, I., Molins, A. & Estivill, X. 1995 Clinical, neuropathologic, and genetic studies of a large spinocerebellar ataxia type 1 (SCA1) kindred: (CAG)n expansion and early premonitory signs and symptoms. *Neurology* **45**, 24–30.

Green, H. 1993 Human genetic diseases due to codon reiteration: relationship to an evolutionary mechanism. *Cell* **74**, 955–956.

Hockermann, G.H., Johnson, B.D., Abbot, M.R., Schewer,T., Catterroll, W.A. 1997 Molecular determinants of high activity phenylalkylamine block of L-cycle calcium channels transmembrane segment IIIS6 and the fore region of the αg subunit. *J. Biol. Chem.* **272**, 18759-18765

Imbert, G. (and 14 others) 1996 Cloning of the gene for spinocerebellar ataxia 2 reveals a locus with high sensitivity to expanded CAG/glutamine repeats. *Nature Genet.* **14**, 285–291.

Jodice, C., Malaspina, P., Persichetti, F., Novelletto, A., Spadaro, M., Giunti, P., Morocutti, C., Terrenato, L., Harding, A. E. & Frontali, M. 1994 Effect of trinucleotide repeat length and parental sex on phenotypic variation in spinocerebellar ataxia I. *Am. J. Hum. Genet.* **54**, 959–965.

Jodice, C., Giovannone, B., Calabresi, V., Bellocchi, M., Terrenato, L. & Novelletto, A. 1997*a* Population variation analysis at nine loci containing expressed trinucleotide repeats. *Ann. Hum. Genet.* **61**, 425–438.

Jodice, C. *et al.* 1997*b* Episodic ataxia type 2 (EA2) and spinocerebellar ataxia type 6 (SCA6) due to CAG repeat expansion in the CACNA1A gene on chromosome 19p. *Hum. Mol. Genet.* **6**, 1973–1978.

Joutel, A. *et al.* 1994 Genetic heterogeneity of familial hemiplegic migraine. *Am. J. Hum. Genet.* **55**, 1166–1172.

Kahlem, P., Terre, C., Green, H. & Djian, P. 1996 Peptides containing glutamine repeats as substrates for transglutaminase-catalyzed cross-linking: relevance to diseases of the nervous system. *Proc. Natl Acad. Sci. USA* **93**, 14580–14585.

Kameya, T., Abe, K., Aoki, M., Sahara, M., Tobita, M., Konno, H. & Itoyama, Y. 1995 Analysis of spinocerebellar ataxia type 1 (SCA1)-related CAG trinucleotide expansion in Japan. *Neurology* **45**, 1587–1594.

Liu, B. *et al.* 1995 Mismatch repair gene defects in sporadic colorectal cancers with microsatellite instability. *Nature Genet.* **9**, 48–55.

McNeil, S. M., Novelletto, A., Srinidhi, J., Barnes, G., Kornbluth, I., Altherr, M. R., Wasmuth, J. J., Gusella, J. F., MacDonald, M. E. & Myers, R. H. 1997 Reduced penetrance of the Huntington's disease mutation. *Hum. Mol. Genet.* **6**, 775–779.

Ophoff, R. A. *et al.* 1996 Familial hemiplegic migraine and episodic ataxia type-2 are caused by mutations in the Ca^{2+} channel gene CACNL1A4. *Cell* **87**, 543–552.
Ordway, J. M. (and 11 others) 1997 Ectopically expressed CAG repeats cause intranuclear inclusions and a progressive late onset neurological phenotype in the mouse. *Cell* **91**, 753–763.
Orr, H. T., Chung, M., Banfi, S., Kwiatkowski, T. J., Servadio, A., Beaudet, A. L., McCall, A. E., Duvick, L. A., Ranum, L. P. W. & Zoghbi, H. Y. 1993 Expansion in an unstable trinucleotide CAG repeat in spinocerebellar ataxia type 1. *Nature Genet.* **4**, 221–226.
Paulson, H. L., Perez, M. K., Trottier, Y., Trojanowski, J. Q., Subramony, S. H., Das, S. S., Vig, P., Mandel, J.-L., Fishbeck, K. H. & Pittmman, R. N. 1997 Intranuclear inclusions of expanded polyglutamine protein in spinocerebellar ataxia type 3. *Neuron* **19**, 333–344.
Pearson, C. E., Eichler, E. E., Lorenzetti, D., Kramer, S. F., Zoghbi, H. Y., Nelson, D. L. & Sinden, R. R. 1998 Interruptions in the triplet repeats of SCA1 and FRAXA reduce the propensity and complexity of slipped strand DNA (S-DNA) formation. *Biochemistry* **37**, 2701–2708.
Perutz, M. F., Johnson, T., Suzuki, M. & Finch, J. T. 1994 Glutamine repeats as polar zippers: their possible role in inherited neurodegenerative diseases. *Proc. Natl Acad. Sci. USA* **91**, 5355–5358.
Quan, F., Janas, J. & Popovich, B. W. 1995 A novel CAG repeat configuration in the SCA1 gene: implications for the molecular diagnostics of spinocerebellar ataxia type 1. *Hum. Mol. Genet.* **4**, 2411–2413.
Quigley, C. A., Friedman, K. J., Johnson, A., Lafreniere, R. G., Silverman, L. M., Lubahn, D. B., Brown, T. R., Wilson, E. M., Willard, H. F. & French, F. S. 1992 Complete deletion of the androgen receptor gene: definition of the null phenotype of the androgen insensitivity syndrome and determination of carrier status. *J. Clin. Endocrinol. Metab.* **74**, 927–933.
Ranum, L. P. *et al.* 1994 Molecular and clinical correlations in spinocerebellar ataxia type I: evidence for familial effects on the age at onset. *Am. J. Hum. Genet.* **55**, 244–252.
Rubinsztein, D. C., Amos, W., Leggo, J., Goodburn, S., Ramesar, R. S., Old, J., Bontrop, R., McMahon, R., Barton, D. E. & Ferguson-Smith, M. A. 1994 Mutational bias provides a model for the evolution of Huntington's disease and predicts a general increase in disease prevalence. *Nature Genet.* **7**, 525–530.
Sasaki, H., Fukazawa, T., Yanagihara, T., Hamada, T., Shima, K., Matsumoto, A., Hashimoto, K., Ito, N., Wakisaka, A. & Tashiro, K. 1996 Clinical features and natural history of spinocerebellar ataxia type 1. *Acta Neurol. Scand.* **93**, 64–71.
Scherzinger, E., Lurz, R., Turmaine, M., Mangiarini, L., Hollenbach, B., Hasenbank, R., Bates, G. P., Davies, S. W., Lehrach, H. & Wanker, E. E. 1997 Huntingtin-encoded polyglutamine expansions form amyloid-like protein aggregates *in vitro* and *in vivo*. *Cell* **90**, 549–558.
Trettel, F., Mantuano, E., Veneziano, L., Sabbadini, G., Olsen, A. S., Ophoff, R. A., Frants, R. R., Jodice, C. & Frontali, M. 1998 Molecular analysis of the gene CACNA1A: refined mapping of the containing region and screening for the mutations in EA2. *Eur. J. Hum. Genet.* **6** (Suppl. 1), 150.
Wei, X., Neely, A., Lacerda, A. E., Olcese, R., Stefani, E., Perez-Reyes, E. & Birnbaumer, L. 1994 Modification of Ca^{2+} channel activity by deletions at the carboxyl terminus of the cardiac alpha 1 subunit. *J. Biol. Chem.* **269**, 1635–1640.
White, J. K., Auerbach, W., Duyao, M. P., Vonsattel, J. P., Gusella, J. F., Joyner, A. L. & MacDonald, M. E. 1997 Huntingtin is required for neurogenesis and is not impaired by the Huntington's disease CAG expansion. *Nature Genet.* **17**, 404–410.
Yue, Q., Jen, J. C., Nelson, S. F. & Baloh, R. W. 1997 Progressive ataxia due a missense mutation in a calcium-channel gene. *Am. J. Hum. Genet.* **61**, 1078–1087.
Yue, Q., Jen, J. C., Thwe, M. M., Nelson, S. F. & Baloh, R. W. 1998 *De novo* muta-

tion in CACNA1A caused acetazolamide-responsive episodic ataxia. *Am. J. Med. Genet.* **77**, 298–301.
Zeitlin, S., Liu, J. P., Chapman, D. L., Papaioannou, V. E. & Efstratiadis, A. 1995 Increased apoptosis and early embryonic lethality in mice nullizygous for the Huntington's disease gene homologue. *Nature Genet.* **11**, 155–163.
Zhuchenko, O., Bailey, J., Bonnen, P., Ashizawa, T., Stockton, D. W., Amos, C., Dobyns, W. B., Subramony, S. H., Zoghbi, H. Y. & Chi Lee, C. 1997 Autosomal dominant cerebellar ataxia (SCA6) associated with small polyglutamine expansion in the α_{1a}-voltage-dependent calcium channel. *Nature Genet.* **15**, 62–69.

15

Microsatellite and trinucleotide repeat evolution: evidence for mutational bias and different rates of evolution in different lineages

David C. Rubinsztein, Bill Amos and Gillian Cooper

Introduction

Microsatellites are stretches of repetitive DNA found in eukaryotic genomes. The repeat units at these loci comprise one to six bases and repeat number frequently varies at a given locus. This high polymorphism rate and the accessibility of these markers to PCR amplification have led to microsatellites being used as major tools for genetic mapping (Weissenbach *et al.* 1992), studies of human (Bowcock *et al.* 1994) and animal diversity (Bruford & Wayne 1993) and for forensic investigations (Jeffreys *et al.* 1992). An understanding of the mutational and evolutionary processes of such loci can aid data interpretation from many of these applications. In addition, trinucleotide repeats are microsatellites, and a study of their mutational features can help to explain the prevalences of different diseases in different populations and how these mutations originate.

At the time we initiated the work reviewed in this chapter, most workers considered that microsatellites mutate symmetrically, with expansion and contraction mutations occurring at equal frequencies (Di Rienzo *et al.* 1994; Goldstein *et al.* 1995*a*; Kruglyak *et al.* 1998). Furthermore, it has been assumed that such loci show similar rates of evolution in related populations and species (Goldstein *et al.* 1995*b*; Pollock *et al.* 1998). These assumptions underly many of the formulae that have been used to compute genetic distances (Goldstein *et al.* 1995*a*). In this review, we will consider evidence which challenges these two important assumptions.

Wild-type Huntington's disease (HD) CAG repeats provide clues about microsatellite behaviour

Huntington's disease (HD), like most other diseases caused by abnormal expansions of trinucleotide repeat tracts, shows anticipation—the age at onset in affected individuals in a family tends to decrease in successive generations (Duyao *et al.* 1993). This can be explained by the correlation between increasing CAG repeat number on disease chromosomes with earlier age at onset of symptoms and the overall tendency for the CAG repeat mutation to increase in size in successive generations, which is particularly marked in male transmissions. The overall mutational bias in favour of expansions seen in HD disease alleles is also a feature of disease alleles at most trinucleotide repeat loci (for a review, see Rubinsztein & Amos 1998).

The mutational processes of mutant trinucleotide repeat disease alleles are comparatively easy to assess by studying disease pedigrees, since disease alleles have high mutation rates. However, we were interested in studying wild-type alleles in order to understand the origins of trinucleotide repeat mutations. Since these do not mutate frequently, family studies were impractical. Thus, we used population genetic approaches.

HD varies in prevalence. For example, it is comparatively common in East Anglia, UK (1/10 000) but rare in Japan (< 1/1 000 000). The range of normal alleles is from eight to 35 CAG repeats and there are no data which suggest that these alleles are associated with variable genetic fitness. Accordingly, we expected that the distributions of normal alleles at the HD locus would reflect mutational processes and random genetic drift. We typed normal HD gene alleles in a panel of human populations and in non-human primates. The human allele distributions showed three features:

(i) There was a relationship between the proportion of long normal alleles in a population and its frequency of HD. For instance, significantly more long normal alleles were seen in East Anglians compared with Japanese. This relationship was confirmed in HD by the Hayden group and has also been described for myotonic dystrophy and dentatorubral–pallidoluysian atrophy, among other trinucleotide repeat diseases (for a review, see Rubinsztein & Amos 1998). This suggests that the majority of new mutations at trinucleotide repeat disease loci originate from the upper end of the normal allele length distribution.

(ii) The allele distributions among all human populations showed a positive skew in that there was an excess of alleles with longer repeat lengths than the modal length (Rubinsztein *et al.* 1994). Such a positive skew was one of the first clues that led us to consider the possibility that microsatellite mutations are not symmetrical with respect to length changes. Indeed, this type of skewed distribution has been found in the hypermutable minisatellites such as CEB1, and these have been shown empirically to have an excess of expansion versus contraction mutations (Vergnaud *et al.* 1991; Jeffreys *et al.* 1994; Monckton *et al.* 1994).

(iii) The HD CAG repeats in a large panel of non-human primates appeared to be shorter than those seen in human populations. Since all primates share a common ancestor, the most parsimonious explanation for this finding was that the repeats have expanded in length in the human lineage.

We performed a series of computer simulation experiments, in order to explore possible mechanisms leading to the positively skewed allele length distributions in all the populations we studied and to account for the expansion of allele lengths in the human lineage. Our empirical data were best explained by a mutational model that incorporated mutational bias in favour of expansions as a key component (Rubinsztein *et al.* 1994).

Microsatellites show mutational bias and different expansion rates in different lineages

We decided to expand the above approach and consider the features of microsatellites in general, in order to see whether the triplet repeats at the HD locus were typical or exceptional. We examined allele frequency distributions of more than 300 microsatellite loci in humans and found that positively skewed length distributions were more common than all other distributions pooled together (J. Swinton & B. Amos, unpublished data). Similar findings have been reported by Farrall & Weeks (1998).

A panel of 44 polymorphic human microsatellite loci were studied in humans, chimpanzees, gorillas, orang-utans, baboons, macaques and marmosets (Rubinsztein *et al.* 1995). These included ‘neutral’ dinucleotide and trinucleotide repeat loci and some trinucleotide disease genes. Allele sizes in 33 of the human loci were significantly larger than those in the chimpanzee homologues; seven of the loci were significantly larger in chimpanzees compared with the humans (33:7 is significantly different from 1:1, $p < 0.0005$). Two loci were of similar size in both species and two did not amplify in chimpanzees. Similar significant trends in favour of larger human loci compared with their homologues in other primates were seen in gorillas, orang-utans, baboons and macaques. In marmosets, the number of loci amplified was too small to detect a significant difference. Similar significant trends were observed when the loci were confined to those which were ‘neutral’, by excluding trinucleotide repeat disease loci and trinucleotide repeats associated with brain cDNAs from the microsatellite panel. For example, 21 neutral loci were longer in humans compared with five which were longer in chimpanzees ($p < 0.005$).

The greater length of human microsatellites can be accounted for by three possibilities. First, selection might favour longer repeats in humans or shorter repeats in primates. This explanation seems unlikely, since the trends were observed with diverse loci and remained even if the analyses were confined to neutral loci. Also, the very large number of independently segregating microsatellites means that the effect on any one locus must be small.

Second, if microsatellites are subject to a mutational bias in favour of expansion, then the greater length of human microsatellites could arise through a genome-wide increase in mutation rate in humans or a decrease in mutation rate in chimpanzees. A mutational bias in favour of expansions is supported by the following empirical observations. Weber and Wong considered the possibility of mutation bias when they observed an excess of expansion mutations in microsatellites typed in CEPH pedigrees (Weber & Wong 1993). Unfortunately, only 22 of the 62 mutations observed occurred in genomic DNA from non-transformed cells. This left the possibility that the trends that they observed were due to mutations occurring in transformed cells. Of the 22 mutations observed in genomic DNA, they detected 14 gains and eight losses in allele lengths ($p = 0.14$, exact binomial probability). However, we later supplemented Weber and Wong's data with microsatellite mutations observed by S. Sawcer and R. Feakes, who had been performing a linkage study on multiple sclerosis using genomic DNA from non-transformed cells (Amos *et al.* 1996). This study yielded a further 15 mutations comprising seven gains, one loss and seven ambiguous mutations. Combined, the two data sets show a significant bias in favour of expansion: 21 gains and nine losses ($p = 0.021$, exact binomial probability).

The third possibility involves an observation bias associated with the fact that the microsatellites used in the human–non-human primate comparison were cloned from human and were sufficiently polymorphic to be useful in linkage studies. Being selected for above average length, some argued that one would expect such markers to be longer than their homologues in related species and hence that the effects that we observed could be due entirely to an ascertainment bias (Ellegren *et al.* 1995, 1997).

This possibility was directly addressed by cloning a series of long polymorphic microsatellites from chimpanzees (Cooper *et al.* 1998). A total of 38 chimpanzee-derived loci were investigated initially, including 24 TG/CA repeats which we identified (maximum 23 repeats, minimum 12 repeats, mean 17.3 repeats), one interrupted locus with 23 repeats and 13 chimpanzee loci described by Takenaka *et al.* (1993). Nine loci were excluded from further analysis as they did not amplify products of the expected size consistently, and one locus did not amplify human DNA. The remaining 28 chimpanzee-derived loci were used for chimpanzee–human length comparisons. These chimpanzee loci had a similar mean repeat number compared with the human-derived loci, which were previously compared with their homologues in non-human primates above ($p = 0.13$).

Fourteen of these chimpanzee loci were significantly shorter than their human homologues, eight were significantly longer in chimpanzees and six loci were similar in length in the two species. If the 33:7 ratio of human-derived loci, which were significantly longer in humans compared with those longer in chimpanzees, was entirely due to an ascertainment bias, then one would expect a similar excess of loci longer in chimpanzees for chimpanzee-derived loci. However, 14:8 (chimpanzee-derived loci longer in humans) does not differ significantly from 33:7 (human-derived loci longer in humans) ($p = 0.18$, Fisher exact test); while 14:8 does differ significantly from 7:33, the ratio expected if the effects we had observed with the human-derived loci were due entirely to ascertainment bias ($p = 0.0007$).

We also considered the possibility that our interpretation may have been confounded by an ascertainment bias resulting from the comparison of polymorphic loci in one species with monomorphic homologues in a related species. Monomorphic loci will tend to have lower mutation rates compared with polymorphic homologues. Thus, monomorphic loci will tend to be shorter than their polymorphic homologues in related species, if mutations are biased in favour of expansions. This source of bias was demonstrated by Crawford *et al.* (1998), when they analysed polymorphic cattle-derived microsatellites in sheep. When the homologues of these loci were polymorphic in sheep (308 loci), then 198 (64%) were longer in sheep. Conversely, if the analyses were confined to the 131 loci monomorphic in sheep, then 109 (83%) were longer in cattle.

We next confined our analyses to loci which were polymorphic in both species, or by further restricting such loci to dinucleotide repeats. In both scenarios, we observed similar trends showing no significant differences between the ratio of human loci longer than chimpanzee loci for human-derived versus chimpanzee-derived loci. In these scenarios, we also observed highly significant differences between the ratios of human loci longer than chimpanzee loci for human-derived microsatellites, compared with the ratio expected from chimpanzee microsatellites, if the data from the human-derived loci were due purely to an ascertainment bias.

These data can be explained by a mutational bias in favour of expansions and a greater expansion rate in the human lineage compared with the chimpanzee lineage. A similar discrepancy in microsatellite allele lengths at homologous loci in related species was reported recently in reciprocal comparisons of sheep and cattle microsatellite loci (Crawford *et al.* 1998). Of 20 polymorphic sheep-derived loci that were polymorphic in both species, 16 were longer in sheep compared with their cattle homologues. However, 198 of the 308 cattle-derived loci, which were polymorphic in both species, were significantly longer in sheep.

Multinational bias in favour of expansions has been recently confirmed by empirical observations of dinucleotide repeat mutations, although no bias was seen in tetranucleotide repeats (Ellegren 2000, Xu *et al.*, 2000). Thus, different types of microsatellites may behave differently. Ellegren (2000) also reported an excess of male-transmitted microsatellite mutations.

The molecular mechanisms underlying these observations are unclear, although the following non-mutually exclusive possibilities may help to explain why microsatellites expand faster in humans compared with chimpanzees:

(i) Human polymerases might be more error-prone compared with those in non-human primates.
(ii) Greater microsatellite mutation rates in humans might reflect the longer lag in humans between the onset of sexual maturity and reproduction. It appears that microsatellite mutation rates are higher in males than in females (Weber & Wong 1993; Amos *et al.* 1996; Brinkmann *et al.*, 1998; Primmer *et al.* 1998), probably reflecting a larger number of cell divisions between zygote and sperm compared with between zygote and ovum. Later reproduction implies more cell divisions and hence, plausibly, greater mutation rates for microsatellites. Interestingly, HD alleles also show greater mutability in males (Duyao *et al.* 1993). This model is consistent with recent single sperm analyses from HD patients which suggest that mutations occur throughout germline mitotic divisions and are not confined to single meiotic events (Leeflang *et al.* 1999).
(iii) Microsatellite mutations may involve inter-chromosomal events. This possibility is supported by data on the trinucleotide repeat disease Machado–Joseph disease, where the mutation rate of the mutant chromosome depends partly on the haplotype of the normal chromosome (Takiyama *et al.* 1997). If interchromosomal processes are involved in the mutations of normal microsatellites, and if mutations are more likely in heterozygote individuals (Amos *et al.* 1996; Amos & Harwood 1998), then humans may show increased rates of microsatellite evolution because of their greater effective population size.

In conclusion, our data suggest that microsatellites show a mutational bias in favour of expansion mutations, a process reminiscent of the mutant alleles at most trinucleotide repeat disease loci. The rate of expansion of microsatellites in related species seems to differ, raising the possibility that these major deviations from molecular clock predictions may be a genome-wide phenomenon not specifically confined to microsatellite loci.

This discussion raises a number of speculative questions. First, are trinucleotide repeat diseases found in man and not in other primates because human microsatellites appear to be generally longer than those in other primates? This question is difficult to address because the chance of ever finding a late-onset, rare (< 1/10 000), neurodegenerative disease in non-human primates is remote. However, chimpanzees do appear to have shorter normal alleles than humans for several triplet diseases (Djian *et al.* 1996). Also, transgenic mice with expanded human trinucleotides from disease genes exhibit relatively low mutation rates and smaller sized mutations than would be expected for the same alleles in humans, suggesting that the mutation process in humans may be unusual (reviewed in Rubinsztein & Amos 1998).

Second, how important are interchromosomal events in microsatellite mutations? The molecular mechanisms underlying microsatellite mutations are still unclear. If interchromosomal events are confirmed, then population size and structure may affect microsatellite mutations. This interesting (almost Lamarckian) concept has a possible precedent, which comes from studies of hybrid populations. Here, the meeting of dissimilar chromosomes usually causes an increase in heterozygosity. In hybrid zones, rare alleles not present in either population occur so frequently that they have been named 'hybrizymes' (Barton *et al.* 1983; Woodruff 1989). Limited DNA sequence data suggest that these are likely to be new mutations rather than the result of recombination events (Hoffman & Brown 1995). Such a pattern is certainly consistent with the notion that heterozygosity may act to modulate mutation rate.

Third, we should ask about the implications of these processes for human populations. If population mixing does act to accelerate expansion, we could speculate that triplet repeat disease incidence would be highest and perhaps rising most rapidly in populations where historical events have brought together peoples with diverse origins. At the same time, over evolutionary time-scales, it is possible that we will see the emergence of new diseases associated with trinucleotide expansions which have yet to reach a disease threshold.

This work was funded by the Huntington's Disease Association, UK and the Leverhulme Trust. D.C.R. is a Glaxo Wellcome Research Fellow.

References

Amos, W. & Harwood, J. 1998 Factors affecting levels of genetic diversity in natural populations. *Phil. Trans. R. Soc. Lond.* B **353**, 177–186.

Amos, W., Sawcer, S. J., Feakes, R. & Rubinsztein, D. C. 1996 Microsatellites show mutational bias and heterozygote instability. *Nature Genet.* **13**, 390–391.

Barton, N. H., Halliday, R. B. & Hewitt, G. M. 1983 Rare electrophoretic variants in a hybrid zone. *Heredity* **50**, 139–146.

Bowcock, A. M., Ruiz Linares, A., Tomfohrde, J., Minch, E., Kidd, J. R. & Cavalli-Sforza, L. L. 1994 High resolution trees with polymorphic microsatellites. *Nature* **368**, 455–457.

Brinkmann, B., Klintschar, M., Neuhuber, F., Huhne, J & Rolf, B. 1998 Mutation rate in human microsatellites: influence of the structure and length of the tandem repeat. *Am J. Hum. Genet.* **62**, 1408–1415.

Bruford, M. W. & Wayne, R. K. 1993 Microsatellites and their application to population genetic studies. *Curr. Opin. Genet. Dev.* **3**, 939–943.

Cooper, G., Rubinsztein, D. C. & Amos, W. 1998 Ascertainment bias does not entirely account for human microsatellites being longer than their chimpanzee homologues. *Hum. Mol. Genet.* **7**, 1425–1429.

Crawford, A., Knappes, S. M., Paterson, K. A., deGotari, M. J., Dodds, K. G., Freking, R. T., Stone, R. T. & Beattie, C. W. 1998 Microsatellite evolution: testing the ascertainment bias hypothesis. *J. Mol. Evol.* **46**, 256–260.

Di Rienzo, A., Peterson, A. C., Garza, J. C., Valdes, A. M. & Slatkin, M. 1994 Mutational processes of simple sequence repeat loci in human populations. *Proc. Natl Acad. Sci. USA* **91**, 3166–3170.

Djian, P., Hancock, J. M. & Chana, H. S. 1996 Codon repeats in genes associated with human diseases: fewer repeats in the genes of nonhuman primates and nucleotide substitutions concentrated at sites of reiteration. *Proc. Natl Acad. Sci. USA* **93**, 417–421.

Duyao, M. (and 12 others) 1993 Trinucleotide repeat length instability and age of onset in Huntington's disease. *Nature Genet.* **4**, 387–392.

Ellegren, H. 2000 Hetrogeneous mutation processes in human microsatellite DNA sequences. *Nature Genet.* **24**, 400–402

Ellegren, H., Primmer, C. R. & Sheldon, B. C. 1995 Microsatellite evolution: directionality or bias in locus selection. *Nature Genet.* **11**, 360–362.

Ellegren, H., Moore, S., Robinson, N., Byrne, K., Ward, W. & Sheldon, B. C. 1997 Microsatellite evolution—a reciprocal study of repeat lengths at homologous loci in cattle and sheep. *Mol. Biol. Evol.* **14**, 854–860.

Farrall, M. & Weeks, D. E. 1998 Mutational mechanisms for generating microsatellite allele frequency distributions: an analysis of 4,558 markers. *Am. J. Hum. Genet.* **62**, 1260–1262.

Goldstein, D. B., Ruiz Linares, A., Cavalli-Sforza, L. L. & Feldman, M. W. 1995*a* An evaluation of genetic distances for use with microsatellite loci. *Genetics* **139**, 463–471.

Goldstein, D. B., Ruiz Linares, A., Cavalli-Sforza, L. L. & Feldman, M. W. 1995*b* Genetic absolute dating based on microsatellites and the origin of modern humans. *Proc. Natl Acad. Sci. USA* **92**, 6723–6727.

Hoffman, S. M. G. & Brown, W. M. 1995 The molecular mechanism underlying the rare allele phenomenon in a subspecific hybrid zone of the California field-mouse, *Peromyscus californicus*. *J. Mol. Evol.* **41**, 1165–1169.

Jeffreys, A. J., Allen, M. J., Hagelberg, E. & Sonnberg, A. 1992 Identification of the skeletal remains of Josef Mengele by DNA analysis. *Forensic Sci. Int.* **56**, 65–76.

Jeffreys, A. J., Tamaki, K., MacLeod, A., Monckton, D. G., Neil, D. L. & Armour, J. A. L. 1994 Complex gene conversion events in germline mutation at human minisatellites. *Nature Genet.* **6**, 136–145.

Kruglyak, S., Durrett, R. T., Schug, M. D. & Aquadro, C. F. 1998 Equilibrium distributions of microsatellite repeat length resulting from a balance between slippage events and point mutations. *Proc. Natl Acad. Sci. USA* **95**, 10774–10778.

Leeflang, E.P. (and 10 others) 1999 Analysis of germline mutation spectra at the Huntington's disease locus supports a mitotic mutation mechanism. *Hum. Mol. Genet.* **8**, 173–183

Monckton, D. G., Neumann, R., Guram, T., Fretwell, N., Tamaki, K., MacLeod, A. & Jeffreys, A. J. 1994 Minisatellite mutation rate variation associated with a flanking DNA sequence polymorphism. *Nature Genet.* **8**, 162–170.

Pollock, D. D., Bergman, A., Feldman, M. W. & Goldstein, D. B. 1998 Microsatellite behaviour with range constraints: parameter estimation and improved distances for use in phylogenetic reconstruction. *Theor. Pop. Biol.* **53**, 256–271.

Primmer, C. R., Saino, N., Møller, A. P. & Ellegren, G. 1998 Unravelling the process of microsatellite evolution through analysis of germ line mutations in barn swallows *Hirundo rustica*. *Mol. Biol. Evol.* **15**, 1047–1054.

Rubinsztein, D. C. & Amos, W. 1998 Trinucleotide repeat mutation processes. In *Analysis of triplet repeat disorders* (ed. D. C. Rubinsztein & M. R. Hayden), pp. 257–268. Oxford: BIOS.

Rubinsztein, D. C., Amos, W., Leggo, J., Goodburn, S., Ramesar, R. S., Old, J., Bontrop, R., McMahon, R., Barton, D. E. & Ferguson-Smith, M. A. 1994 Mutational bias provides a model for the evolution of Huntington's disease and predicts a general increase in disease prevalence. *Nature Genet.* **7**, 525–530.

Rubinsztein, D. C., Amos, W., Leggo, J., Goodburn, S., Jain, S., Li, S. H., Margolis, R. L., Ross, C. A. & Ferguson-Smith, M. 1995 Microsatellites are

generally longer in humans compared to their homologues in non-human primates: evidence for directional evolution at microsatellite loci. *Nature Genet.* **10**, 337–343.
Takenaka, O., Takasaki, H., Kawamoto, S., Arakawa, M. & Takenaka, A. 1993 Polymorphic microsatellite DNA amplification customised for chimpanzee paternity testing. *Primates* **34**, 27–35.
Takiyama Y (and 10 others) 1997 Single sperm analysis of the CAG repeats in the gene for Machado–Joseph disease (MJD1): evidence for non-Mendelian transmission of the MJD1 gene and for the effect of the intragenic CGG/GGG polymorphism on the intergenerational instability. *Hum Mol Genet* **7**, 525–530.
Vergnaud, G., Mariat, D., Apion, F., Aurias, A., Lathrop, M. & Lauthier, V. 1991 The use of synthetic tandem repeats to isolate new VNTR loci—cloning of a human hypermutable sequence. *Genomics* **11**, 135–144.
Weber, J. L. & Wong, C. 1993 Mutation of human short tandem repeats. *Hum. Mol. Genet.* **2**, 1123–1128.
Weissenbach, J., Gyapay, G., Dib, C., Vignal, A., Morissette, J., Millasseau, P., Vaysseix, G. & Lathrop, M. 1992 A second-generation linkage map of the human genome. *Nature* **359**, 794–801.
Woodruff, R. C. 1989 Genetic anomalies associated with *Cerion* hybrid zones: the origin and maintenance of new electrophoretic variants called hybrizymes. *Biol. J. Linn. Soc.* **36**, 281–294.
Xu, X., Peng, H., Fang, Z., & Xu, X. 2000 The direction of microsatellite mutations is dependent upon allele length. *Nature Genet.* **24**, 396–399

Pathology caused by other proteins with expanded glutamine repeats

16

Genotype–phenotype correlation in the spinocerebellar ataxias

Paul F. Worth, Alexis Brice and Nicholas W. Wood

Introduction

The spinocerebellar ataxias constitute a group of neurological conditions which are both clinically and genetically heterogeneous. The word 'ataxia' derives from the Greek word *taxis*, literally meaning 'order'. Hence 'ataxia' can be translated as 'lack of order' or 'disorder', and is usually applied to describe the clinical syndrome consisting of an unsteady, wide-based gait, incoordination of the limbs, slurred, sometimes 'staccato' speech, and disordered eye movements which may be jerky or slow. These features are usually attributable to dysfunction of the cerebellum or its connections.

The most common hereditary ataxia is that first described by Friedreich in a series of papers between 1863 and 1877. It usually has a juvenile onset and is inherited in an autosomal recessive fashion. In this condition, ataxia is due to dysfunction of various neuronal pathways including some which supply the cerebellum with sensory information, and is therefore not strictly a cerebellar ataxia. This chapter will concentrate on the autosomal dominant cerebellar ataxias (ADCAs).

In 1893, Pierre Marie (Marie 1893) observed that certain individuals with hereditary ataxia were distinct from those described by Friedreich in that the age of onset was later, tendon reflexes were increased, abnormal eye movements were frequent and inheritance was autosomal dominant. However, it is now apparent that his cases were heterogeneous, and many other clinical and pathological studies over the next 90 years sought to distinguish reliably between the various subtypes. It was not until 1982 that Harding proposed a simple classification of the ADCAs based on the different clinical features of 11 families (Harding 1982) (table 16.1). All subtypes display cerebellar ataxia, but the presence of other clinical features such as spasticity, extrapyramidal signs, ophthalmoplegia and maculopathy determines the ADCA type. The exact prevalence of ADCA is not known but it appears to be less than 1/10 000.

Although Harding's classification remains a useful clinical tool, she recognized the possibility that a given clinical subtype would be found to be

Table 16.1 Autosomal dominant cerebellar ataxia: clinico-genetic classification

ADCA type	*clinical features*	*genetic loci and chromosomal location*	*normal allele*	*pathological allele*
ADCA I	cerebellar syndrome plus:			
	pyramidal signs	**SCA1 6p22–23**	6–44	39–83
	supranuclear ophthalmoplegia	**SCA2 12q23–24.1**	13–33	32–77
	extrapyramidal signs	**SCA3 14q32.1**	12–40	54–89
	peripheral neuropathy	SCA4 16q24-ter	–	–
	dementia	**SCA 8 13q21**	16–91	110–130
ADCA II	cerebellar syndrome plus:			
	pigmentary maculopathy	**SCA7 3p12–21.1**	4–35	36–306
	other signs as ADCA I			
ADCA III	'pure' cerebellar syndrome	SCA5 cent 11	–	–
	mild pyramidal signs	**SCA6 19p13**	4–18	20–33
		SCA10 22q	–	–
		SCA11 15q14–21.3	–	–

The SCA genes which have been cloned are shown in bold type. Size ranges for normal and pathological alleles observed in the different SCAs are given. Note that for SCA8, combined CTG/CTA repeat sizes are given (see text).

genetically heterogeneous. Indeed, there are at least 11 loci which have been shown to account for the three ADCA subtypes (table 16.1). These loci have been assigned the notation SCA (spinocerebellar ataxia) followed by a number referring to the order in which the loci were identified. The genes for six of these loci have been cloned to date, and in five of these, SCA1 (Orr *et al.* 1993), SCA2 (Imbert *et al.* 1996; Pulst *et al.* 1996; Sanpei *et al.* 1996), SCA3 (Kawaguchi *et al.* 1994; Cancel *et al.* 1995; Schols *et al.* 1995), SCA6 (Zhuchenko *et al.* 1997) and SCA7 (David *et al.* 1997; Del-Favero *et al.* 1998; Koob *et al.* 1998), affected individuals have expansions in a CAG trinucleotide repeat in the coding region of at least one copy of the relevant gene, analogous to the mutational mechanism in Huntington's disease (HD) (Huntington's Disease Collaborative Research Group 1993), Kennedy's syndrome (X-linked spino-bulbar muscular atrophy) (La Spada *et al.* 1991, 1992) and dentatorubro–pallidoluysian atrophy (DRPLA) (Nagafuchi *et al.* 1994). However, these five genes together account for only around 25–60% of all ADCA families (A. Dürr, personal communication), and as more SCA loci and genes are identified, the complexity of the clinico-genetic classification of the dominant ataxias will continue to increase.

The relative prevalence of the SCA mutations varies throughout the world; for example, in Portugal, SCA3 accounts for 80% of ADCA families (Silveira *et al.* 1998), but only 43% in Japan (Takano *et al.* 1998), and is virtually unheard of in Italy. Similarly, SCA6 is relatively frequent in Japan (30%) and the UK, but rare in France and Portugal.

This chapter will describe how the study of the molecular characteristics of the CAG repeats, and genotype–phenotype correlation in the SCAs has led to important advances in the understanding of the molecular pathology of trinucleotide repeat disorders in general.

Molecular characteristics of the CAG repeats in the spinocerebellar ataxias

Normal and pathological alleles

The CAG repeat in normal SCA1, SCA2, SCA3, SCA6 and SCA7 alleles encodes a polyglutamine domain in the specified protein. In abnormal SCA genes, this domain is expanded or lengthened. The normal function of the proteins (referred to as ataxins) encoded by SCA1, SCA2, SCA3 and SCA7 is obscure; they have no sequence homology with each other or with other known proteins. However, the function of the protein product of the SCA6 gene (ataxin-6) is known. The SCA6 gene contains 47 exons which encode the α_{1A}-subunit of the voltage-gated P/Q type calcium channel CACNA1A (Zhuchenko *et al.* 1997). The primary transcript undergoes differential splicing such that the cerebellum preferentially expresses a subunit containing the polyglutamine domain in the C-terminal region (Ishikawa *et al.* 1999), although, in common with the other ataxins, CACNA1A is expressed throughout the nervous system and all peripheral tissues. The molecular properties of SCA8, the sixth SCA gene to be cloned, are rather different from those of the other SCA genes, in that affected individuals in the only family described to date have expansions in an untranslated CTG/CTA repeat on chromosome 13q21 (Koob *et al.* 1999). As no protein or polyglutamine tract is translated, it has been suggested that the transcript may exert its presumed pathogenic effect at an RNA level, perhaps as an antisense RNA, but this has not been proved. Useful analogy with the other SCA genes cannot be made at the time of writing, and therefore SCA8 will not be discussed further in this text.

The number of CAG repeats which represents the pathological threshold, and the range of the number of CAG repeats found in both normal and pathological alleles, varies between the different SCAs (table 16.1). For example, in SCA3, the range of normal is 12–40 repeats, while pathological alleles contain 54–89 repeats. In contrast, in SCA6, the normal range is 4–18, while abnormal is 20–31. The gap between the ranges of repeat size of normal and pathological alleles also varies from two repeats in SCA6, to 14 in SCA3. The implication of these differences and why, for example, a repeat size of less than 30 is pathological in SCA6 but not in any of the other SCAs, is unclear.

An important feature of CAG trinucleotide repeats disorders is their tendency to exhibit both meiotic and mitotic instability, which is observed to varying degrees in SCA1, SCA2, SCA3 and SCA7. Meiotic instability is the phenomenon whereby the CAG repeat undergoes either an increase or a decrease in size during vertical, i.e. parent-to-child, transmission. This may occur as a result of slippage during DNA replication or from the formation of stable hairpin structures. This latter mechanism would be predicted to result in large expansions or contractions depending on their location on the leading or lagging strand, respectively (Eichler *et al.* 1994; Kang *et al.* 1995; Wells *et al.* 1998).

Normal alleles show little or no meiotic instability, whereas expanded alleles tend to be unstable. In normal SCA1 and most normal SCA2 alleles, the CAG repeat is interrupted by CAT and CAA, respectively. These interruptions, which also code for glutamines, are usually absent in the expanded allele, and it has been suggested that they may confer stability on the normal allele. Alleles with repeat numbers which fall within the expected pathological range, but which are apparently not pathological, have been found in both SCA1 and 2; sequencing has identified the presence of interruptions in these 'normal' alleles (Chung *et al.* 1993; Quan *et al.* 1995; Cancel *et al.* 1997). Normal alleles in the other SCAs have no such interruption, and therefore the length of the repeat alone may be the critical factor.

Increases in repeat size tend to occur more frequently in paternal transmissions than in maternal transmissions. In addition, instability is much lower in SCA3 than in SCA7, in which the most dramatic meiotic instability is observed. Paternally transmitted SCA7 alleles have been shown to undergo large increases (figure 16.1) of up to 166 repeats (David *et al.* 1998; Giunti 1999, #18). The reason for this apparent paternal bias toward instability is thought to be related to the fact that oogenesis stops early in prenatal development, but spermatogenesis is a lifelong process, involving many more cell divisions, and therefore more opportunities for 'mistakes' in replication to occur. In contrast to SCA7, SCA6 expansions show little or no meiotic instability, there being only two reports in which paternally derived, 20- and 24-repeat alleles underwent expansion by five repeats and two repeats, respectively (Jodice *et al.* 1997; Matsuyama *et al.* 1997). Single sperm analyses of patients with SCA1 (Chong *et al.* 1995; Koefoed *et al.* 1998), SCA2 (Cancel *et al.* 1997), SCA3 (Watanabe *et al.* 1996), SCA7 (David *et al.* 1998) and DRPLA (Takiyama *et al.* 1999) have shown varying degrees of gonadal mosaicism. For example, in the sperm of one SCA7 patient, alleles with repeats of 42 to more than 155 were found. In contrast, single sperm analysis of SCA6 patients revealed no gonadal mosaicism. There is good evidence that this meiotic instability accounts for the clinical phenomenon of anticipation (see below).

Mitotic instability results in somatic mosaicism, such that different somatic cells including neurons contain expanded alleles with varying numbers of repeats. For a given SCA, meiotic and mitotic instability tend to be correlated,

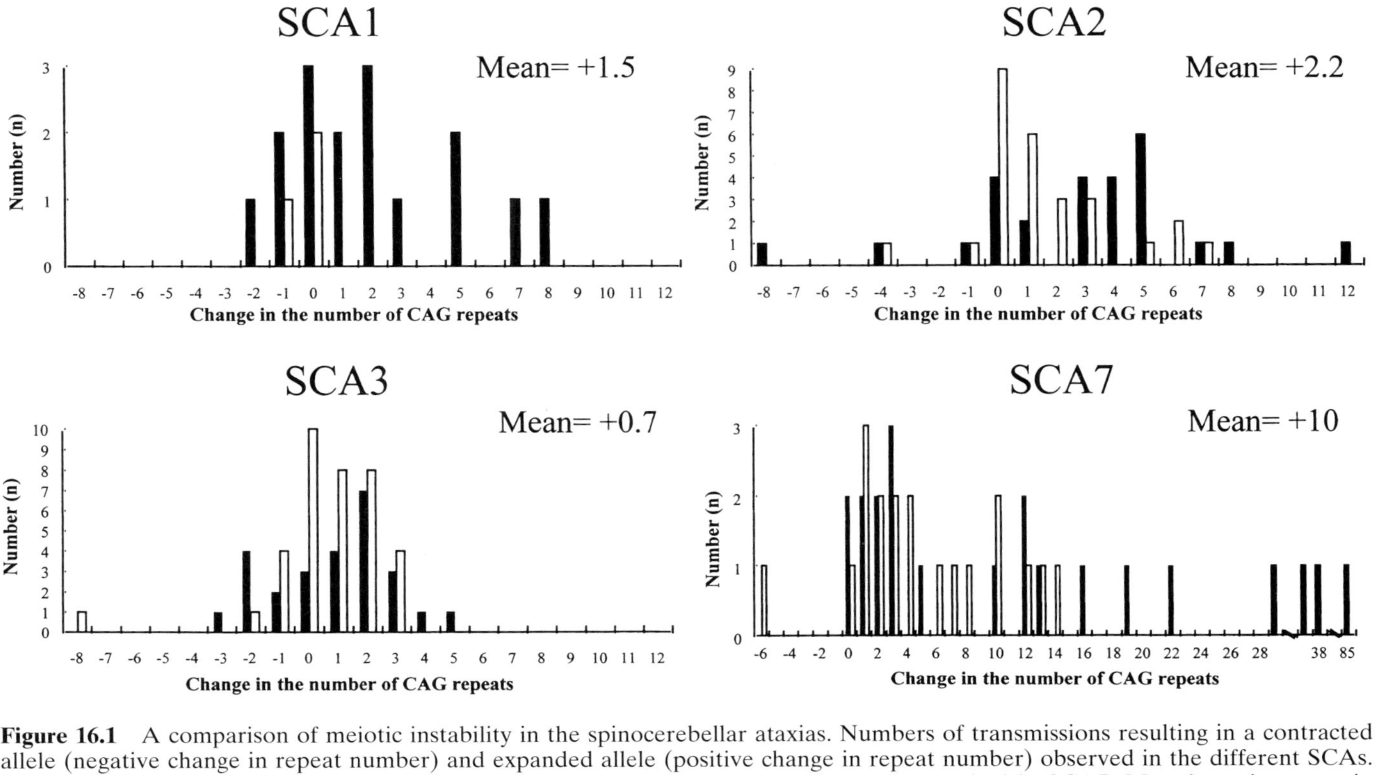

Figure 16.1 A comparison of meiotic instability in the spinocerebellar ataxias. Numbers of transmissions resulting in a contracted allele (negative change in repeat number) and expanded allele (positive change in repeat number) observed in the different SCAs. Mean overall change in repeat number is given. Instability is low in SCA3, and most marked in SCA7. Most large increases in repeat number occur during paternal transmissions.

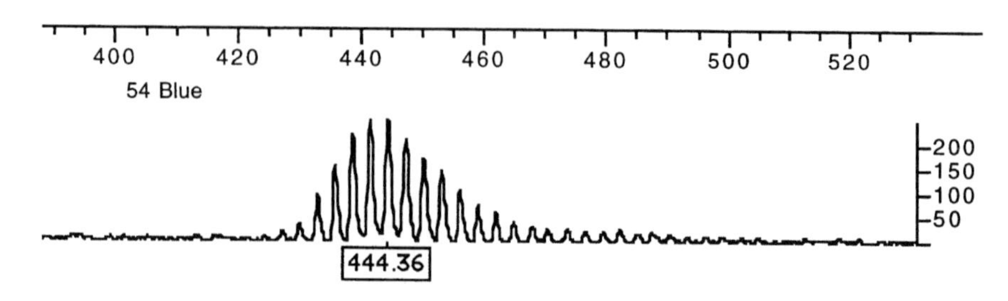

Figure 16.2 An electropherogram of a patient with an expansion in the SCA7 gene. Mitotic instability present in peripheral blood leucocytes produces this characteristic 'hedgehog' appearance of the electropherogram, with the multiple peaks each representing alleles with different numbers of repeats, widely scattered around the mode, in this case *ca.* 445 bp. Normal, unexpanded alleles produce a single peak.

such that in SCA7, a high degree of somatic mosaicism is observed in blood leucocytes (figure 16.2), whereas in SCA6, no such mosaicism has been detected. Somatic mosaicism is, however, always less marked than gonadal mosaicism for a given SCA. To date, there is no firm evidence of a significant effect of somatic mosaicism on the clinical phenotype.

While pathological alleles which are already expanded may undergo further expansion during meiosis, it has been shown that normal alleles at the upper limit of the normal range, or in the range between normal and pathological, so-called intermediate alleles (IAs), may also undergo expansion into the abnormal, pathological range during meiosis, even though this may occur infrequently. In so doing, they would provide a plausible explanation for the occurrence of alleged *de novo*, or new mutations. Evidence for *de novo* pathological expansion has been shown in HD (Goldberg *et al.* 1995), and now also in SCA7 (Stevanin *et al.* 1998; Giunti *et al.* 1999) where, in two separate pedigrees, a paternally derived IA has been observed to undergo expansion into the pathological range (figure 16.3). One study (Takano *et al.* 1998) recently showed that the different prevalence of the dominantly inherited ataxias and DRPLA in separate Caucasian and Japanese populations was related to the frequency of the larger normal alleles of the respective gene. They found that the relative prevalence of SCA1 and SCA2 was higher in 177 Caucasian pedigrees (15% and 14%, respectively) than in 202 Japanese pedigrees (3% and 5%, respectively), and that, accordingly, the frequencies of the large normal SCA1 alleles (> 30 repeats) and SCA2 alleles (> 22 repeats) were greater in Caucasians. Conversely, SCA3, SCA6 and DRPLA were more common in Japanese pedigrees (43%, 11% and 20%) than in Caucasian pedigrees (30%, 5% and 0%), and the frequencies of the large normal SCA3 (> 27 repeats), SCA6 (> 13 repeats) and DRPLA (> 17 repeats) alleles were higher in the Japanese. The close correlation observed was taken by the authors to imply that a number of large normal alleles provide a reservoir for the appearance of *de novo* cases by undergoing pathological expansion.

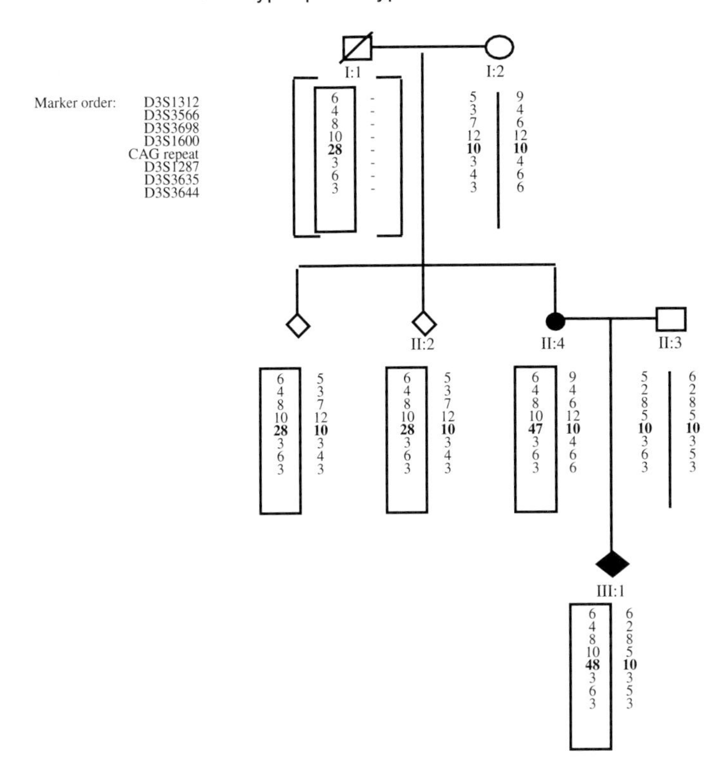

Figure 16.3 A SCA7 pedigree demonstrating a *de novo* mutation. The birth order has been changed and the sex of three individuals, indicated by diamonds, has been concealed to preserve anonymity. The haplotype of individual I:1 has been reconstructed from that of his offspring. Paternal transmission of a 28 repeat SCA7 allele from I:1, who was unaffected, to two of his children resulted in an unchanged number of repeats, but this same allele underwent expansion into the pathological range (47 repeats) when transmitted to individual II:4, who was affected. This allele was then relatively stable, undergoing expansion by only one repeat during maternal trans mission to her child, III:1, who was also affected.

Further circumstantial evidence for the occurrence of new mutations comes from the expectation that, at least in SCA7 where meiotic instability is high, the prevalence of the disease might be expected to fall over time. As pathological alleles undergo progressive expansion with successive meioses, resulting in larger repeat numbers, higher infantile or juvenile incidence would be predicted, with subsequent pre-reproductive death. Conversely, if the disease prevalence is to remain in equilibrium, *de novo* mutations must occur to replace these untransmitted pathological alleles. There is no evidence that SCA7 prevalence is falling, and, consequently, credence must be given to this latter hypothesis. The corollary of this argument is aptly demonstrated with reference to SCA6. Here, meiotic instability is low or zero, and as strong linkage disequilibrium is observed in unrelated German kindreds (Dichgans *et al.* 1999), suggesting a single ancestral founder, a low *de novo* mutation rate in this disorder can be inferred.

Genotype–phenotype correlation

While there are important differences between both the molecular characteristics and the clinical phenotypes of the different SCAs, there are also striking similarities. To varying degrees, patients with SCA1, 2, 3, 6 and 7 all display the phenomenon of anticipation, i.e. the tendency towards earlier age at onset, increased disease severity and increased rate of disease progression with successive generations. In general, and for a given type of SCA, individuals may be expected to develop the disease earlier, and exhibit a more severe phenotype, including a faster rate of disease progression, with increasing repeat number. Figure 16.4*a*,*b* shows plots of age at onset against repeat number for the different SCAs, and a significant inverse correlation has been demonstrated in each case. It will be noted that the slope of the correlation curve is different for each SCA; the slope of the SCA1 curve is steeper than that of SCA7, suggesting that the size of the polyglutamine domain has a greater effect on age at onset in SCA7 than in SCA1. A plot of disease progression rate against repeat number would produce a similar pattern. Despite this close correlation, it is not possible to give an accurate prediction of the age at onset from knowledge of the repeat number, owing to considerable data scatter. Compound heterozygotes for SCA2 (Sanpei *et al.* 1996), SCA3 (Kawakami *et al.* 1995; Lerer *et al.* 1996; Sobue *et al.* 1996) and SCA6 (Geschwind *et al.* 1997; Ikeuchi *et al.* 1997; Matsumura *et al.* 1997) are reported to have an earlier age at onset, implying a dosage effect. As discussed above, in all but SCA6, the size of the CAG repeat in the pathological allele tends to increase with successive meioses, especially during paternal transmission, and this accounts neatly for these clinical observations. Although anticipation has been described for SCA6, it is not particularly dramatic, and no consistent meiotic instability is observed. Hence the alleged finding of anticipation in SCA6 may be the result of observer or ascertainment bias, i.e. the tendency for investigators to 'look' harder for evidence of the disease in offspring of affected individuals, and thus detect signs of the disease earlier. If anticipation is a real phenomenon in this condition, then another mechanism must be responsible, but this remains obscure.

Clinico-pathological correlation

As discussed above, although there are important similarities in the molecular characteristics of the different SCA mutations, there is considerable variation in the clinical manifestations. In addition to cerebellar ataxia, attributable to dysfunction of the cerebellum or its connections, signs and symptoms which result from degeneration of different parts of the central and/or peripheral nervous system are characteristic of the different SCAs (table 16.1). However, with the exception of pigmentary retinopathy in SCA7, no single clinical sign is found exclusively in, or is diagnostic of, any one form of SCA.

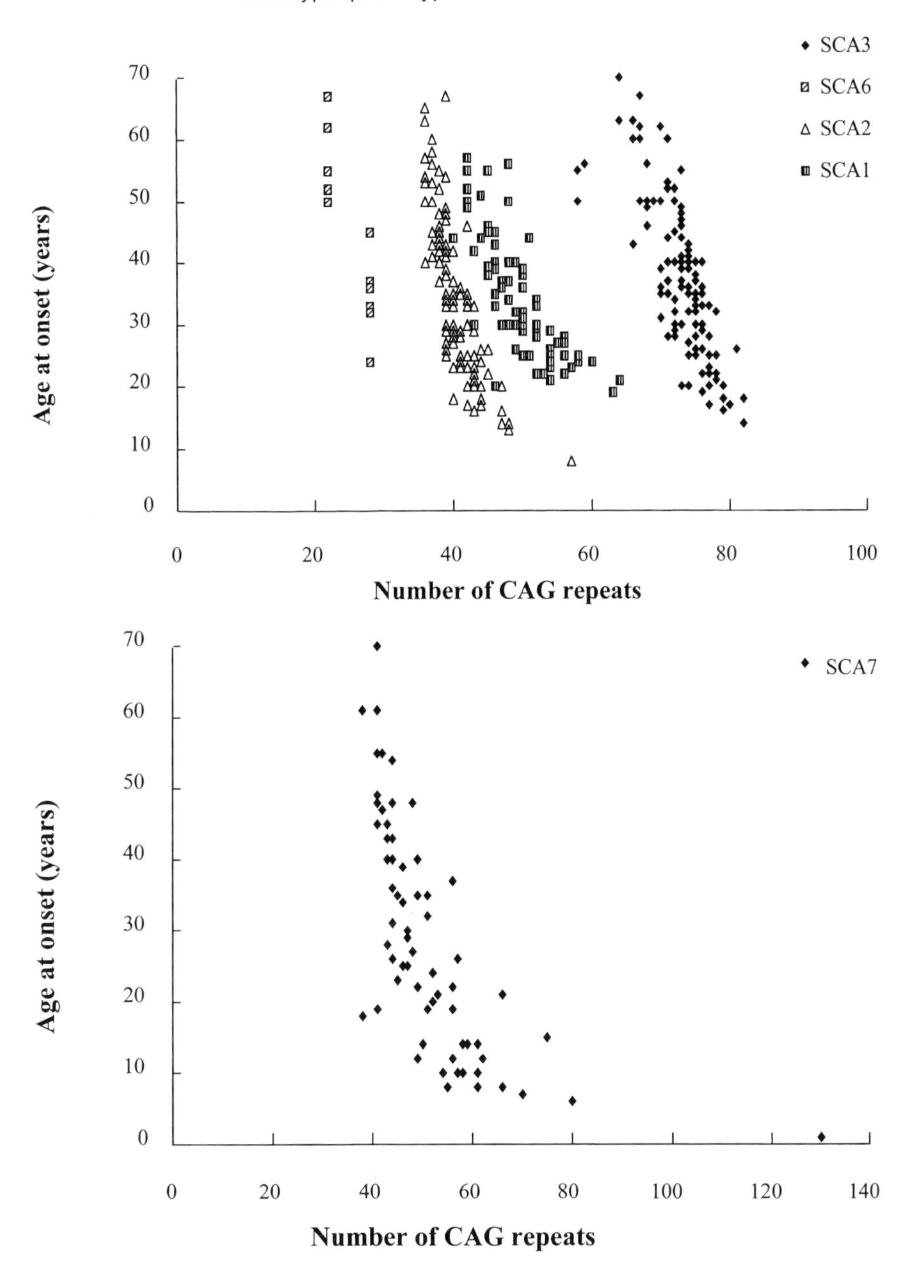

Figure 16.4 Age at onset versus number of CAG repeats for the different SCAs. A significant inverse correlation between repeat number and age at onset has been shown for each SCA. Note the different pathological thresholds and different slopes; the SCA1 and SCA3 (*a*) correlation curves are appreciably steeper than that of SCA7 (*b*), indicating a greater effect of repeat number on phenotypic expression in SCA7. Data scatter precludes prediction of age at onset from knowledge of repeat number alone. Note the SCA7 individual with 130 repeats whose age at onset was one year.

Therefore, because of this phenotypic overlap, prediction of the genotype of an individual patient from the clinical phenotype is not usually possible. In spite of this important principle, neuropathological studies (Martin *et al.* 1994; Durr *et al.* 1995, 1996; Robitaille *et al.* 1995; Gilman *et al.* 1996; Tsuchiya *et al.* 1998) have shown that the clinical phenotype of a given SCA corresponds quite well to the degree of involvement of the relevant part of the nervous system (table 16.2). For example, in SCA3, extrapyramidal signs such as parkinsonism and dystonia occur as a result of basal ganglia involvement, and amyotrophy due to anterior horn cell loss. In SCA2, pontine nuclear involvement is manifested by slowed saccadic eye movements. Similarly, visual loss in SCA7 is usual, and is due to a pigmentary macular dystrophy; degeneration of the retinal ganglion cells is seen in pathological material from affected individuals. In the case of SCA6, which usually manifests as a relatively 'pure' ataxia, with no other major clinical features, the Purkinje cells of the cerebellum and the inferior olives are the only structures which show significant abnormalities at autopsy.

Neuroimaging studies have also shown a correlation between clinical features and magnetic resonance imaging (MRI) abnormalities (Watanabe *et al.* 1996; Gomez *et al.* 1997; Stevanin *et al.* 1997; Klockgether *et al.* 1998; Schols *et al.* 1998). Brainstem and cerebellar atrophy is seen in SCA1, 2 and 3, though this may be most marked in SCA2. In contrast, SCA6 patients usually show cerebellar atrophy alone, with sparing of the brainstem. Atrophy of the basal ganglia, specifically the caudate and putamen, is most marked in SCA3, in which extrapyramidal (Parkinsonian) features are often observed.

Table 16.2 A comparison of the major pathological changes in the different SCAs

areas or neurons affected	*SCA1*	*SCA2*	*SCA3*	*SCA6*	*SCA7*
Purkinje cells	+	++	+/–	+++	+
inferior olives	+++	+++	–	±	+++
globus pallidus	+ext	+	++int	–	+
retina	–	–	–	–	+
substantia nigra pars compacta	+	++	++	–	++
nucleus pontis	+	+++	+	–	+
dentate nucleus	++	–	++	+/–	++
anterior horn cells	+	+	+	–	+
axonal neuropathy	+	++	++	++	–
dorsal columns	+	+++	+	–	+
spinocerebellar tracts	+	–	++	–	+

(Key: –, not affected; +/–, sometimes affected; +, mildly affected; ++, moderately affected; +++, severely affected. Ext = globus pallidus, external part; int = internal part.)

In summary, the expanded trinucleotide repeat results in region-specific neurodegeneration the pattern of which is different for each of the SCAs. The reasons for this region specificity are unclear.

Conclusions

During the last ten years, much has been learnt about the molecular pathology of the trinucleotide repeat disorders; the instability of the expanded repeat and genotype–phenotype correlation. Recently we have gained insight into the complex cell biology which underlies these conditions. Many of these insights are discussed elsewhere in this book. Predictably, many findings seem to raise more questions than they answer.

The variable instability of the mutant alleles of the different SCA genes during replication (figure 16.1) suggests that the DNA context of the gene may exert a stabilizing (or destabilizing) effect. Genetic and epigenetic factors such as the position and orientation of the repeat, with respect to the origin of replication, can modify instability in bacterial (Kang *et al.* 1995) and yeast (Maurer *et al.* 1996; Freudenreich *et al.* 1997; Schweitzer & Livingston 1998) models. Stable hairpin formation by the repeat may also contribute to replication stability.

One of the key questions is how the region-specific neurodegeneration observed in the different SCAs occurs. The mutant and wild-type proteins are expressed throughout the brain, and small variations in the levels of expression do not account for the striking neurospecificity. Similarly, the degree of somatic mosaicism in DRPLA detected in the nervous system does not account for the selectivity of neuronal death (Takano *et al.* 1996). These observations have led to attempts to identify proteins which have greater affinity for the expanded, mutant ataxin, and which themselves exhibit region specificity. One protein of interest is the leucine-rich acidic nuclear protein (LANP), which interacts with ataxin-1 and co-localizes with nuclear inclusions (Matilla *et al.* 1997) (see below). However, while polyglutamine-containing proteins appear to have many potential protein partners, no single protein which is expressed appropriately has yet been identified.

The exact mechanism by which expanded polyglutamine domains in the respective proteins cause certain populations of neurons to die is also far from clear. It is thought that an expanded polyglutamine domain confers on the protein a novel but deleterious, toxic property or 'gain of function'. In addition to interactions with other normal proteins, polyglutamine polypeptides have been shown to associate *in vitro* (Hollenbach *et al.*, Chapter 11; Scherzinger *et al.* 1997, 1999), and intranuclear inclusions containing ubiquitinated aggregates of mutant protein have been found in the affected brain regions in SCA1 (Duyckaerts *et al.* 1999), SCA3 (Paulson *et al.* 1997) and SCA7 (Holmberg *et al.* 1998) patients, similar to those found in the brains of DRPLA patients (Hayashi *et al.* 1998) and of adult-onset HD patients

(Gourfinkel-An *et al.* 1998) and transgenic HD mice (Davies *et al.* 1997). Interestingly, inclusions in SCA2 have not been reported. It may be that intranuclear inclusions are toxic to neurons, but whether the development of inclusions is either necessary or sufficient to cause cell death remains controversial, as does the question of whether nuclear localization is necessary for the inclusions to exert their toxic effect. It was predicted that intranuclear inclusions would not be found in SCA6, as the protein is expressed in the cell membrane. However, a recent study (Ishikawa *et al.* 1999) has shown that cytoplasmic inclusions containing aggregated calcium channel subunits are found in this condition. Are these cytoplasmic inclusions toxic to the neurons, or is cell death in SCA6 the result of altered channel permeability, leading to progressive ionic damage to the neuron? Some authors have proposed that 'full blown' inclusions may represent a protective response to toxicity rather than being the toxic agent themselves (Sisodia 1998). Hence, the initial stages of protein aggregation, which may precede the development of identifiable inclusions, may be one important step in the process of cell death.

A related problem is what determines the point at which clinical symptoms become apparent in a given SCA patient. Conventional wisdom would suggest that a certain critical number of neurons in a given region must die before symptoms attributable to dysfunction of that area become manifest. The time-course of protein aggregation is measured in seconds and minutes, whereas that of these diseases from birth to presenting symptom is measured in years in all but those with the largest repeat expansions which may render the foetus non-viable. Squaring this circle is also likely to prove problematic.

Factors other than the repeat number, which may be both intrinsic and extrinsic to the protein, seem to be of great importance in determining phenotypic expression, and the disease duration until death, in the different SCAs. The protein context of the polyglutamine domain appears to be one such factor, as evidenced by the variable repeat number representing the pathological threshold (table 16.1), and the variable influence of repeat number on age at onset (figure 16.4*a*,*b*).

In summary, there are many crucial questions which remain unanswered with respect to the molecular instability and pathophysiological consequences of the expanded trinucleotide repeat. While study and elucidation of a single aspect of one of the CAG repeat disorders has relevance to this whole group of conditions, important differences between the SCAs suggest that no single model will provide a unifying explanation. We have probably only scratched the surface of the integrated mechanism which leads to phenotypic expression of the spinocerebellar ataxias.

The authors would like to thank Dr P. Giunti, Dr A. Dürr and Dr G. Stevanin for contributing to this work, including unpublished data and figures. P.F.W. is a UK Medical Research Council Clinical Training Fellow. A.B. is supported by the VERUM Foundation, the Association Française contre les Myopathies and Biomed (No. CEE BMH4-CT960244).

References

Cancel, G., Abbas, N., Stevanin, G., Durr, A., Chneiweiss, H., Neri, C., Duyckaerts, C., Penet, C., Cann, H. M., Agid, Y., *et al.* 1995 Marked phenotypic heterogeneity associated with expansion of a CAG repeat sequence at the spinocerebellar ataxia 3/Machado–Joseph disease locus. *Am. J. Hum. Genet.* **57**, 809–816.

Cancel, G. (and 23 others) 1997 Molecular and clinical correlations in spinocerebellar ataxia 2: a study of 32 families. *Hum. Mol. Genet.* **6**, 709–7015.

Chong, S. S., McCall, A. E., Cota, J., Subramony, S. H., Orr, H. T., Hughes, M. R. & Zoghbi, H. Y. 1995 Gametic and somatic tissue-specific heterogeneity of the expanded SCA1 CAG repeat in spinocerebellar ataxia type 1. *Nature Genet.* **10**, 344–350.

Chung, M. Y., Ranum, L. P., Duvick, L. A., Servadio, A., Zoghbi, H. Y. & Orr, H. T. 1993 Evidence for a mechanism predisposing to intergenerational CAG repeat instability in spinocerebellar ataxia type I. *Nature Genet.* **5**, 254–258.

David, G. (and 18 others) 1997 Cloning of the SCA7 gene reveals a highly unstable CAG repeat expansion. *Nature Genet.*, **17**, 65–70.

David, G. (and 15 others) 1998 Molecular and clinical correlations in autosomal dominant cerebellar ataxia with progressive macular dystrophy (SCA7). *Hum. Mol. Genet.* **7**, 165–170.

Davies, S. W., Turmaine, M., Cozens, B. A., DiFiglia, M., Sharp, A. H., Ross, C. A., Scherzinger, E., Wanker, E. E., Mangiarini, L. & Bates, G. P. 1997 Formation of neuronal intranuclear inclusions underlies the neurological dysfunction in mice transgenic for the HD mutation. *Cell* **90**, 537–548.

Del-Favero, J. (and 12 others) 1998 Molecular genetic analysis of autosomal dominant cerebellar ataxia with retinal degeneration (ADCA type II) caused by CAG triplet repeat expansion. *Hum. Mol. Genet.* **7**, 177–186.

Dichgans, M. (and 11 others) 1999 Spinocerebellar ataxia type 6: evidence for a strong founder effect among German families. *Neurology* 52, 849–51.

Durr, A., Smadja, D., Cancel, G., Lezin, A., Stevanin, G., Mikol, J., Bellance, R., Buisson, G. G., Chneiweiss, H., Dellanave, J., *et al.* 1995 Autosomal dominant cerebellar ataxia type I in Martinique (French West Indies). Clinical and neuropathological analysis of 53 patients from three unrelated SCA2 families. *Brain* **118**, 1573–1581.

Durr, A. (and 13 others) 1996 Spinocerebellar ataxia 3 and Machado–Joseph disease: clinical, molecular, and neuropathological features. *Ann. Neurol.* **39**, 490–499.

Duyckaerts, C., Durr, A., Cancel, G. & Brice, A. 1999 Nuclear inclusions in spinocerebellar ataxia type 1. *Acta Neuropathol.* **97**, 201–207.

Eichler, E. E., Holden, J. J., Popovich, B. W., Reiss, A. L., Snow, K., Thibodeau, S. N., Richards, C. S., Ward, P. A. & Nelson, D. L. 1994 Length of uninterrupted CGG repeats determines instability in the FMR1 gene. *Nature Genet.* **8**, 88–94.

Freudenreich, C. H., Stavenhagen, J. B. & Zakian, V. A. 1997 Stability of a CTG/CAG trinucleotide repeat in yeast is dependent on its orientation in the genome. *Mol. Cell. Biol.* **17**, 2090–2098.

Geschwind, D. H., Perlman, S., Figueroa, K. P., Karrim, J., Baloh, R. W. & Pulst, S. M. 1997 Spinocerebellar ataxia type 6. Frequency of the mutation and genotype–phenotype correlations. *Neurology* **49**, 1247–1251.

Gilman, S., Sima, A. A., Junck, L., Kluin, K. J., Koeppe, R. A., Lohman, M. E. & Little, R. 1996 Spinocerebellar ataxia type 1 with multiple system degeneration and glial cytoplasmic inclusions. *Ann. Neurol.* **39**, 241–255.

Giunti, P., Stevanin, G., Worth, P. F., David, G., Brice, A. & Wood, N. W. 1999 Molecular and clinical study of 18 families with ADCA type II: evidence for genetic heterogeneity and *de novo* mutation. *Am. J. Hum. Genet.* **64**, 1594–603.

Goldberg, Y. P., McMurray, C. T., Zeisler, J., Almqvist, E., Sillence, D., Richards, F., Gacy, A. M., Buchanan, J., Telenius, H. & Hayden, M. R. 1995 Increased instability of intermediate alleles in families with sporadic Huntington disease compared to similar sized intermediate alleles in the general population. *Hum. Mol. Genet.* **4**, 1911–1918.

Gomez, C. M., Thompson, R. M., Gammack, J. T., Perlman, S. L., Dobyns, W. B., Truwit, C. L., Zee, D. S., Clark, H. B. & Anderson, J. H. 1997 Spinocerebellar ataxia type 6: gaze-evoked and vertical nystagmus, Purkinje cell degeneration, and variable age of onset. *Ann. Neurol.* **42**, 933–950.

Gourfinkel-An, I., Cancel, G., Duyckaerts, C., Faucheux, B., Hauw, J. J., Trottier, Y., Brice, A., Agid, Y. & Hirsch, E. C. 1998 Neuronal distribution of intranuclear inclusions in Huntington's disease with adult onset. *NeuroReport* **9**, 1823–1826.

Harding, A. E. 1982 The clinical features and classification of the late onset autosomal dominant cerebellar ataxias. A study of 11 families, including descendants of the 'the Drew family of Walworth'. *Brain* **105**, 1–28.

Hayashi, Y. (and 10 others) 1998 Hereditary dentatorubral–pallidoluysian atrophy: detection of widespread ubiquitinated neuronal and glial intranuclear inclusions in the brain. *Acta Neuropathol.* **96**, 547–552.

Holmberg, M. (and 10 others) 1998 Spinocerebellar ataxia type 7 (SCA7): a neurodegenerative disorder with neuronal intranuclear inclusions. *Hum. Mol. Genet.* **7**, 913–918.

Huntington's Disease Collaborative Research Group 1993 A novel gene containing a trinucleotide repeat that is expanded and unstable on Huntington's disease chromosomes. *Cell* **72**, 971–983.

Ikeuchi, T. (and 13 others) 1997 Spinocerebellar ataxia type 6: CAG repeat expansion in alpha1A voltage- dependent calcium channel gene and clinical variations in Japanese population. *Ann. Neurol.* **42**, 879–884

Imbert, G. (and 14 others) 1996 Cloning of the gene for spinocerebellar ataxia 2 reveals a locus with high sensitivity to expanded CAG/glutamine repeats. *Nature Genet.* **14**, 285–291.

Ishikawa, K. (and 14 others) 1999 Abundant expression and cytoplasmic aggregations of α_{1A} voltage- dependent calcium channel protein associated with neurodegeneration in spinocerebellar ataxia type 6. *Hum. Mol. Genet.* **8**, 1185–93.

Jodice, C. (and 12 others) 1997 Episodic ataxia type 2 (EA2) and spinocerebellar ataxia type 6 (SCA6) due to CAG repeat expansion in the CACNA1A gene on chromosome 19p. *Hum. Mol. Genet.* **6**, 1973–1978.

Kang, S., Jaworski, A., Ohshima, K. & Wells, R. D. 1995 Expansion and deletion of CTG repeats from human disease genes are determined by the direction of replication in *E. coli. Nature Genet.* **10**, 213–218.

Kawaguchi, Y. (and 12 others) 1994 CAG expansions in a novel gene for Machado–Joseph disease at chromosome 14q32.1. *Nature Genet.* **8**, 221–228.

Kawakami, H., Maruyama, H., Nakamura, S., Kawaguchi, Y., Kakizuka, A., Doyu, M. & Sobue, G. 1995 Unique features of the CAG repeats in Machado–Joseph disease. *Nature Genet.* **9**, 344–345.

Klockgether, T. (and 10 others) 1998 Autosomal dominant cerebellar ataxia type I. MRI-based volumetry of posterior fossa structures and basal ganglia in spinocerebellar ataxia types 1, 2 and 3. *Brain* **121**, 1687–1693.

Koefoed, P., Hasholt, L., Fenger, K., Nielsen, J. E., Eiberg, H., Buschard, K. & Sorensen, S. A. 1998 Mitotic and meiotic instability of the CAG trinucleotide repeat in spinocerebellar ataxia type 1. *Hum. Genet.* **103**, 564–569.

Koob, M. D., Benzow, K. A., Bird, T. D., Day, J. W., Moseley, M. L. & Ranum, L. P. 1998 Rapid cloning of expanded trinucleotide repeat sequences from genomic DNA. *Nature Genet.* **18**, 72–75.

Koob, M. D., Moseley, M. L., Schut, L. J., Benzow, K. A., Bird, T. D., Day, J. W. & Ranum, L. P. 1999 An untranslated CTG expansion causes a novel form of spinocerebellar ataxia (SCA8). *Nature Genet.* **21**, 379–384.

La Spada, A. R., Wilson, E. M., Lubahn, D. B., Harding, A. E. & Fischbeck, K. H. 1991 Androgen receptor gene mutations in X-linked spinal and bulbar muscular atrophy. *Nature* **352**, 77–79.

La Spada, A. R., Roling, D. B., Harding, A. E., Warner, C. L., Spiegel, R., Hausmanowa-Petrusewicz, I., Yee, W. C. & Fischbeck, K. H. 1992 Meiotic stability and genotype–phenotype correlation of the trinucleotide repeat in X-linked spinal and bulbar muscular atrophy. *Nature Genet.* **2**, 301–304.

Lerer, I., Merims, D., Abeliovich, D., Zlotogora, J. & Gadoth, N. 1996 Machado–Joseph disease: correlation between the clinical features, the CAG repeat length and homozygosity for the mutation. *Eur. J. Hum. Genet.* **4**, 3–7.

Marie, P. 1893 Sur l'heredoataxie cerebelleuse. *Sem. Med. Paris* **13**, 444–447.

Martin, J. J., Van Regemorter, N., Krols, L., Brucher, J. M., de Barsy, T., Szliwowski, H., Evrard, P., Ceuterick, C., Tassignon, M. J., Smet-Dieleman, H., *et al.* 1994 On an autosomal dominant form of retinal-cerebellar degeneration: an autopsy study of five patients in one family. *Acta Neuropathol.* **88**, 277–286.

Matilla, A., Koshy, B. T., Cummings, C. J., Isobe, T., Orr, H. T. & Zoghbi, H. Y. 1997 The cerebellar leucine-rich acidic nuclear protein interacts with ataxin-1. *Nature* **389**, 974–978.

Matsumura, R., Futamura, N., Fujimoto, Y., Yanagimoto, S., Horikawa, H., Suzumura, A. & Takayanagi, T. 1997 Spinocerebellar ataxia type 6. Molecular and clinical features of 35 Japanese patients including one homozygous for the CAG repeat expansion. *Neurology* **49**, 1238–1243.

Matsuyama, Z. (and 10 others) 1997 Molecular features of the CAG repeats of spinocerebellar ataxia 6 (SCA6). *Hum. Mol. Genet.* **6**, 1283–1287.

Maurer, D. J., O'Callaghan, B. L. & Livingston, D. M. 1996 Orientation dependence of trinucleotide CAG repeat instability in *Saccharomyces cerevisiae*. *Mol. Cell. Biol.* **16**, 6617–6622.

Nagafuchi, S. (and 22 others) 1994 Dentatorubral and pallidoluysian atrophy expansion of an unstable CAG trinucleotide on chromosome 12p. *Nature Genet.* **6**, 14–18.

Orr, H. T., Chung, M. Y., Banfi, S., Kwiatkowski Jr, T. J., Servadio, A., Beaudet, A. L., McCall, A. E., Duvick, L. A., Ranum, L. P. & Zoghbi, H. Y. 1993 Expansion of an unstable trinucleotide CAG repeat in spinocerebellar ataxia type 1. *Nature Genet.* **4**, 221–226.

Paulson, H. L., Perez, M. K., Trottier, Y., Trojanowski, J. Q., Subramony, S. H., Das, S. S., Vig, P., Mandel, J. L., Fischbeck, K. H. & Pittman, R. N. 1997 Intranuclear inclusions of expanded polyglutamine protein in spinocerebellar ataxia type 3. *Neuron* **19**, 333–344.

Pulst, S. M. (and 15 others) 1996 Moderate expansion of a normally biallelic trinucleotide repeat in spinocerebellar ataxia type 2. *Nature Genet.* **14**, 269–276.

Quan, F., Janas, J. & Popovich, B. W. 1995 A novel CAG repeat configuration in the SCA1 gene: implications for the molecular diagnostics of spinocerebellar ataxia type 1. *Hum. Mol. Genet.* **4**, 2411–2413.

Robitaille, Y., Schut, L. & Kish, S. J. 1995 Structural and immunocytochemical features of olivopontocerebellar atrophy caused by the spinocerebellar ataxia type 1 (SCA-1) mutation define a unique phenotype. *Acta Neuropathol.* **90**, 572–581.

Sanpei, K. (and 24 others) 1996 Identification of the spinocerebellar ataxia type 2 gene using a direct identification of repeat expansion and cloning technique, DIRECT. *Nature Genet.* **14**, 277–284.

Scherzinger, E., Lurz, R., Turmaine, M., Mangiarini, L., Hollenbach, B., Hasenbank, R., Bates, G. P., Davies, S. W., Lehrach, H. & Wanker, E. E. 1997 Huntingtin-encoded polyglutamine expansions form amyloid-like protein aggregates *in vitro* and *in vivo*. *Cell* **90**, 549–558.

Scherzinger, E., Sittler, A., Schweiger, K., Heiser, V., Lurz, R., Hasenbank, R., Bates, G. P., Lehrach, H. & Wanker, E. E. 1999 Self-assembly of polyglutamine-containing huntingtin fragments into amyloid-like fibrils: implications for Huntington's disease pathology. *Proc. Natl Acad. Sci. USA* **96**, 4604–4609.

Schols, L., Vieira-Saecker, A. M., Schols, S., Przuntek, H., Epplen, J. T. & Riess, O. 1995 Trinucleotide expansion within the MJD1 gene presents clinically as spinocerebellar ataxia and occurs most frequently in German SCA patients. *Hum. Mol. Genet.* **4**, 1001–1005.

Schols, L., Kruger, R., Amoiridis, G., Przuntek, H., Epplen, J. T. & Riess, O. 1998 Spinocerebellar ataxia type 6: genotype and phenotype in German kindreds. *J. Neurol. Neurosurg. Psychiatry* **64**, 67–73.

Schweitzer, J. K. & Livingston, D. M. 1998 Expansions of CAG repeat tracts are frequent in a yeast mutant defective in Okazaki fragment maturation. *Hum. Mol. Genet.* **7**, 69–74.

Silveira, I., Coutinho, P., Maciel, P., Gaspar, C., Hayes, S., Dias, A., Guimaraes, J., Loureiro, L., Sequeiros, J. & Rouleau, G. A. 1998 Analysis of SCA1, DRPLA, MJD, SCA2, and SCA6 CAG repeats in 48 Portuguese ataxia families. *Am. J. Med. Genet.* **81**, 134–138.

Sisodia, S. S. 1998 Nuclear inclusions in glutamine repeat disorders: are they pernicious, coincidental, or beneficial? *Cell* **95**, 1–4.

Sobue, G., Doyu, M., Nakao, N., Shimada, N., Mitsuma, T., Maruyama, H., Kawakami, S. & Nakamura, S. 1996 Homozygosity for Machado–Joseph disease gene enhances phenotypic severity. *J. Neurol. Neurosurg. Psychiatry* **60**, 354–356.

Stevanin, G., Durr, A., David, G., Didierjean, O., Cancel, G., Rivaud, S., Tourbah, A., Warter, J. M., Agid, Y. & Brice, A. 1997 Clinical and molecular features of spinocerebellar ataxia type 6. *Neurology* **49**, 1243–124

Stevanin, G., Giunti, P., Belal, G. D. S., Durr, A., Ruberg, M., Wood, N. & Brice, A. 1998 *De novo* expansion of intermediate alleles in spinocerebellar ataxia 7. *Hum. Mol. Genet.* **7**, 1809–18.

Takano, H. (and 10 others) 1996 Somatic mosaicism of expanded CAG repeats in brains of patients with dentatorubral–pallidoluysian atrophy: cellular population-dependent dynamics of mitotic instability. *Am. J. Hum. Genet.* **58**, 1212–1222.

Takano, H. (and 19 others) 1998 Close aassociations between prevalences of dominantly inherited spinocerebellar ataxias with CAG-repeat expansions and frequencies of large normal CAG alleles in Japanese and Caucasian populations. *Am. J. Hum. Genet.* **63**, 1060–1066.

Takiyama, Y., Sakoe, K., Amaike, M., Soutome, M., Ogawa, T., Nakano, I. & Nishizawa, M. 1999 Single sperm analysis of the CAG repeats in the gene for dentatorubral– pallidoluysian atrophy (DRPLA): the instability of the CAG repeats in the DRPLA gene is prominent among the CAG repeat diseases. *Hum. Mol. Genet.* **8**, 453–457.

Tsuchiya, K. (and 10 others) 1998 A clinical, genetic, neuropathological study in a Japanese family with SCA6 and a review of Japanese autopsy cases of autosomal dominant cortical cerebellar atrophy. *J. Neurol. Sci.* **160**, 54–59.

Watanabe, M. (and 11 others) 1996 Analysis of CAG trinucleotide expansion associated with Machado–Joseph disease. *J. Neurol Sci.* **136**, 101–107.

Wells, R. D., Parniewski, P., Pluciennik, A., Bacolla, A., Gellibolian, R., and Jaworski, A. (1998). Small slipped register genetic instabilities in *Escherichia coli* in triplet repeat sequences associated with hereditary neurological diseases. *J. Biol. Chem.* **273**, 19532–19541.

Zhuchenko, O., Bailey, J., Bonnen, P., Ashizawa, T., Stockton, D. W., Amos, C., Dobyns, W. B., Subramony, S. H., Zoghbi, H. Y. & Lee, C. C. 1997 Autosomal dominant cerebellar ataxia (SCA6) associated with small polyglutamine expansions in the alpha 1A-voltage-dependent calcium channel. *Nature Genet.* **15**, 62–69.

17

CAG-polyglutamine1-repeat mutations: independence from gene context

Jared M. Ordway, Jamie A. Cearley and Peter J. Detloff

Introduction

Inheritance of a long CAG–polyglutamine repeat underlies several human neurological disorders including Huntington's disease (HD), dentato-rubral–pallido-luysian atrophy (DRPLA), spinal and bulbar muscular atrophy (SBMA), and several spinocerebellar ataxias (SCA1, SCA2, SCA3 and SCA7) (Saudou *et al.* 1996; David *et al.* 1997; Reddy & Housman 1997). In each disease the mutation is located in a different gene and different sets of neurons are affected (Ross 1995). Nevertheless, these disorders share several phenotypic and molecular similarities. For each of these progressive late-onset neurological disorders, the CAG repeat codes for a polyglutamine stretch. The inheritance of a report more than 36 units in length is necessary for pathology; longer repeat lengths are correlated with an earlier age of onset (Ross 1995). Furthermore, neuronal intranuclear inclusions (NIIs) that contain the polyglutamine portion of the disease protein are found in patient material and transgenic models of the disorders (Davies *et al.* 1998). These similarities suggest that a common molecular mechanism underlies these disorders (Cha & Dure 1994). These mutations are probably acting through a toxic gain-of-function mechanism, because in at least two disorders (HD and SBMA), loss-of-function mutations cause phenotypes that differ from the disease phenotype (La Spada *et al.* 1991; Ambrose *et al.* 1994).

Several molecular mechanisms have been proposed to explain the pathologies of the CAG–polyglutamine1- repeat disorders. The proposed mechanisms can be placed into two distinct categories. In the first are molecular mechanisms in which the toxic effect of the CAG–polyglutamine repeat is absolutely dependent on the context of the gene in which it resides. This category would include, for example, mechanisms in which the binding of a heterologous protein at a site unique to the disease protein (rather than the polyglutamine stretch alone) is a necessary step towards toxicity (Strittmatter *et al.* 1997). The second category includes mechanisms in which the CAG–polyglutamine mutation is not absolutely dependent on the context of the gene in which it resides. Examples of this category of proposed mechanisms include those in which polyglutamine stretches bind heterologous proteins or cause self-aggregation, which in turn leads to neuronal toxicity (Perutz *et al.* 1994; Stott *et al.* 1995; Burke *et al.* 1996). Genetically this second category is unusual because, by the classical view, a mutation acts through the context of the locus in which it resides. As an illustration, take the example of the mutation that causes sickle-cell anaemia. This glutamine to valine substitution near the N-terminus of β-globin acts by a gain-of-function mechanism, a similarity shared with the CAG–polyglutamine 1-repeat mutations (Ingram 1959; Bunn & Forget 1986; Ross 1995). The sickle-cell mutation induces haemoglobin to form fibres that distort erythrocytes, which then occlude capillaries (Bunn & Forget 1986). Therefore, sickle-cell anaemia is also a good example of a mutation causing a physical intermediate that contributes directly to the disease process. Analogously, NIIs might also act as a toxic physical intermediate in the pathology of the CAG repeat disorders (Davies *et al.* 1998). The sickle-cell mutation, however, is totally dependent on its genetic context. Inserting this same mutation into a different gene would not cause sickle-cell anaemia. For example, a glutamine to valine substitution is found in variants of the α_1-antitrypsin protein in humans who are not affected by sickle-cell anaemia or sickle-cell trait (Long *et al.* 1984).

To assess the role of genetic context in the CAG–polyglutamine 1- repeat disorders, we expressed a CAG–polyglutamine repeat in a novel genetic context. Mice were made with a gene-targeted insertion of a 146-unit CAG repeat into the X-linked murine *Hprt* locus. These mice showed several common features of the CAG–polyglutamine 1- repeat disorders of humans, including late-onset progressive neurological symptoms and NIIs (Ordway *et al.* 1997). Thus, this work has shown that CAG–polyglutamine repeats need not reside in one of the classic repeat disorder genes to have a neurotoxic effect. Here we expand our report on the *Hprt*-CAG mice, including a description of the effects of mice with a 70-unit CAG repeat and mice heterozygous for the 146-unit CAG–polyglutamine repeat.

Rationale of targeting the *Hprt* locus

To control for chromosomal context, gene targeting in murine embryonic stem cells was used to insert CAG repeats into the *Hprt* locus (Ordway *et al.* 1997). This strategy permits the comparison of lines of mice having different alleles in the same chromosomal location. Four characteristics of the *Hprt* locus make it a good gene for studying the effects of long CAG repeats. First, the function of the gene product has been well characterized (Stout & Caskey 1985). Second, selection with 6-thioguanine permits the efficient insertion of different repeat variants. Third, this housekeeping gene is expressed ubiquitously throughout development (Stout & Caskey 1985). *In situ* hybridization studies of Hprt mRNA in mouse brain show staining in all neurons (Jinnah *et al.* 1992). Widespread expression is also found for several of the human CAG repeat disorder genes (Reddy & Housman 1997). Thus, this system permits the detection of a pathology in a specific set of neurons in the background of ubiquitous expression. Fourth, the loss of *Hprt* function in the mouse results in no spontaneous behavioural phenotype (Williamson *et al.* 1992). *Hprt*-deficient mice, however, have some mild neurochemical alterations (Finger *et al.* 1988; Dunnett *et al.* 1989; Jinnah *et al.* 1994). Therefore, *Hprt* deletion mice were used in addition to wild-type controls (Ordway *et al.* 1997). Here we describe the lack of abnormalities characteristic of the $Hprt^{(CAG)146}$ phenotype in mice made *Hprt*-deficient by the insertion of a shorter CAG–polyglutamine repeat.

Old $Hprt^{(CAG)70}$ mice lack abnormalities found in $Hprt^{(CAG)146}$ mice

For all of the CAG–polyglutamine 1- repeat disorders, the length of repeat determines whether an individual will be affected (Ross 1995). With the exception of SCA6, which is thought to act by a molecular mechanism that differs from the other CAG–polyglutamine repeat disorders, the threshold of length needed before an individual will become affected is approximately 38 CAG repeats (Ross 1995; Zhuchenko *et al.* 1997). The length of repeat also influences the age of onset of these disorders: longer repeats cause earlier ages of onset (Ross 1995). Onsets for humans, however, are usually well after 20 years of age (Gusella *et al.* 1996). We had previously shown that hemizygous male or homozygous female $Hprt^{(CAG)146}$ mice develop neurological abnormalities at 18 weeks of age (Ordway *et al.* 1997). This contrasts with hemizygous and homozygous $Hprt^{(CAG)70}$ mice of the same age, which showed no overt behavioural abnormalities. Since 70 repeats is in the pathological range for the human disorders, we followed 25 hemizygous and homozygous $Hprt^{(CAG)70}$ mice up to 65 weeks of age to determine whether the onset of these abnormalities occurred at an great age than in the $Hprt^{(CAG)146}$ mice. Mice with the $Hprt^{(CAG)146}$ mice have died (53 weeks). Therefore, we compared the characteristics of old $Hprt^{(CAG)70}$ mice with the abnormalities present in $Hprt^{(CAG)146}$ mice between their age of onset (18 weeks) and their death (45 weeks). The $Hprt^{(CAG)70}$ mice were subject to the same battery of neurological tests described previously for the $Hprt^{(CAG)146}$ mice (Ordway *et al.* 1997). We suspended the mice 1 cm from the base of the tail at a height of 35 cm for 1 min to record clasping of front or rear paws and whether the mouse could escape from the trial by climbing onto the observer's fingers. This result shows that $Hprt^{(CAG)70}$ mice up to 65 weeks of age do not exhibit the increased tendency to clasp that was previously shown in $Hprt^{(CAG)146}$ mice (figure 17.1*a*). Old $Hprt^{(CAG)70}$ mice usually escape from tail suspension, which is an ability that the $Hprt^{(CAG)146}$ mice lose by 18 weeks of age (figure 17.1*b*). Mice with the $Hprt^{(CAG)146}$ allele are

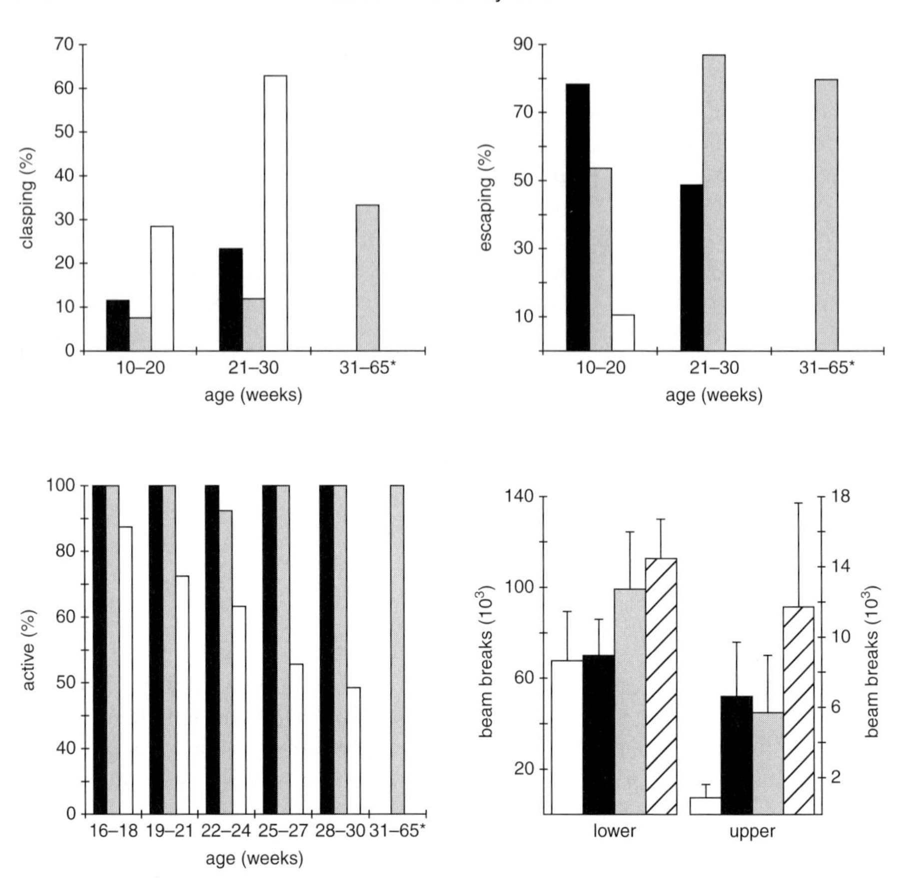

Figure 17.1 Comparison of behavioural characteristics between *Hprt*$^{(CAG)70}$ and *Hprt*$^{(CAG)146}$ mice. (*a*) Percentage of wild-type (filled bars, n = 16), *Hprt*$^{(CAG)70}$ (grey bars, n = 25) and *Hprt*$^{(CAG)146}$ (open bars, n = 30) mice clasping during 1 min tail-suspension trials. Each bar represents at least three trials for each of at least five mice. (*b*) Percentage of wild-type (filled bars, n = 16), *Hprt*$^{(CAG)70}$ (grey bars, n = 25) and *Hprt*$^{(CAG)146}$ (open bars, n = 33) mice escaping during 1 min tail-suspension trials. Each bar represents at least three trials for each of at least five mice. (*c*) Percentage of wild-type (filled bars, n = 16, 106 trials), *Hprt*$^{(CAG)70}$ (grey bars, n = 20, 200 trials) and *Hprt*$^{(CAG)146}$ (open bars, n = 31, 228 trials) mice active after removal of cage tops. (*d*) Number of lower and upper infrared beams broken by *Hprt*$^{(CAG)146}$ (open bars, n = 18), wild-type bars, n = 14), *Hprt*$^{(CAG)70}$ (grey bars, n = 9) and *Hprt* deletion mice, *Hprt*$^{b\text{-}m3}$ (hatched bars, n = 20) mice during experiments monitoring the activity cage. Results are the average number of beam breaks over a continuous five-day–five-night trial period. *The 31–65-week age group includes data for *Hprt*$^{(CAG)70}$ mice only.

susceptible to seizures during tail suspension; however, no seizures were observed in *Hprt*$^{(CAG)70}$ mice in 307 trials. We previously reported that *Hprt*$^{(CAG)146}$ mice show a progressive loss in exploratory activity as they aged (Ordway *et al.* 1997). This was monitored by recording whether mice roamed during the first 10 s after removal of the cage cover in a well-lit laminar-flow hood. As shown in figure 17.1*c*, *Hprt*$^{(CAG)70}$ mice retain this roaming exploration activity until 65 weeks of age. Undisturbed activity was measured in an infrared-beam activity monitor described in Ordway *et al.* (1997). The number of infrared beam breaks caused by old *Hprt*$^{(CAG)70}$ mice during a five-day and five-night trial did not differ significantly from wild-type or *Hprt* deletion controls (figure 17.1*d*). Furthermore, the severe deficit in vertical activity shown by a decrease in the number of upper-beam breaks by *Hprt*$^{(CAG)146}$ mice was not shown for old *Hprt*$^{(CAG)70}$ mice. Statistical analyses of the tail suspension and activity trials described above were performed as described in Ordway *et al.* (1997). These analyses showed no significant differences between *Hprt*$^{(CAG)70}$ mice and wild-type controls. In contrast, we found large and significant differences between *Hprt*$^{(CAG)70}$ and *Hprt*$^{(CAG)146}$ mice ($p<0.0155$). Nine hemizygous and homozygous *Hprt*$^{(CAG)70}$ mice between six and 70 weeks of age were stained for the immunohistochemical detection of NIIs. In contrast with *Hprt*$^{(CAG)146}$ mice, NIIs were not detected in *Hprt*$^{(CAG)70}$ mice (figure 17.2). Therefore, *Hprt*$^{(CAG)70}$ mice up to 65 weeks of age do not develop the behavioural or molecular abnormalities of the *Hprt*$^{(CAG)146}$ mice.

Repeats with a length of 70 units in the *Hprt* locus might cause toxicity much later than 65 weeks, or this length of repeat might be insufficient to cause a phenotype during the life span of the mouse. In contrast, 70 repeats in one of the classical CAG–polyglutamine 1- repeat disorder genes would generally cause an onset before middle age in humans (Gusella *et al.* 1996). There are several factors that might explain this difference. First, the expression levels of the *Hprt* locus might not be sufficient for toxicity. Second, the *Hprt* gene context might decrease the toxicity of the repeats. Ikeda *et al.* (1996) have shown that SCA3 gene sequences decrease (CAG–polyglutamine 1- repeat toxicity in transgenic mice. Third, CAG–polyglutamine 1-repeat toxicity might reflect a process that is dependent on absolute time rather than ageing of the organism. Therefore, given the onset of the 70 repeat alleles in the human, expression of the *Hprt*$^{(CAG)70}$ allele might require decades to cause toxicity in a transgenic animal. Fourth, the threshold length for CAG repeat toxicity in the mouse might differ from the threshold of 40 found in the human disorders. This human threshold is thought to be special because some evidence indicates a physical change in polyglutamine structure at 40 units in length. For example, a change was shown with antibody 1C2, which recognizes polyglutamine tracts longer than 40 units, but fails to recognize shorter polyglutamine tracts (Trottier *et al.* 1995). This antibody recognizes Hprt protein with 70 glutamine residues (results not shown) from the *Hprt*$^{(CAG)70}$ allele, suggesting that this non-pathogenic gene product retains the 1C2 epitope of the pathogenic human proteins. Our results suggest repeats greater than 70 units in length in the murine *Hprt* locus are needed to cause a CAG–polyglutamine 1-repeat toxicity in the mouse.

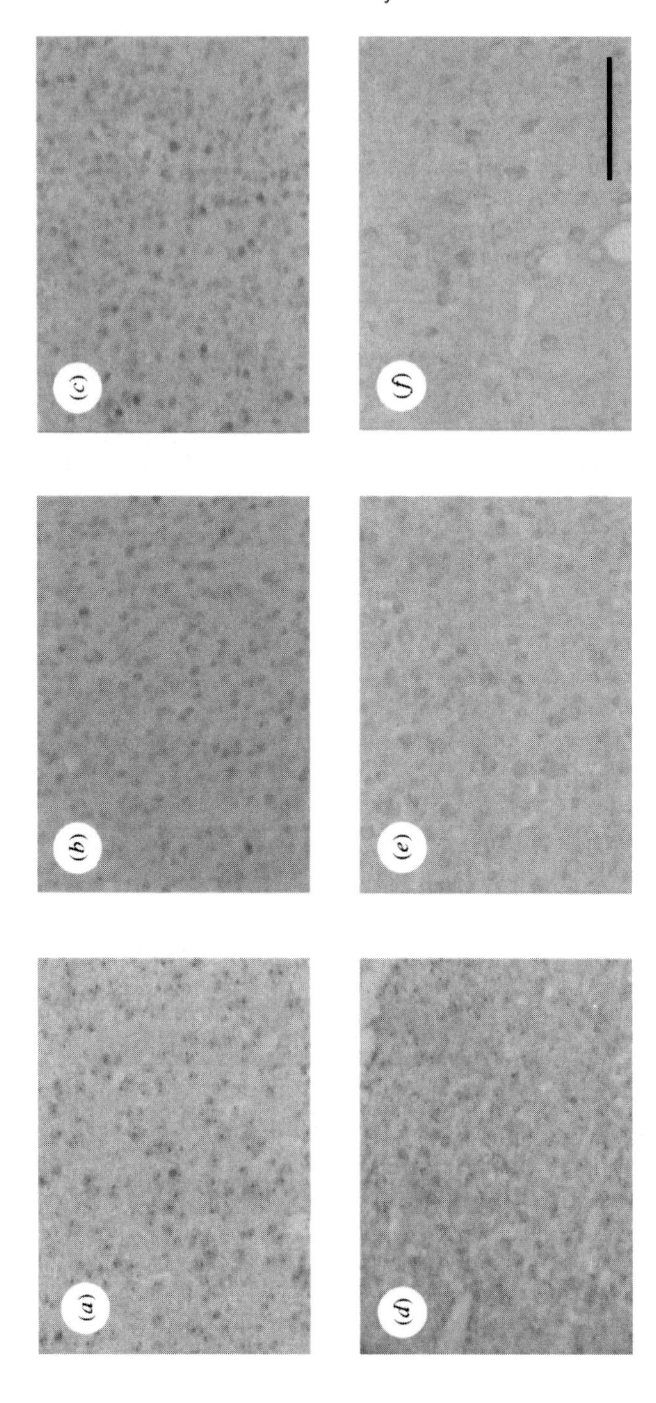
(a)
(b)
(c)
(d)
(e)
(f)

Heterozygosity delays the phenotypic effects of the X-linked *Hprt*$^{(cag)146}$ allele

The murine *Hprt* locus is subject to X-chromosome inactivation (Melton *et al.* 1984). It is therefore likely that every cell expressing this locus in a heterozygous mouse would express only one of the two alleles present. Heterozygous *Hprt*$^{(CAG)146}$ females might therefore possess a substantial number of neurons that express the wild-type *Hprt* allele. In this case heterozygous females might be expected to exhibit one of the following: variability in their expression of abnormalities, expression of some but not all of the abnormalities, variability in age of onset, or a delay in onset. Given the many redundancies thought to be present in neuronal circuitry, it is also possible that these mosaic expressors would have no abnormalities. To explore these possibilities we analysed heterozygous females at two different ages. Six heterozygous *Hprt*$^{(CAG)146}$ mice 18–23 weeks old were analysed for abnormalities described previously for homozygous and hemizygous *Hprt*$^{(CAG)146}$ mice (Ordway *et al.* 1997). These heterozygotes showed none of the abnormalities described for age-matched hemizygous and homozygous mice and were indistinguishable from age-matched wild-type controls. For example, these heterozygotes showed no seizures in 108 tail-suspension trials. This contrasts with 23 seizures in 223 tail-suspension trials for the 35 age-matched hemizygous and homozygous *Hprt*$^{(CAG)146}$ mice studied. Furthermore, heterozygotes were less likely to clasp their paws during tail suspension and were more likely to escape from the trial by climbing onto the observer's hand than age-matched hemizygous and homozygous *Hprt*$^{(CAG)146}$ mice. Mice that clasped or escaped in 20% or more of the trials (minimum five trials per mouse) were classified as claspers and escapers respectively. Zero of six 18–23-week-old heterozygotes were claspers and six of six were escapers. This contrasts with eight of 11 hemizygotes and homozygotes classified as claspers and zero of 13 as escapers. These values are significantly different (clasping, $p = 0.009$; escaping $p < 0.0001$ (Fisher exact)). The most profound effect of heterozygosity, however, was increased lifespan. All of the 15 hemizygous and homozygous *Hprt*$^{(CAG)146}$ mice in our previous study died by 53 weeks of age (Ordway *et al.* 1997). In contrast, none of the six heterozygous mice died before 53 weeks of age ($p < 0.0001$, Fisher exact). The median age of death for heterozygotes has yet to be established. These comparisons between mice that express only the mutant *Hprt*$^{(CAG)146}$ allele

Figure 17.2 Immunohistochemical analysis to detect the presence of neuronal intranuclear inclusions. Cerebral cortex of a 40-week-old *Hprt*$^{(CAG)146}$ mouse (*a,d*), a 30-week-old wild-type mouse (*b,e*) and a 70-week-old *Hprt*$^{(CAG)70}$ mouse (*c,f*) are shown. Brain sections were stained by immunoperoxidase labelling with polyclonal antiserum against HPRT (*a–c*) or ubiquitin (*d–f*) as described previously (Ordway *et al.* 1997). Scale bar, 100 μm.

and heterozygotes show a normalizing effect of heterozygosity. Heterozygosity of the X-linked CAG–polyglutamine alleles responsible for SBMA also has a normalizing effect in humans. Female carriers of SBMA alleles are normal, or exhibit mild abnormalities that are usually considered subclinical (Sobue *et al.* 1993; Guidetti *et al.* 1996).

To determine whether abnormalities could arise in heterozygotes, we studied older mice. Three of the six heterozygotes at 62–67 weeks of age showed no abnormalities. The remaining three showed many of the abnormalities described previously for affected hemizygous and homozygous *Hprt*$^{(CAG)146}$ mice (Ordway *et al.* 1997). The three affected mice showed a noticeable ataxia, which was recorded by painting ink on the footpads of mice and having the mice walk on paper (figure 17.3). The three affected heterozygous *Hprt*$^{(CAG)146}$ mice also showed an inability to improve performance during successive trials on the rotarod. Once a day for ten days, mice were placed on the rotarod (a cylindar 4.0cm in diameter rotating slowly (2.5 rev min^{-1}) while suspended 30 cm above a padded bench) for a 1 min trial. In contrast with unaffected age-matched heterozygotes (and old *Hprt* deletion controls), which did not fall in their last six trials, the affected heterozygotes fell from the rotarod during every trial. Two of the affected

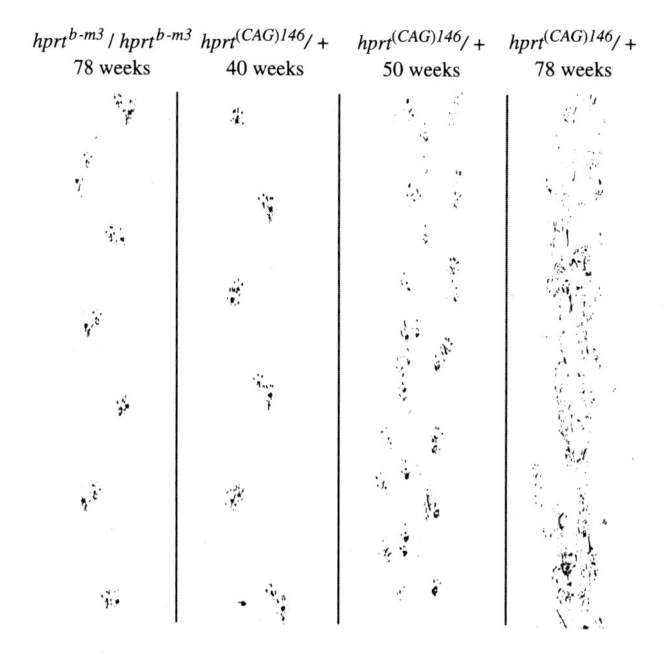

Figure 17.3 Late-onset gait abnormality of heterozygous *Hprt*$^{(CAG)146}$/+ mice. To analyse gait, footpads of mice were painted with ink and the mice were allowed to walk through a narrow pathway. Genotypes and ages of mice analysed are indicated above the lanes. The deletion allele *Hprt*$^{b\text{–}m3}$ results produces no functional *Hprt* gene product (Williamson *et al.* 1992).

heterozygotes exhibited a resting tremor and one of these mice had handling-induced seizures. In summary, the heterozygous mice between 62 and 67 weeks of age were either unaffected or profoundly affected. Affected heterozygotes, however, had a later onset of abnormalities than homozygotes. These results raise the possibility that female carriers of SBMA, also heterozygotes for an X-linked CAG–polyglutamine mutation, have a high probability of developing SBMA-like symptoms at a later date than hemizygotes with the same alleles.

The incomplete penetrance described here for old heterozygous *Hprt*$^{(CAG)146}$ mice is expected for a locus that is subject to random X-chromosome inactivation. We are developing immunocytochemical markers to study the extent and pattern of *Hprt*$^{(CAG)146}$ expression and X-chromosome inactivation in these heterozygotes. These studies will provide us with an estimate of the numbers of wild-type neurons needed to postpone or eliminate a CAG–polyglutamine phenotype. Neural transplants have been suggested as a treatment for HD (Shannon & Kordower 1996). Because inactivation of the disease allele in *Hprt*$^{(CAG)146}$ heterozygotes provides wild-type neurons that have developed in their proper location with proper functional connections, studies of these heterozygotes might represent the best possible outcome for transplants.

The degree of independence of CAG-polyglutamine 1-repeat mutations

Our results and those of others indicate that the expression of CAG–polyglutamine repeats causes neurological abnormalities (Ikeda *et al.* 1996; Mangiarini *et al.* 1996; Ordway *et al.* 1997). Furthermore, these results suggest a certain degree of independence from gene context. Many lines of evidence show that this independence is not absolute. For example, there is probably a threshold level of expression that a carrier gene needs if it is to cause CAG–polyglutamine toxicity. These thresholds are shown in experiments with transgenic mice in which lines with poor expression have no phenotype or a less severe phenotype than lines with robust CAG–polyglutamine 1-repeat expression (Burright *et al.* 1995; Mangiarini *et al.* 1996). Despite widespread expression of CAG–polyglutamine repeats in the human disorders, each disorder affects different types of neurons (Ross 1995). Therefore, gene context probably has an effect beyond the level of expression. This view is supported by experiments in which alterations of transgenes outside the CAG–polyglutamine 1- repeat region have been shown to alter the phenotype of mice. For example, a mutation that inactivates a nuclear localization signal eliminates the CAG–polyglutamine toxicity of an SCA1 transgene (Klement *et al.* 1998).

Extrapolation of our results would suggest that CAG–polyglutamine repeats in any carrier gene expressed properly would cause neurotoxicity. There is at least one example in humans of a ubiquitously expressed nuclear gene product with a long polyglutamine repeat that is not known to be associated with a disorder. Individuals with a 42-unit polyglutamine repeat in the TATA-binding protein (TBP) are apparently healthy (Rubinsztein *et al.* 1996). There are several possible reasons that these long repeats are not associated with a pathology. First, the length of repeat in TBP needed to cause disease might be longer than 42. Second, the polyglutamine in TBP might be otherwise engaged or bound to a TBP-specific protein that masks a toxic effect. Third, long-repeat TBP alleles might cause a very-late-onset disorder that has not yet been described.

For each disorder, repeat toxicity might be only one component of a complex pathology. Other gene-specific toxicities might contribute to each of the disorders. For example, proteins have been identified whose binding to a disease protein is influenced by the length of the polyglutamine tract (Strittmatter *et al.* 1997). Such alteration in binding might, in turn, create a toxicity specific to a particular disease. Our results show that CAG–polyglutamine repeats can cause toxicity in a foreign gene whose gene-specific interactions are not likely to mimic those of the repeat disorder genes. This work also suggests that CAG–polyglutamine 1- repeat mutations do not fit the classical view that a mutation acts through the context of the gene in which it resides. It is more likely that the gene context modulates a toxic property of the CAG–polyglutamine 1- repeat mutation. Understanding the factors that alter this toxicity should provide strategies designed to interrupt the pathogenic process.

We thank S. Tallaksen-Greene and Roger Albin for assistance with neuroanatomic analyses. This work was supported by grants from the Hereditary Disease Foundation's Cure HD Initiative, and National Institutes of Health grant NS34492.

References

Ambrose, C. M. (and 20 others) 1994 Structure and expression of the Huntington's disease gene: evidence against simple inactivation due to an expanded CAG repeat. *Somat. Cell Mol. Genet.* **20**, 27–38.

Bunn, H. F. & Forget, B. G. 1986 *Hemoglobin: molecular, genetic and clinical aspects.* Philadelphia, PA: W. B. Saunders Co.

Burke, J. R., Enghild, J. J., Martin, M. E., Jou, Y.-S., Myers, R. M., Roses, A. D., Vance, J. M. & Strittmatter, W. J. 1996 Huntingtin and DRPLA proteins selectively interact with the enzyme GAPDH. *Nature Med.* **2**, 347–350.

Burright, E. N., Clark, H. B., Servadio, A., Matilla, T., Feddersen, R. M., Yunis, W. S., Duvick, L. A., Zoghbi, H. Y. & Orr, H. T. 1995 SCA1 transgenic mice: a model for neuro-degeneration caused by an expanded CAG trinucleotide repeat. *Cell* **82**, 937–948.

Cha, J.-H. J. & Dure IV, L. S. 1994 Trinucleotide repeats in neurologic diseases: an hypothesis concerning the pathogenesis of Huntington's disease, Kennedy's disease, and spinocerebellar ataxia type I. *Life Sci.* **54**, 1459–1464.

David, G. (and 18 others) 1997 Cloning of the SCA7 gene reveals a highly unstable CAG repeat expansion. *Nature Genet.* **17**, 65–70.

Davies, S. W., Beardsall, K., Turmaine, M., Difiglia, M., Aronin, N. & Bates, G. P. 1998 Are neuronal intranuclear inclusions the common neuropathology of triplet-repeat disorders with poly-glutamine-repeat expansions? *Lancet* **351**, 131–133.

Dunnett, S. B., Sirinathsinghji, D. J., Heavens, R., Rogers, D. C. & Kuehn, M. R. 1989 Monoamine deficiency in a transgenic (Hprt-) mouse model of Lesch–Nyhan syndrome. *Brain Res.* **501**, 401–406.

Finger, S., Heavens, R. P., Sirinathsinghji, D. J., Kuehn, M. R. & Dunnett, S. B. 1988 Behavioural and neurochemical evaluation of a transgenic mouse model of Lesch–Nyhan syndrome. *J. Neurol. Sci.* **86**, 203–213.

Guidetti, D., Vescovini, E., Motti, L., Ghidoni, E., Gemignani, F., Marbini, A., Patrosso, M. C., Ferlini, A. & Solime, F. 1996 X-linked bulbar and spinal muscular atrophy, or Kennedy disease: clinical, neurophysiological, neuropathological, neuropsychological and molecular study of a large family. *J. Neurol. Sci.* **135**, 140–148.

Gusella, J. F., McNeil, S., Persichetti, F., Srinidhi, J., Nevelletto, A., Bird, E., Faber, P., Vansattel, J.-P., Myers, R. H. & MacDonald, M. E. 1996 Huntington's disease. *Cold Spring Harb. Symp. Quant. Biol.* **61**, 615–626.

Ikeda, H., Yamaguchi, M., Sugai, S., Aze, Y., Narumiya, S. & Kakizuka, A. 1996 Expanded polyglutamine in the Machado–Joseph disease protein induces cell death *in vitro* and *in vivo*. *Nature Genet.* **13**, 196–202.

Ingram, V. M. 1959 Abnormal human haemoglobins. III. The chemical difference between normal and sickle cell haemoglobins. *Biochim. Biophys. Acta* **36**, 402–411.

Jinnah, H. A., Hess, E. J., Wilson, M. C., Gage, G. H. & Friedmann, T. 1992 Localization of hypoxanthine–guanine phosphoribosyltransferase mRNA in the mouse brain by *in situ* hybridization. *Mol. Cell. Neurosci.* **3**, 64–78.

Jinnah, H. A., Wojcik, B. E., Hunt, M., Narang, N., Lee, K. Y., Goldstein, M., Wamsley, J. K., Langlais, P. J. & Friedmann, T. 1994 Dopamine deficiency in a genetic mouse model of Lesch–Nyhan disease. *J. Neurosci.* **14**, 1164–1175.

Klement, I. A., Skinner, P. J., Kaytor, M. D., Yi, H., Hersch, S. M., Clark, H. B., Zoghbi, H. Y. & Orr, H. T. 1998 Ataxin-1 nuclear localization and aggregation—role in polyglutamine-induced disease in SCA1 transgenic mice. *Cell* **95**, 41–53.

La Spada, A., Wilson, E. M., Lubahn, D. B., Harding, A. E. & Fischbeck, K. H. 1991 Androgen receptor gene mutations in X-linked spinal and bulbar muscular atrophy. *Nature* **352**, 77–79.

Long, G. L., Chandra, T., Woo, S. L., Davie, E. W. & Kurachi, K. 1984 Complete sequence of the cDNA for human alpha-1-antitrypsin and the gene for the S variant. *Biochemistry* **23**, 4828–4837.

Mangiarini, L. (and 10 others) 1996 Exon 1 of the HD gene with an expanded CAG repeat is sufficient to cause a progressive neurological phenotype in transgenic mice. *Cell* **87**, 493–506.

Melton, D. W., Konecki, D. S., Brennand, J. & Caskey, C. T. 1984 Structure, expression, and mutation of the hypoxanthine phosphoribosyltranferase gene. *Proc. Natl Acad. Sci. USA* **81**, 2147–2151.

Ordway, J. M. (and 11 others) 1997 Ectopically expressed CAG repeats cause intranuclear inclusions and a progressive late onset neurological phenotype in the mouse. *Cell* **91**, 753–763.

Perutz, M. F., Johnson, T., Suzuki, M. & Finch, J. T. 1994 Glutamine repeats as polar zippers: their possible role in inherited neurodegenerative diseases. *Proc. Natl Acad. Sci. USA* **91**, 5355–5358.

Reddy, R. S. & Housman, D. E. 1997 The complex pathology of trinucleotide repeats. *Curr. Opin. Cell Biol.* **9**, 364–372.

Ross, C. A. 1995 When more is less: pathogenesis of glutamine repeat neurodegenerative diseases. *Neuron* **15**, 493–496.

Rubinsztein, D. C. Leggo, J., Crow, T. J., DeLisi, L. E., Walsh, C., Jain, S. & Paykel, E. S. 1996 Analysis of polyglutamine-coding repeats in the TATA-binding protein in different human populations and in patients with schizophrenia and bipolar affective disorder. *Am. J. med. Genet.* **67**, 495–498.

Saudou, F., Devys, D., Trottier, Y., Imbert, G., Stoeckel, M.-E., Brice, A. & Mandel, J.-L. 1996 Polyglutamine expansions and neurodegenerative diseases. *Cold Spring Harb. Symp. Quant. Biol.* **61**, 639–647.

Shannon, K. M. & Kordower, J. H. 1996 Neural transplantation for Huntington's disease: experimental rationale and recommendations for clinical trials. *Cell Transplant.* **5**, 339–352.

Sobue, G., Doyu, M., Kachi, T., Yasuda, T., Mukai, E., Kumagai, T. & Mitsuma, T. 1993 Subclinical phenotypic expressions in heterozygous females of X-linked recessive bulbospinal neuronopathy. *J. Neurol. Sci.* **117**. 74–78.

Stott, K., Blackburn, J. M., Butler, P. J. & Perutz, M. 1995 Incorporation of glutamine repeats makes protein oligomerize: implications for neurodegenerative diseases. *Proc. Natl Acad. Sci. USA* **92**, 6509–6513.

Stout, J. T. & Caskey, C. T. 1985 HPRT: gene structure, expression, and mutation. *A Rev. Genet.* **19**, 127–148.

Strittmatter, W. J., Burke, J. R., DeSerrano, V. S., Huane, D. Y., Matthew, W., Saunders, A. M., Scott, B. L., Vance, J. M., Weigraber, K. H. & Roses, A. D. 1997 Protein:protein interactions in Alzheimer's disease and the CAG triplet repeat diseases. *Cold Spring Harb. Symp. Quant. Biol.* **61**, 597–605.

Trottier, Y. (and 12 others) 1995 Polyglutamine expansion as a pathological epitope in Huntington's disease and four dominant cerebellar ataxias. *Nature* **378**, 403–406.

Williamson, D. J., Hooper, M. L. & Melton, D. W. 1992 Mouse models of hypoxanthine phosphoribosyltransferase deficiency. *J. Inherit. Metab. Dis.* **15**, 665–673.

Zhuchenko, O., Bailey, J., Bonnen, P., Ashizawa, T., Stockton, D. W., Amos, C., Dobyns, W. B., Subramony, S. H., Zoghbi, H. Y. & Lee, C. C. 1997 Autosomal dominant cerebellar ataxia (SCA6) associated with small polyglutamine expansions in the alpha(la)-voltage-dependent calcium channel. *Nature Genet.* **15**, 62–69.

18

Molecular pathology of dentatorubral–pallidoluysian atrophy

Ichiro Kanazawa

Introduction

Dentatorubral–pallidoluysian atrophy (DRPLA) is an autosomal dominant neurodegenerative disorder characterized clinically by myoclonus, epileptic seizures, cerebellar ataxia, choreoathetotic movements, personality change and dementia. The cardinal clinical features depend on the age of onset. The dentate nucleus of the cerebellum is the most severely affected site, followed by the pallidum. Recently, the causative gene of DRPLA was assigned to the short arm of chromosome 12, and the underlying abnormality was identified as an expansion of a CAG repeat within its coding region. The function of the gene product, atrophin 1, has not yet been determined. Among the CAG repeat diseases, DRPLA exhibits the most prominent instability in the number of CAG repeats; they expand significantly in following generations. Furthermore, DRPLA shows a strong ethnic predilection for Asian, particularly Japanese, populations. A short history of DRPLA, its clinical and pathological features, classical and molecular genetics, and molecular pathology will be summarized here in order to contribute to a better understanding of DRPLA.

How DRPLA was found

The original presumed case of hereditary DRPLA was reported by Titica & Van Bogaert (1946), describing unique pathological features of the pallidoluysian atrophy combined with an atrophy of the dentatorubral system. This is an important contribution, which is revisited later. Naito *et al.* (1972) made a clinical report on two families who suffered from progressive myoclonus epilepsy with autosomal dominant transmission, pathological studies of which were performed later suggesting the combined degeneration of the pallidal and dentatal systems. Responding to these findings, Oyanagi *et al.* (1978) examined 15 autopsied brains based 'purely on pathological criteria', i.e. the presence of degenerative changes in efferent systems of both the pallidum and the dentate. Surprisingly, the clinical diagnosis of these cases fell into three groups: (i) progressive myoclonus epilepsy; (ii) Huntington's disease; and (iii) not carrying a specific diagnosis (combination of myoclonus, cerebellar ataxia and choreoathetosis). In addition to this, Naito & Oyanagi (1982) noticed a strong heritability of DRPLA with the pattern of an autosomal dominant trait. Although the first report of DRPLA appeared in a Western country, the establishment of this disease was achieved mostly in Japan.

Clinico-pathological features of DRPLA

In DRPLA patients of every stage overall, cerebellar ataxia and dementia appeared in 100%, choreoathetoid movements in 74%, epileptic seizures in 65% and myoclonus in 56% (Komure *et al.* 1995). The clear relationship between the clinical picture and the age of onset was noted. Indeed, the juvenile type of DRPLA (age of onset less than 20 years old) predominantly exhibits myoclonus and epileptic seizures and is often diagnosed as progressive myoclonus epilepsy, whereas the late adult type (age of onset more than 40 years) exhibits choreoathetosis, cerebellar ataxia and dementia with no myoclonus or epileptic seizures (Naito 1990). Almost all patients of the adult type show personality changes, i.e. mood changes swinging from euphoria to anger, childish behaviour and severe attention deficit. This personality change is extremely important for the clinical diagnosis.

Electroencephalograph abnormalities are frequently observed, especially in early onset DRPLA patients.

Abnormalities revealed by brain magnetic resonance imaging (MRI) are more clear than those found in computed tomography scans. A cerebellar 'atrophy' with VIth ventricular dilatation (figure 18.1*a*), and tegmental atrophy of the midbrain with aqueductal dilatation are the cardinal features of brain imaging. A T2-weighted MRI image of late adult-onset DRPLA patients frequently demonstrates diffuse high intensity in the cerebral white matter (figure 18.1*b*). In addition to this, high-signal lesions in the T2-weighted MRI image are found in the pons (figure 18.1*c*).

Macroscopic pathology revealed that the cerebellum and the brainstem are 'just small in size' or 'hypoplastic', rather than atrophic, because the contours of the cerebellum and pons are well preserved. Microscopically, combined degeneration of the dentatorubral (dentatofugal) and pallidoluysian (pallidofugal) systems is a characteristic feature of this disease. Neurons in the dentate nucleus are constantly, sometimes most severely, affected (figure 18.2) (Takahashi *et al.* 1995). The red nucleus neurons are least frequently affected, with mild gliosis, if any. On the other hand, neuronal loss in the pallidum is almost always present. It is noteworthy that the neuronal loss in the external segments of the globus pallidus is usually more pronounced than in the internal segments (Takahata *et al.* 1978). Gliosis is clearly demonstrated by the Holzer staining in the external segments (figure 18.3).

Besides the lesions mentioned above, there is another characteristic feature of DRPLA in the cerebral white matter in which myelin is diffusely lost, corresponding to the high-intensity area in T2-weighted MRI images and being regarded as the lesion responsible for the dementia associated with DRPLA (Yagishita & Inoue 1997).

Classical and molecular genetics of DRPLA

Autosomal dominant inheritance of DRPLA has been reported repeatedly (Naito *et al.* 1972; Takahata *et al.* 1978). Although seemingly isolated sporadic cases have sometimes been reported in the literature based on clinical and/or pathological findings, most of those patients were revealed later to have an abnormality in the gene. DRPLA families clearly show anticipation, where in later generations, symptoms began earlier than in preceding generations. Indeed, the age of onset is 10–30 years earlier in the following generation (Sano *et al.* 1994).

Using the CTG-B37 clone reported by Li *et al.* (1993), two Japanese groups (Koide *et al.* 1994; Nagafuchi *et al.* 1994) independently showed that CAG repeats in the gene were expanded exclusively in patients with DRPLA, and the DRPLA gene was assigned to 12q13.31. The consensus DRPLA cDNA sequence was revealed to encode 1185 amino acids.

Polymerase chain reaction using a primer set reported by Li *et al.* (1993) amplified products encompassing 7–34 CAG repeats in normal DNA samples and 53–88 CAG repeats in DRPLA patients (Ikeuchi *et al.* 1995; Komure *et al.* 1995). No overlap was found in the number of CAG repeats between normal and DRPLA chromosomes (figure 18.4). When the relationship between DRPLA gene CAG repeat length on the DRPLA chromosomes and the age of onset of disease in the patients is examined, there is a statistically significant negative correlation between the two; the longer the CAG repeat length, the earlier the age of onset (figure 18.5).

The acceleration of the age of onset based on a larger increase of the expanded CAG repeats is prominent when the disease is paternally transmitted compared with maternal transmission. Instability of the CAG repeat is considered to result from 'slippage', especially in paternal transmission, i.e. during spermatogenesis. Concerning the correlation between the numbers of CAG repeats and disease phenotype, clinical features are strongly influenced by the age of onset of DRPLA. Since the age of onset is directly related to CAG repeat length, clinical features should correlate with the number of CAG repeats. Indeed, the mean value of the CAG repeat lengths was much longer for the juvenile type (< 20 years old before onset) and shorter for the adult type (> 20 years old before onset). Moreover, the age of manifestation of each clinical symptom such as myoclonus, epilepsy, ataxia,

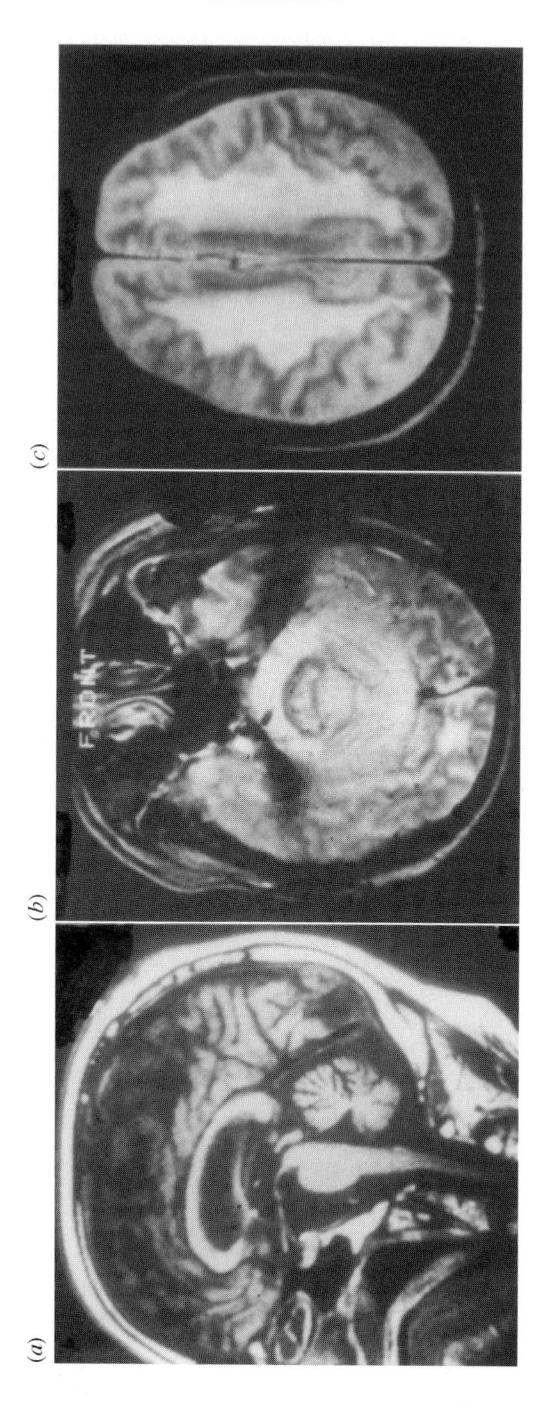
(a)
(b)
FRONT
(c)

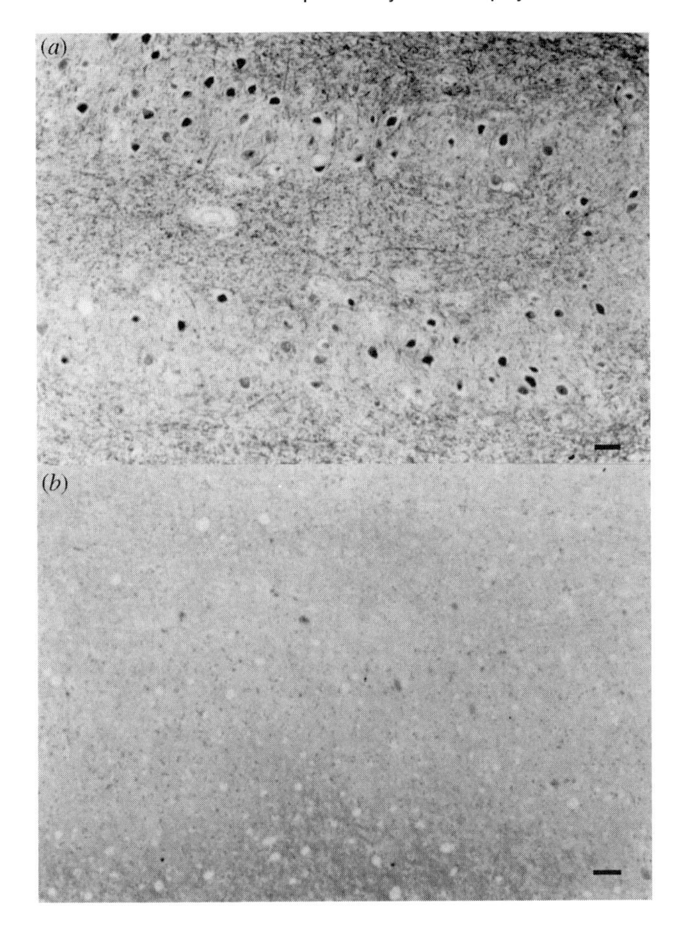

Figure 18.2 Neuropathological features of the dentate nucleus. (*a*) The normal control dentate nucleus (Kluver–Barrera stain, bar = 100 μm). (*b*) Severe neuronal loss of the dentate nucleus (Kluver–Barrera stain, bar = 100 μm).

choreoathetosis and dementia strongly correlates with the length of the CAG repeat (Ikeuchi *et al.* 1995).

The frequency in normal populations of alleles possessing more than 18 or 19 CAG repeats in the DRPLA gene was clearly higher in Asians

Figure 18.1 MRI (1.5 T) findings in an adult-onset DRPLA patient (54-year-old male). (*a*) The T1-weighted (TR/TE 400/15) midsagittal MRI image reveals proportional but 'small in size' brainstem and cerebellum with VIth ventricular dilatation. (*b*) The T2-weighted (TR/TE 3000/45) axial MRI image reveals diffuse high-intensity signals in the middle pontine region. (*c*) The T2-weighted (TR/TE 3000/45) axial MRI image reveals symmetrical, diffuse high-intensity signals in the periventricular white matter.

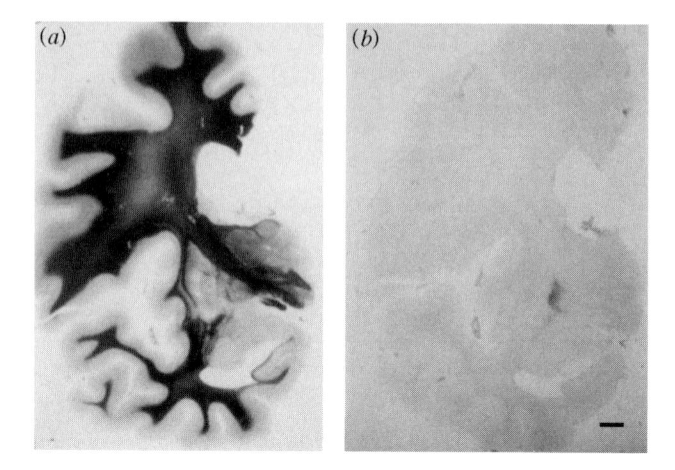

Figure 18.3 Neuropathological features of the globus pallidus (bar = 1 cm). (*a*) Myelin stain. Note a slight atrophic change of the globus pallidus especially of the external segment of the globus pallidus. (*b*) Holzer stain. Note a strong gliosis restricted to the external segment.

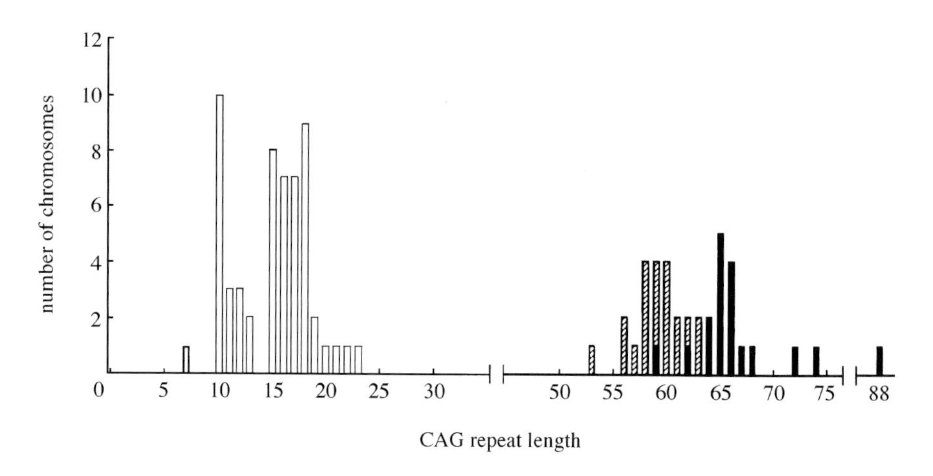

Figure 18.4 Distribution of CAG repeat length in normal chromosomes (white columns) and the larger of the two alleles (black or shaded columns) in DRPLA patients. Black columns indicate DRPLA chromosomes of juvenile-onset patients and shaded columns those of adult-onset patients (from Komure *et al.* 1995).

(Japanese, Korean and Chinese) than in Caucasians. Therefore, these populations representing longer CAG repeat alleles would provide the basis for a high frequency of DRPLA in Asian populations (Yanagisawa *et al.* 1996).

Haplotypes associated with DRPLA have been analysed extensively (Yanagisawa *et al.* 1996). In normal individuals from Asian populations, the frequency of the A1-B1 haplotype (base 1010 is adenine and base 1885 is

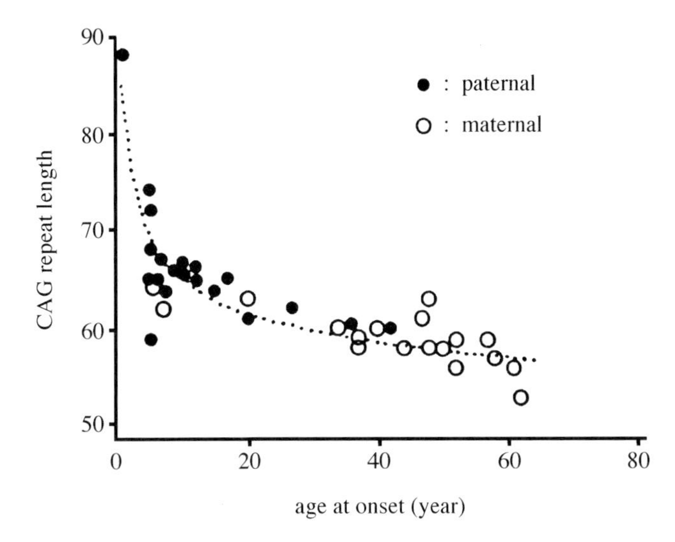

Figure 18.5 Relationship between CAG repeat length in DRPLA chromosomes and age of onset. Open circles represent alleles of maternal origin, and closed circles alleles of paternal origin. Note the longer the repeat, the earlier the onset (from Komure *et al.* 1995).

thymine) is the highest. On the other hand, in Caucasians the frequency of this allele is the lowest. All Japanese DRPLA patients were found to have a haplotype of A1-B1. This would be the reason why DRPLA is more frequent in Asians than Caucasians.

Molecular pathology of DRPLA

Somatic instability or mosaicism

Gonadal mosaicism in CAG repeat diseases was first investigated in HD. However, somatic cell mosaicism was not prominent in HD (Duyao *et al.* 1993). On the other hand, in the DRPLA cerebellum, the expanded allele is consistently smaller than in other tissues (Takano *et al.* 1996; Hashida *et al.* 1997). Interestingly, the degree of relative reduction of expansion of the CAG repeat in the cerebellum is positively correlated with the age of onset: individuals with an onset of disease at a younger age show smaller differences (figure 18.6). Since DNA in the cerebellar tissue comes largely from granule cells, it is reasonable to suppose that post-mitotic neuronal cells show a smaller degree of expansion of the CAG repeat compared with mitotic cells such as glia, endothelial cells or other non-neural tissues.

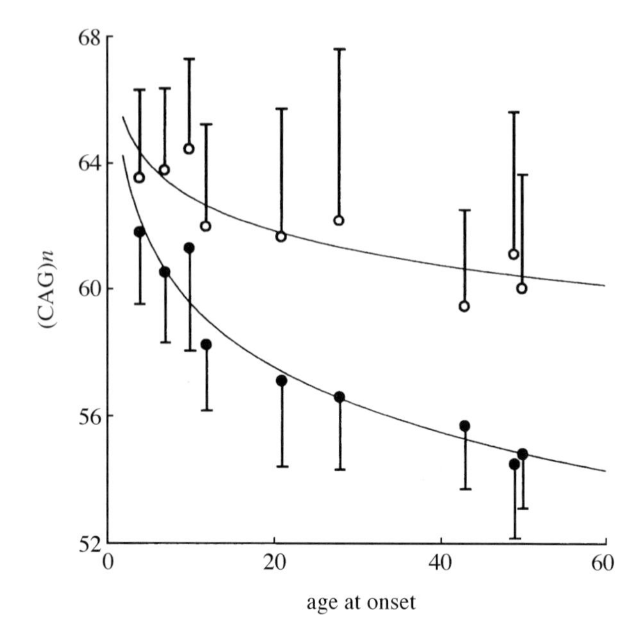

Figure 18.6 Relationship between the age of onset of DRPLA in patients and the mean size of the expanded allele in frontal cortex (open circles) and cerebellar cortex (closed circles). Bar = 1 s.d. Note the smaller expansion in the cerebellar cortex compared with those in the cerebral cortex (from Hashida *et al.* 1997).

Expression of mRNA and protein, atrophin 1

Northern blot analyses of mRNA isolated from various tissues revealed that the DRPLA gene product, atrophin 1, is ubiquitously expressed in all normal human tissues examined. However, preferential expression of atrophin 1 mRNA in the cerebellum was also demonstrated using *in situ* hybridization methods (Margolis *et al.* 1996; Nishiyama *et al.* 1997) and antisense riboprobes, especially in the granular cell layer. Moreover, the gene was found to be expressed predominantly in neurons. It is noted, however, that atrophin 1 expression in DRPLA brain was not different from that in normal brain (Nishiyama *et al.* 1997).

We raised rabbit polyclonal antisera against the C-terminal peptide of atrophin 1 (Yazawa *et al.* 1995). Employing Western blot analysis of human brain tissues using the antisera, the DRPLA gene product in normal brain was identified as a *ca.* 190 kDa protein. In DRPLA brain, an additional protein of *ca.* 205 kDa was found, corresponding to an expanded CAG repeat allele exclusively present in the DRPLA tissues (figure 18.7).

Using C-terminal antisera and newly raised rabbit polyclonal antisera against the N-terminal peptide of atrophin 1, we confirmed that the atrophin 1 is untruncated in tissues (Yazawa *et al.* 1997). In addition, we found that

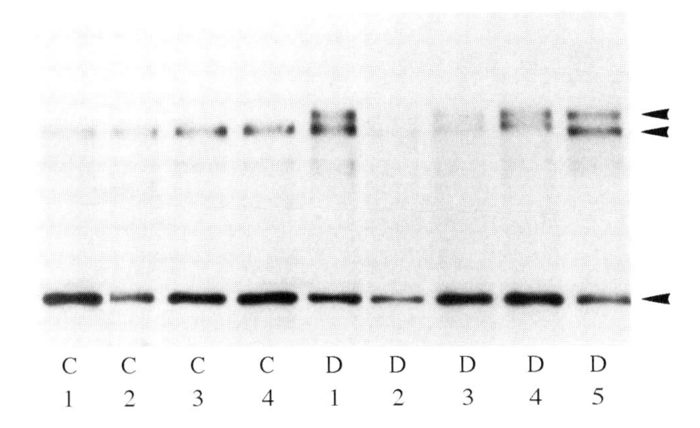

Figure 18.7 Western blot of DRPLA protein, atrophin 1, in brain tissues, using specific anti-atrophin 1 antiserum detecting the C-terminus peptides. C, control brains (1–4); D, DRPLA brains (1–5). Arrowheads correspond to molecular masses of 205, 190 and 100 kDa, respectively. The bands of 205 kDa indicate the protein containing an abnormally long polyglutamine stretch (from Yazawa *et al.* 1995).

under non-reducing conditions human atrophin 1 tends to aggregate and to form a large molecular complex (> 250 kDa) at the top of the stacking gel (Yazawa *et al.* 1998). Since DRPLA is transmitted in an autosomal dominant fashion, the underlying mechanism for neuronal cell death should be the result of a 'gain of function' due to the presence of protein(s) with abnormal structures and toxic properties.

Intranuclear inclusion bodies

Perutz *et al.* (1994) proposed a possible mechanism for neuronal cell death caused by polyglutamine stretches, demonstrating that these molecules can form polar zippers based on their β-sheet structure, which could form the possible intracellular precipitation of such aggregated protein and could cause neurons to die. In fact, Bates and her colleagues (Mangiarini *et al.* 1996; Davies *et al.* 1997) succeeded in revealing intranuclear inclusions in neurons of transgenic mice expressing exon 1 of the HD gene and carrying extremely expanded CAG repeats. Although obvious neuronal cell death was absent, the inclusions are thought to bring about cellular dysfunction.

DRPLA patients also exhibit neuronal inclusions (Igarashi *et al.* 1998). Intranuclear inclusions in neurons of affected patients are located near the nucleolus with their size ranging from 1 to 3 μm. These inclusions can be immunostained with antibodies raised against the expanded polyglutamine stretch or against ubiquitin. It is worth noting that transgenic mice exhibit neurological phenotypes. It is possible, therefore, to suppose that the intranuclear inclusions are somehow related to the pathogenesis of expanding CAG repeat diseases. Neuronal cell loss in DRPLA is prominent in the dentate nucleus, globus pallidus and other regions. Therefore, one should attempt to elucidate the mechanism underlying this selective neuronal cell death.

Recently, using a DRPLA gene expressed in COS cells, Igarashi *et al.* (1998) provided some evidence on the *in vitro* formation of intranuclear inclusions and the induction of apoptosis. They found that only truncated DRPLA protein containing an expanded polyglutamine tract forms perinuclear and/or intranuclear aggregates when expressed in cultured COS-7 cells, which subsequently undergo apoptosis.

Although the above-mentioned 'intranuclear inclusion body hypothesis' is most plausible for explaining neuronal cell death at the moment, a problem yet to be solved is that we cannot fully explain the mechanism(s) for the 'selectivity' of neuronal cell death, since atrophin 1 and other abnormally long polyglutamine stretch-containing proteins are expressed widely throughout the nervous system, i.e. there is no evidence for their predominant presence in specific neuronal populations.

Conclusion

Clinical diagnoses of DRPLA patients are sometimes difficult due to the diversity of the clinical features. However, it is not impossible to reach the correct diagnosis, if (i) the clear relationship between the age of onset and the clinical types, (ii) autosomal dominant inheritance with clear anticipation and (iii) MRI imaging showing small-sized cerebellum and brainstem, and a diffuse high-intensity lesion in the T2-weighted condition in adult patients, are taken into consideration. After the identification of the gene responsible for DRPLA, the above concept is strongly supported by the clear correlation between the length of the CAG repeat and the age of onset. As one of the members of the CAG repeat diseases, neurons in the nervous system of DRPLA patients contain inclusion bodies, which provide the most plausible marker of neuronal cell death at present.

I wish to thank my colleagues, Drs N. Nukina, J. Goto, S. Hashida, I. Yazawa, S. Murayama, K. Nakamura and N. Hazeki, Department of Neurology, University of Tokyo, for their participation in DRPLA research in my laboratory, and my collaborators, Dr M. Yamada, National Institute for Paediatrics, Professor I. Kondo, Ehime University and Professor S. Tsuji, Niigata University, for their constant encouragement.

References

Davies, S. W., Turmaine, M., Cozens, B. A., DiFiglia, M., Sharp, A. H., Ross, C. A., Scherzinger, E., Wanker, E. E., Mangiarini, L. & Bates, G. P. 1997 Formation of

neuronal intranuclear inclusions underlies the neurological dysfunction in mice transgenic for the HD mutation. *Cell* **90**, 537–548.
Duyao, M. (and 41 others) 1993 Trinucleotide repeat length instability and age of onset in Huntington's disease. *Nature Genet.* **4**, 387–392.
Hashida, S., Goto, J., Kurisaki, H., Mizusawa, H. & Kanazawa, I. 1997 Brain regional differences in the expansion of a CAG repeat in the spinocerebellar ataxias: dentatorubral–pallidoluysian atrophy, Machado–Joseph disease, and spinocerebellar ataxia type 1. *Ann. Neurol.* **41**, 505–511.
Igarashi, S. (and 18 others) 1998 Suppression of aggregate formation and apoptosis by transglutaminase inhibitors in cells expressing truncated DRPLA protein with an expanded polyglutamine stretch. *Nature Genet.* **18**, 111–117.
Ikeuchi, T. (and 20 others) 1995 Dentatorubral–pallidoluysian atrophy: clinical features are closely related to unstable expansions of trinucleotide (CAG) repeat. *Ann. Neurol.* **37**, 769–775.
Koide, R. (and 14 others) 1994 Unstable expansion of CAG repeat in hereditary dentatorubral–pallidoluysian atrophy (DRPLA). *Nature Genet.* **6**, 9–13.
Komure, O. (and 12 others) 1995 DNA analysis in hereditary dentatorubral–pallidoluysian atrophy: correlation between CAG repeat length and phenotypic variation and the molecular basis of anticipation. *Neurology* **45**, 143–149.
Li, S. H., McInnis, M. G., Margolis, R. L., Antonarakis, S. E. & Ross, C. A. 1993 Novel triplet repeat containing genes in human brain: cloning, expression, and length polymorphisms. *Genomics* **16**, 572–579.
Mangiarini, L. (and 10 others) 1996 Exon 1 of the HD gene with an expanded CAG repeat is sufficient to cause a progressive neurological phenotype in transgenic mice. *Cell* **87**, 493–506.
Margolis, R. L., Li, S. H., Young, W. S., Wagster, M. V., Stine, O. C., Kidwai, A. S., Ashworth, R. G. & Ross, C. A. 1996 DRPLA gene (atrophin 1) sequence and mRNA expression in human brain. *J. Neurol. Sci.* **36**, 219–226.
Nagafuchi, S. (and 22 others) 1994 Dentatorubral and pallidoluysian atrophy expansion of an unstable CAG trinucleotide on chromosome 12p. *Nature Genet.* **6**, 14–18.
Naito, H. 1990 The clinical picture and classification of dentatorubral–pallidoluysian atrophy (DRPLA). *Shinkei-Naika* **32**, 450–456. (In Japanese.)
Naito, H. & Oyanagi, S. 1982 Familial myoclonus epilepsy and choreoathetosis: hereditary dentatorubral–pallidoluysian atrophy. *Neurology* **32**, 798–807.
Naito, H., Izawa, K., Kurosaki, T., Kaji, S. & Sawa, M. 1972 Progressive myoclonus epilepsy with Mendelian dominant heredity. *Psychiat. Neurol. Japan.* **74**, 871–897. (In Japanese.)
Nishiyama, K., Nakamura, K., Murayama, S., Yamada, M. & Kanazawa, I. 1997 Regional and cellular expression of the dentatorubral–pallidoluysian atrophy gene in brains of normal and affected individuals. *Ann. Neurol.* **41**, 599–605.
Oyanagi, S. 1978 On pathology of the chorea—with reference to 'hereditary pallidal and dentate system atrophy proposed by us. *Saishin-Igaku* **33**, 236–242. (In Japanese.)
Perutz, M. F., Johnson, T., Suzuki, M. & Finch, J. T. 1994 Glutamine repeats as polar zippers: their possible role in inherited neurodegenerative diseases. *Proc. Natl Acad. Sci. USA* **91**, 5355–5358.
Sano, A., Yamauchi, N., Kakimoto, Y., Komure, O., Kawai, J., Hazama, F., Kuzume, K., Sano, N. & Kondo, I. 1994 Anticipation in hereditary dentatorubral–pallidoluysian atrophy. *Hum. Genet.* **93**, 699–702.
Takahashi, H., Yamada, M. & Takeda, S. 1995 Neuropathology of dentatorubral–pallidoluysian atrophy and Machado–Joseph disease. *No-to-Shinkei* **47**, 947–953. (In Japanese.)
Takahata, N., Ito, K., Yoshimura, Y., Nishihori, K. & Suzuki, H. 1978 Familial chorea and myoclonus epilepsy. *Neurology* **28**, 913–919.

Takano, H. (and 10 others) 1996 Somatic mosaicism of expanded CAG repeats in brains of patients with dentatorubral–pallidoluysian atrophy: cellular population-dependent dynamics of mitotic instability. *Am. J. Hum. Genet.* **58**, 1212–1222.

Titica, J. & Van Bogaert, L. 1946 Heredo-degenerative hemiballismus. *Brain* **69**, 251–263.

Yagishita, S. & Inoue, M. 1997 Clinicopathology of spinocerebellar degeneration: its correlation to the unstable CAG repeat of the affected gene. *Pathol. Int.* **47**, 1–15.

Yanagisawa, H. (and 22 others) 1996 A unique origin and multistep process for the generation of expanded DRPLA triplet repeats. *Hum. Mol. Genet.* **5**, 373–379.

Yazawa, I., Nukina, N., Hashida, S., Goto, J., Yamada, M. & Kanazawa, I. 1995 Abnormal gene product identified in hereditary dentatorubral–pallidoluysian atrophy (DRPLA) brain. *Nature Genet.* **10**, 99–103.

Yazawa, I., Haseki, N. & Kanazawa, I. 1998 Expanded glutamine repeat enhances complex formation of dentatorubral–pallidoluysian atrophy (DRPLA) protein in human brains. *Biochem. Biophys. Res. Commun.* (In the press.)

19

Androgen receptor mutation in Kennedy's disease

Kenneth H. Fischbeck, Andrew Lieberman, Christine K. Bailey, Annette Abel and Diane E. Merry

Introduction

X-linked spinal and bulbar muscular atrophy (SBMA) was the first repeat expansion disease gene to be discovered. This review covers what we have come to learn of the disease mechanism of SBMA, and how it relates to the other polyglutamine expansion neurodegenerative diseases.

Kennedy's disease

Although there were earlier reports, particularly in the Japanese literature, SBMA often goes by the name 'Kennedy's disease' after a description of the X-linked pattern of inheritance published by William Kennedy and his colleagues 30 years ago (Kennedy *et al.* 1968). SBMA is a chronic, progressive neuromuscular disorder, characterized by proximal muscle weakness, atrophy and fasciculations. Affected males may show signs of androgen insensitivity, including gynaecomastia, reduced fertility and testicular atrophy. The cause of the disease is expansion of a trinucleotide repeat in the androgen receptor gene on the X chromosome at Xq11–12 (La Spada *et al.* 1991).

The principal pathological manifestation of SBMA is the loss of motor neurons in the spinal cord and brainstem (Sobue *et al.* 1989). There is also a subclinical loss of sensory neurons in the dorsal root ganglia.

The causative defect is an expanded CAG repeat in the first exon of the androgen receptor gene, near the 5' end of the gene. It encodes a run of glutamine residues near the amino-terminus of the protein, separate from the DNA- and hormone-binding domains and close to the transcriptional activation domain. In normal individuals, the repeat averages about 20 CAGs, with a range of 11–33 CAGs. In patients with SBMA, the repeat is two to three times its normal length, about 38–62 CAGs. As with the other repeat expansion diseases, the longer the repeat the earlier the onset of the disease (La Spada *et al.* 1992).

Androgen receptor function

The androgen receptor is a nuclear receptor, a member of the steroid and thyroid hormone receptor family. The members of this family are intracellular receptors with well defined interactions. The androgen receptor is produced in the cytoplasm, where it is phosphorylated, binds stoichiometrically to heat shock proteins and is transported to the nucleus, where it is actively taken up through nuclear pores. In the presence of ligand (testosterone or dihydrotestosterone), it dissociates from the heat shock proteins and is free to dimerize and bind DNA at specific regulatory elements. It then functions as a ligand-dependent transcription factor; that is, it functions by up- or down-regulating target genes, through interactions with other proteins in the transcriptional activation complex.

We sought to determine whether the normal function and interactions of the androgen receptor protein are altered by the polyglutamine expansion (Brooks *et al.* 1997). We found that the receptor binds ligand normally. The full-length protein also has normal intracellular localization in cultured motor neuron–neuroblastoma hybrid cells. In the absence of ligand, the localization is primarily cytoplasmic, and in the presence of ligand primarily nuclear. This is a good indication that most of the normal interactions of the androgen receptor protein are unaffected by the polyglutamine expansion. We did find, as have others (Mhatre *et al.* 1993), that target gene transactivation by the mutant receptor is reduced, probably because of decreased levels of the receptor protein (Choong *et al.* 1996). This may account for the signs of androgen insensitivity often seen in patients with SBMA, and may also contribute to the motor neuron degeneration that occurs in the disease.

Androgens have long been known to affect muscle strength, and at least some of this effect may take place at the level of the motor neuron rather than the muscle. Androgens promote the survival and dendritic arborization of sexually dimorphic motor neurons, and they have effects on non-dimorphic motor neurons, as well. An early study of ligand binding in the central nervous system showed the highest levels of binding to spinal and bulbar motor neurons (Sar & Stumpf 1977), the same cells that are prone to degeneration in SBMA. Also, androgens lead to increased survival of brainstem motor neurons after cranial nerve section (Yu 1989). Thus it is possible that the loss of androgen receptor function contributes to the motor neuron degeneration and progressive weakness of SBMA. This is consistent with anecdotal clinical reports and the results of one controlled study (Mendell *et al.* 1996), which showed increased strength in SBMA patients treated with exogenous androgen. It is not clear, however, that this effect is specific to the disease. It may be that anyone treated with androgens will get stronger.

We are currently attempting to identify the targets of androgen receptor action that are responsible for the trophic effects of androgens in motor neurons. For this purpose, we are using a cell culture system that recapitulates some of the effects of androgens *in vivo* (Brooks *et al.* 1998). In this system, androgen treatment leads to altered morphology and increased cell survival. Preliminary results indicate that among the genes induced by androgen treatment in this system is the mammalian homologue of *tra-2*, a factor responsible for sexual differentiation in *Drosophila*. Thus the mechanism for development of sexual dimorphism, including sexual differentiation of motor neurons, could be broadly conserved across species.

Toxic gain of function

While loss of androgen receptor function may be a factor in the motor neuron dysfunction and degeneration of SBMA, this is not likely to be the primary effect of polyglutamine expansion. Other patients are known to have mutations that cause a loss of androgen receptor function, and these patients have a different phenotype (androgen insensitivity or testicular feminization syndrome), with feminization but no weakness due to motor neuron degeneration. Since loss of androgen receptor function has a different effect, we have long suspected that polyglutamine expansion instead produces a gain of function; that is, it alters the structure of the receptor protein so that it becomes toxic to motor neurons.

Further evidence that androgen receptor polyglutamine expansion causes a toxic gain of function in the gene product comes from the finding of similar mutations in other disorders. The other polyglutamine expansion diseases are all dominantly inherited neurodegenerative diseases with similar age of onset and rate of progression. They are all caused by the same kind of mutation in widely expressed genes, and in each case there is a correlation between repeat length and age of onset. That these disorders involve a toxic gain of function mechanism is indicated by the finding that for several of them loss of gene function leads to a different phenotype, yet transgenic expression of the mutant protein in mice recapitulates features of the human disease.

Transgenic models

Over the past several years, we have made various lines of transgenic mice with mutant versions of the androgen receptor gene (see table 19.1). Mouse lines with expression of full-length androgen receptor driven by the interferon-inducible Mx, myosin light chain, neurofilament light chain, neuron-specific enolase and human androgen receptor promoters all failed to show an abnormal phenotype, despite repeat expansions of up to 66 glutamines (longer than the longest repeat observed in SBMA patients) and up to twice endogenous expression levels in the spinal cord (Bingham *et al.* 1995; Merry *et al.* 1996; La Spada *et al.* 1998). Recently, we produced mice with a truncated androgen receptor containing 112 glutamines driven by the prion protein promoter, and these animals have a striking neurological phenotype,

Table 19.1 Androgen receptor transgenic mice

promoter:	*AR construct*	*repeat length*	*phenotype*
Mx	full-length	45 gln	normal
NSE			
MLC			
NSE	full-length	66 gln	normal
NFL			
PRP	truncated	112 gln	gait difficulty, tremor, circling, foot clasping

(Abbreviations: AR, androgen receptor; gln, glutamine; Mx, interferon-inducible Mx promoter; NSE, neuron-specific enolase; MLC, myosin light chain; NFL, neurofilament light chain; PRP, prion protein promoter.)

with progressive gait difficulty, circling behaviour, tremor and seizures (Abel *et al.* 1998). This finding is consistent with results reported in huntingtin and ataxin-3 transgenics, where truncated protein has a particularly pronounced effect (Ikeda *et al.* 1996; Mangiarini *et al.* 1996). It is not clear at this point whether the severe phenotype we find in transgenic mice with the truncated expanded protein is due to the truncation, the expression level or the extent of repeat expansion in the construct used to make these, as compared to the earlier lines of androgen receptor transgenic mice. Further transgenic experiments with full-length protein containing 112 repeats and driven by the same promoter should help to resolve this issue. It may be noted that the truncated version of the androgen receptor expressed in the transgenic mice with the neurological phenotype is similar (although not identical) to the fragment produced by caspase cleavage (Kobayashi *et al.* 1998; Ellerby *et al.* 1999), consistent with the hypothesis that an androgen receptor cleavage product is more toxic than the full-length protein *in vivo*.

Recently, polyglutamine neurotoxicity has been reproduced in *Drosophila* (Warrick *et al.* 1998). As with other polyglutamine transgenes, truncated, expanded androgen receptor produces a neurodegenerative phenotype in flies (N. Bonini, Chapter 6), indicating a similar mechanism of action. Available transgenic mouse and fly models of SBMA and the other polyglutamine expansion diseases should allow us to answer the outstanding questions in this field: What accounts for the similarities among these diseases? What accounts for the differences? And most importantly, what can be done to treat them?

Inclusions and aggregates

In 1997, a common pathological feature was discovered to go with the presumed common mechanism of the polyglutamine expansion diseases:

nuclear inclusions of polyglutamine-containing protein. These inclusions, which are ubiquitinated and present in neurons that become dysfunctional and die, have been found in most of the polyglutamine diseases and animal models where they have been sought, including the transgenic flies (Davies *et al.* 1997; DiFiglia *et al.* 1997; Paulson *et al.* 1997; Warrick *et al.* 1998). Nuclear inclusions are present in motor neurons in SBMA (Li *et al.* 1998*a*), as well as in the transgenic mice we have produced (Abel *et al.* 1998). Interestingly, nuclear inclusions of ubiquitinated androgen receptor protein are also present in non-neural tissues in SBMA patients (Li *et al.* 1998*b*).

We have been able to reproduce nuclear inclusions with truncated, expanded androgen receptor protein in transiently transfected cells in culture (Merry *et al.* 1998). In Cos cells, the inclusions are primarily cytoplasmic, while in motor neuron-like MN-1 cells the inclusions are nuclear, perhaps indicating a cell-specific mechanism for inclusion formation. The Western blot correlate of the nuclear inclusions is aggregated protein that barely enters the gel. With increasing repeat length, more of the protein becomes aggregated. At the same time, a specific cleavage product appears, indicating that aberrant androgen receptor proteolysis may lead to, or result from, the protein aggregation. The protein cleavage and aggregation are also associated with repeat-length-dependent cellular toxicity. Further experiments to characterize this process in stably transfected, inducible cell lines are in progress.

If protein aggregation and nuclear inclusions are important to the pathogenesis of SBMA and the other polyglutamine expansion diseases, then this suggests several approaches to treatment. One could attempt to block expression of the mutant protein, or inhibit its processing and nuclear uptake, or block the aggregation and downstream effects, including apoptotic cell death. In any event, cell culture and animal models that are now becoming available can serve as useful systems for further elucidation of the disease mechanism and pharmacological screening. In the case of SBMA, it remains to be determined whether ligand effects, which can alter the subcellular distribution and processing of the receptor protein, could provide a handle on effective treatment of the disease.

This work was supported in part by grants from the Muscular Dystrophy Association and the National Institutes of Health (intramural funds and grant number NS32214).

References

Abel, A., Taye, A. A., Fischbeck, K. H. & Merry, D. E. 1998 Truncated androgen receptor gene with an expanded CAG repeat causes phenotypic changes in transgenic mice (abstract). *Am. J. Hum. Genet.* **63**, A319.

Bingham, P. M., Scott, M. O., Wang, S., McPhaul, M. J., Wilson, E. M., Garbern, J. Y., Merry, D. E. & Fischbeck, K. H. 1995 Stability of an expanded trinucleotide repeat in the androgen receptor gene in transgenic mice. *Nature Genet.* **9**, 191–196.

Brooks, B. P., Paulson, H. L., Merry, D. E., Salazar-Grueso, E. F., Brinkmann, A. O., Wilson, E. M. & Fischbeck, K. H. 1997 Characterization of an expanded glutamine repeat androgen receptor in a neuronal cell culture system. *Neurobiol. Dis.* **4**, 313–323.

Brooks, B. P., Merry, D. E., Paulson, H. L., Lieberman, A., Kolson, D. & Fischbeck, K. H. 1998 A cell culture model for androgen effects in motor neurons. *J. Neurochem.* **70**, 1054–1060.

Choong, C. S., Kemppainen, J. A., Zhou, Z. & Wilson, E. M. 1996 Reduced androgen receptor gene expression with first exon CAG repeat expansion. *Mol. Endocrinol.* **10**, 1527–1535.

Davies, S. W., Turmaine, M., Cozens, B. A., DiFiglia, M., Sharp, A. H., Ross, C. A., Scherzinger, E., Wanker, E. E., Mangiarini, L. & Bates, G. P. 1997 Formation of neuronal intranuclear inclusions underlies the neurological dysfunction in mice transgenic for the HD mutation. *Cell* **90**, 537–548.

DiFiglia, M., Sapp, E., Chase, K. O., Davies, S. W., Bates, G. P., Vonsattel, J. P. & Aronin, N. 1997 Aggregation of huntingtin in neuronal intranuclear inclusions and dystrophic neurites in brain. *Science* **277**, 1990–1993.

Ellerby, L. M. (and 11 others) 1999 Kennedy's disease: caspase cleavage of the androgen receptor is a crucial event in cytotoxicity. *J. Neurochem.* **72**, 185–195.

Ikeda, H., Yamaguchi, M., Sugai, S., Aze, Y., Narumiya, S. & Kakizuka, A. 1996 Expanded polyglutamine in the Machado–Joseph disease protein induces cell death *in vitro* and *in vivo*. *Nature Genet.* **13**, 196–202.

Kennedy, W. R., Alter, M. & Sung, J. H. 1968 Progressive proximal spinal and bulbar muscular atrophy of late onset: a sex-linked recessive trait. *Neurology* **18**, 671–680.

Kobayashi, Y., Miwa, S., Merry, D. E., Kume, A., Mei, L., Doyu, M. & Sobue, G. 1998 Caspase-3 cleaves the expanded androgen receptor protein of spinal and bulbar muscular atrophy in a polyglutamine length-dependent manner. *Biochem. Biophys. Res. Commun.***252**, 145–150.

La Spada, A., Wilson, E. M., Lubahn, D. B., Harding, A. E. & Fischbeck, K. H. 1991 Androgen receptor gene mutations in X-linked spinal and bulbar muscular atrophy. *Nature* **352**, 77–79.

La Spada, A. R., Roling, D., Harding, A. E., Warner, C. L., Spiegel, R., Hausmanowa-Petrusewicz, I., Yee, W. C. & Fischbeck, K. H. 1992 Meiotic stability and genotype–phenotype correlation of the trinucleotide repeat in X-linked spinal and bulbar muscular atrophy. *Nature Genet.* **2**, 301–304.

La Spada, A. R. (and 13 others) 1998 Androgen receptor YAC transgenic mice carrying CAG 45 alleles show trinucleotide repeat instability. *Hum. Mol. Genet.* **7**, 959–967.

Li, M., Miwa, S., Kobayashi, Y., Merry, D. E., Yamamoto, M., Tanaka, F., Doyu, M., Hashizume, Y., Fischbeck, K. H. & Sobue, G. 1998*a* Nuclear inclusions of the androgen receptor protein in spinal and bulbar muscular atrophy. *Ann. Neurol.* **44**, 249–254.

Li, M., Nakagomi, Y., Kobayashi, Y., Merry, D. E., Tanaka, F., Doyu, M., Fischbeck, K. H. & Sobue, G. 1998*b* Non-neural inclusions of androgen receptor protein in spinal and bulbar muscular atrophy. *Am. J. Pathol.* **153**, 695–701.

Mangiarini, L. (and 10 others) 1996 Exon 1 of the HD gene with an expanded CAG repeat is sufficient to cause a progressive neurological phenotype in transgenic mice. *Cell* **87**, 493–506.

Mendell, J. R., Freimer, M. & Kissel, J. T. 1996 Randomized, double-blind crossover trial of androgen hormone deficiency and replacement in X-linked bulbar spinal muscular atrophy (abstract). *Neurology* **46**, A469.

Merry, D. E., McCampbell, A., Taye, A. A., Winston, R. L. & Fischbeck, K. H. 1996 Toward a mouse model for spinal and bulbar muscular atrophy: effect of neuronal expression of androgen receptor in transgenic mice (abstract). *Am. J. Hum. Genet.* **59**, A271.

Merry, D. E., Kobayashi, Y., Bailey, C. K., Taye, A. A. & Fischbeck, K. H. 1998 Cleavage, aggregation and toxicity of the expanded androgen receptor in spinal and bulbar muscular atrophy. *Hum. Mol. Genet.* **7**, 693–701.

Mhatre, A. N., Trifiro, M. A., Kaufman, M., Kazemi-Esfarjani, P., Figlewicz, D., Rouleau, G. & Pinsky, L. 1993 Reduced transcriptional regulatory competence of the androgen receptor in X-linked spinal and bulbar muscular atrophy. *Nature Genet.* **5**, 184–187.

Paulson, H. L., Perez, M. K., Trottier, Y., Trojanowski, J. Q., Subramony, S. H., Das, S. S., Vig, P., Mandel, J. L., Fischbeck, K. H. & Pittman, R. N. 1997 Intranuclear inclusions of expanded polyglutamine protein in spinocerebellar ataxia type 3. *Neuron* **19**, 333–344.

Sar, M. & Stumpf, W. E. 1997 Androgen concentration in motor neurons of cranial nerves and spinal cord. *Science* **197**, 77–79.

Sobue, G., Hashizume, Y., Mukai, E., Hirayama, M., Mitsuma, T. & Takahashi, A. 1989 X-linked recessive bulbospinal neuronopathy. A clinicopathological study. *Brain* **112**, 209–232.

Warrick, J. M., Paulson, H. L., Gray-Board, G. L., Bui, Q. T., Fischbeck, K. H., Pittman, R. N. & Bonini, N. M. 1998 Expanded polyglutamine protein forms nuclear inclusions and causes neural degeneration in *Drosophila*. *Cell* **93**, 939–949.

Yu, W. H. 1989 Administration of testosterone attenuates neuronal loss following axotomy in the brain-stem motor nuclei of female rats. *J. Neurosci.* **9**, 3908–3914.

20

Progress in pathogenesis studies of spinocerebellar ataxia type 1

C. J. Cummings, H. T. Orr and H. Y. Zoghbi

The dominantly inherited spinocerebellar ataxias (SCAs) are a heterogeneous group of neurological disorders characterized by variable degrees of degeneration of the cerebellum, spinal tracts and brainstem (Greenfield 1954; Koeppen & Barron 1984). Clinical and pathological classification of the SCAs has been very difficult because of intra- and inter-familial variability. The decade of the brain has witnessed an incredible amount of progress in genetic studies of SCAs. So far, the loci for ten ataxias have been mapped and the genes and/or mutations have been identified for eight. The challenge for the next decade is to decipher the functions of the protein products and to unravel disease pathogenesis. The following is a summary of gene function and disease pathogenesis studies in SCA type 1 (SCA1).

SCA1 is characterized by progressive ataxia, dysarthria, amyotrophy and bulbar dysfunction. The typical age of onset is in the third or fourth decade, but early onset in the first decade has been documented in some families. Increase in the severity of the phenotype in later generations, a phenomenon known as 'anticipation', has been observed in at least two large SCA1 kindreds (Schut 1950; Zoghbi *et al.* 1988). The disease typically progresses over ten to 15 years, but a more rapidly progressive course has been described in juvenile-onset cases (Zoghbi *et al.* 1988). Pathologically, SCA1 is characterized by the degeneration of cerebellar Purkinje cells, inferior olive neurons and neurons within brainstem cranial nerve nuclei.

The *SCA1* gene was identified in 1993 and the mutation was demonstrated to be an expansion of a translated CAG repeat (Orr *et al.* 1993). Normal alleles contain 6–44 repeats and are always interrupted with one to three CAT nucleotides when the CAG tract contains 21 or more repeats (Chung *et al.* 1993). In contrast, expanded disease alleles contain 39–82 uninterrupted CAG repeats. The *SCA1* gene has a wide pattern of expression in the nervous system and peripheral tissues. The gene product, ataxin-1, is a novel protein that has nuclear localization in neurons and cytoplasmic distribution in peripheral tissues. Within the central nervous system it is abundantly expressed in neurons that are affected by the disease (Purkinje cells and brainstem neurons) and those that are spared (hippocampal and cortical neurons) (Servadio *et al.* 1995).

To gain insight into the normal function of ataxin-1, mice with a targeted deletion in the *Sca1* gene were generated. These mice are viable, fertile and show no evidence of ataxia or neurodegeneration. *Sca1*-null mice do, however, demonstrate decreased exploratory behaviour, pronounced deficits in spatial memory and impaired performance on the rotating-rod apparatus. At the neurophysiological level, studies on area CA1 of the hippocampus revealed decreased paired-pulse facilitation (PPF) but normal long-term potentiation (LTP) and post-tetanic potentiation (PTP) (Matilla *et al.* 1998). These findings prove incontrovertibly that SCA1 is not caused by a loss of function of ataxin-1 and suggest that the protein has some role in learning and memory.

Overexpression of a mutant *Sca1* allele (82 glutamine residues, line BO5) in mice with the use of the Purkinje cell promoter (*Pcp2*) resulted in progressive ataxia and Purkinje cell degeneration (Burright *et al.* 1995). The phenotype was similar when the mutant allele was expressed on a wild-type background or on an *Sca1*-null background, confirming that the SCA1 mutation causes disease via a toxic gain-of-function mechanism. Ataxin-1 aggregates and localizes to a single nuclear inclusion (NI) in the Purkinje cells of SCA1 transgenic mice. This finding prompted the careful evaluation of ataxin-1 distribution in the tissue of SCA1 patients. Ataxin-1 aggregates are detected in brainstem neurons that typically degenerate in this disease (Skinner *et al.* 1997). Intranuclear protein aggregation has also been observed for huntingtin, ataxin-3, ataxin-7, atrophin-1 and the androgen receptor (Davies *et al.* 1997; Paulson *et al.* 1997; Scherzinger *et al.* 1997; Hayashi *et al.* 1998; Holmberg *et al.* 1998; Li *et al.* 1998). An important issue is whether these intranuclear aggregates have a role in initiating pathogenesis or whether they represent a late downstream effect in the pathogenetic process. To address this question, transgenic mice were generated by using a form of mutant ataxin-1 that does not aggregate in transfected cells. This form of ataxin-1 lacks the self-association domain that is necessary for ataxin-1 to interact with itself in yeast two-hybrid studies (Burright *et al.* 1997). Overexpression of ataxin-1 containing 77 CAG repeats but lacking the self-association domain (ataxin-1{77}Δ) in mice with the use of the *Pcp2* promoter resulted in the same phenotype of ataxia and Purkinje cell pathology as in the original BO5 SCA1 transgenic mice. However, the nuclear inclusions were not detected (Klement *et al.* 1998). These results suggest that ataxin-1 aggregates are not necessary to initiate SCA1 pathogenesis.

Immunohistological studies reveal that the SCA1 NIs are ubiquitin-positive and that various components of the proteasome are redistributed to their site, suggesting that the cell's proteolytic machinery is attempting to degrade mutant ataxin-1 (Cummings *et al.* 1998). Even more intriguing is the finding that aggregates of ataxin-1 also stain positively for the molecular chaperone HDJ-2/HSDJ. These results suggest that protein misfolding is responsible for the nuclear aggregates seen in SCA1. This hypothesis is supported by the finding that chaperone overexpression in cell culture subdues ataxin-1 aggregation, perhaps by promoting the recognition of the aberrant polyglutamine repeat protein and allowing its refolding and/or ubiquitin-dependent degradation (Cummings *et al.* 1998).

The issue of selective neuronal degeneration in polyglutamine diseases remains an important aspect of the pathogenesis given that, like ataxin-1, many of the mutant proteins are abundantly expressed in cells spared by the disease. One hypothesis proposes that mutant ataxin-1 might interact with one or more proteins that are more abundantly expressed and/or are essential in the vulnerable neurons. The leucine-rich acidic nuclear protein (LANP) is a candidate for the mediation of SCA1 pathogenesis on the basis of its interactions with mutant ataxin-1 and its patterns of expression and subcellular distribution (Matilla *et al.* 1997). It is interesting to note that LANP interacts with ataxin-1{77}Δ, which is consistent with a role of LANP–ataxin-1 interactions in the early pathogenesis of disease. Present work is focusing on determining the role of LANP in SCA1 pathogenesis with the use of mouse models *in vivo* and on overexpressing the HDJ-2/HSDJ chaperone in mice to assess its role in modulating SCA1 pathogenesis.

This work was supported by grants from the NIH (NS27699 NS22920, NS35255). H.Y.Z. is a Howard Hughes Medical Institute Investigator.

References

Burright, E. N., Clark, H. B., Servadio, A., Matilla, T., Feddersen, R. M., Yunis, W. S., Duvick, L. A., Zoghbi, H. Y. & Orr, H. T. 1995 SCA1 transgenic mice: a model for neurodegeneration caused by an expanded CAG trinucleotide repeat. *Cell* **82**, 937–948.

Burright, E. N., Davidson, J. D., Duvick, L. A., Koshy, B., Zoghbi, H. Y. & Orr, H. T. 1997 Identification of a self-association region within the *SCA1* gene product, ataxin-1. *Hum. Mol. Genet.* **6**, 513–518.

Chung, M.-y., Ranum, L. P. W., Duvick, L., Servadio, A., Zoghbi, H. Y. & Orr, H. T. 1993 Analysis of the CAG repeat expansion in spinocerebellar ataxia type I: evidence for a possible mechanism predisposing to instability. *Nature Genet.* **5**, 254–258.

Cummings, C. J., Mancini, M. A., Antalffy, B., DeFranco, D. B., Orr, H. T. & Zoghbi, H. Y. 1998 Chaperone suppression of aggregation and altered subcellular proteasome localization imply protein misfolding in SCA1. *Nature Genet.* **19**, 148–154.

Davies, S. W., Turmaine, M., Cozens, B. A., DiFiglia, M., Sharp, A. H., Ross, C. A., Scherzinger, E., Wanker, E. E., Mangiarini, L. & Bates, G. P. 1997 Formation of neuronal intranuclear inclusions underlies the neurological dysfunction in mice transgenic for the HD mutation. *Cell* **90**, 537–548.

Greenfield, J. G. 1954 *The spino-cerebellar degenerations*. Springfield, IL: C. C. Thomas.

Hayashi, Y., Yamada, M., Egawa, S., Oyanagi, S., Naito, H., Tsuji, S. & Takahashi, H. 1998 Hereditary dentatorubral–pallidoluysian atrophy: ubiquitinated filamentous inclusions in the cerebellar dentate nucleus neurons. *Acta Neuropathol.* **95**, 479–482.

Holmberg, M. (and 10 others) 1998 Spinocerebellar ataxia type 7 (SCA7): a neurodegenerative disorder with neuronal intranuclear inclusions. *Hum. Mol. Genet.* **7**, 913–918.

Klement, I. A., Skinner, P. J., Kaytor, M. D., Yi, H., Hersch, S. M., Clark, H. B. & Zoghbi, H. Y. 1998 Ataxin-1 nuclear localization and aggregation: role in polyglutamine-induced disease in SCA1 transgenic mice. *Cell* **95**, 41–53.

Koeppen, A. H. & Barron, K. D. 1984 The neuropathology of olivopontocerebellar atrophy. In *The olivopontocerebellar atrophies* (ed. R. C. Duvoisin & A. Plaitakis), pp. 13–38. New York: Raven.

Li, M. (and 10 others) 1998 Nuclear inclusions of the androgen receptor in spinal and bulbar muscular atrophy. *Ann. Neurol.* **44**, 249–254.

Matilla, T., Koshy, B., Cummings, C. J., Isobe, T., Orr, H. T. & Zoghbi, H. Y. 1997 The cerebellar leucine-rich acidic nuclear protein interacts with ataxin-1. *Nature* **389**, 974–978.

Matilla, A., Roberson, E. D., Banfi, S., Morales, J., Armstrong, D. L., Burright, E. N., Orr, H. T., Sweatt, J. D., Zoghbi, H. Y. & Matzuk, M. M. 1998 Mice lacking ataxin-1 display learning deficits and decreased hippocampal paired-pulse facilitation. *J. Neurosci.* **18**, 5508–5516.

Orr, H., Chung, M.-y., Banfi, S., Kwiatkowski Jr, T. J., Servadio, A., Beaudet, A. L., McCall, A. E., Duvick, L. A., Ranum, L. P. W. & Zoghbi, H. Y. 1993 Expansion of an unstable trinucleotide (CAG) repeat in spinocerebellar ataxia type 1. *Nature Genet.* **4**, 221–226.

Paulson, H. L., Perez, M. K., Trottier, Y., Trojanowsk, J. Q., Subramony, S. H., Das, S. S., Vig, P., Mandel, J.-L., Fischbeck, K. H. & Pittman, R. N. 1997 Intranuclear inclusions of expanded polyglutamine protein in spinocerebellar ataxia type 3. *Neuron* **19**, 333–334.

Scherzinger, E., Lurz, R., Turmaine, M., Mangiarini, L., Hollenbach, B., Hasenbank, R., Bates, G. P., Davies, S. W. & Wanker, E. E. 1997 Huntingtin-encoded polyglutamine expansions form amyloid-like protein aggregates *in vitro* and *in vivo*. *Cell* **90**, 549–558.

Schut, J. W. 1950 Hereditary ataxia: clinical study through six generations. *Arch. Neurol. Psychiatr.* **63**, 535–568.

Servadio, A., Koshy, B., Armstrong, D., Antalfy, B., Orr, H. T. & Zoghbi, H. Y. 1995 Expression analysis of the ataxin-1 protein in tissues from normal and spinocerebellar ataxia type 1 individuals. *Nature Genet.* **10**, 94–98.

Skinner, P. J., Koshy, B., Cummings, C., Klement, I. A., Helin, K., Servadio, A., Zoghbi, H. Y. & Orr, H. T. 1997 Ataxin-1 with extra glutamines induces alterations in nuclear matrix-associated structures. *Nature* **389**, 971–974.

Zoghbi, H. Y., Pollack, M. S., Lyons, L. A., Ferell, R. E., Daiger, S. P. & Beaudet, A. L. 1988 Spinocerebellar ataxia: variable age of onset and linkage to human leukocyte antigen in a large kindred. *Ann. Neurol.* **23**, 580–584.

21

Filamentous nerve cell inclusions in neurodegenerative diseases: tauopathies and α-synucleinopathies

Michel Goedert

Introduction

Neurodegenerative diseases of the human brain are characterized by the degeneration of specific populations of nerve cells. Alzheimer's disease, a dementing condition, is the most common of these diseases. It affects 20–25 million people worldwide and is the fourth leading cause of death in the industrialized world. Alzheimer's disease is defined by the presence of two neuropathological abnormalities made of filamentous deposits: neuritic plaques in the extracellular space and neurofibrillary lesions inside nerve cells. Frontotemporal dementias such as Pick's disease account for 5–10% of dementias. They frequently show neuropathological features similar to the neurofibrillary lesions of Alzheimer's disease. Parkinson's disease, a movement disorder, is the second most common neurodegenerative disease. It affects six to seven million individuals worldwide. Neuropathologically, Parkinson's disease is defined by the presence of intracytoplasmic filamentous inclusions in the form of Lewy bodies and Lewy neurites. Dementia with Lewy bodies is a common late-life dementia that shares pathological features with Parkinson's disease and is clinically similar to Alzheimer's disease. Finally, multiple system atrophy, a less common neurodegenerative disease that is often clinically mistaken for Parkinson's disease, is characterized by filamentous inclusions in glial cells. All these diseases exist as rare genetic forms and as much more common sporadic forms. Taken together, they account for the majority of late-onset neurodegenerative diseases in man (Goedert *et al.* 1998*a*).

At the beginning of this century, Alzheimer and Lewy described the characteristic light-microscopic neuropathological features of Alzheimer's disease, Pick's disease and Parkinson's disease (Alzheimer 1907, 1911; Lewy 1912). In the 1960s these lesions were shown to be made of abnormal filamentous material (Kidd 1963; Duffy & Tennyson 1965; Rewcastle & Ball 1968). Over the past 15 years, the molecular components of the filamentous lesions of Alzheimer's disease and Pick's disease have been identified and much continues to be learnt about their formation. The intracellular deposits are made of the microtubule-associated protein tau (Brion *et al.* 1985; Pollock *et al.* 1986; Goedert *et al.* 1988; Kondo *et al.* 1988; Wischik *et al.* 1988; Lee *et al.* 1991), whereas the extracellular deposits of Alzheimer's disease are made of the β-amyloid protein Aβ (Glenner & Wong 1984; Masters *et al.* 1995). Over the past two years, the biochemical nature of the filamentous material of Lewy bodies and Lewy neurites of Parkinson's disease and dementia with Lewy bodies has been discovered (Spillantini *et al.* 1997*a*, 1998*a*). In 1998, the biochemical nature of the filaments of multiple system atrophy was revealed (Arima *et al.* 1998*a*; Spillantini *et al.* 1998*b*; Tu *et al.* 1998; Wakabayashi *et al.* 1998*a*,*b*). In all three diseases, the intracellular filaments are made of the protein α-synuclein.

Recent progress has led to the classification of late-onset neurodegenerative diseases according to the biochemical composition of their intracellular filamentous deposits (Goedert *et al.* 1998*a*; Hardy & Gwinn Hardy 1998). The three major classes of disease are the tauopathies, the α-synucleinopathies and the glutamine repeat diseases (table 21.1). This chapter deals exclusively with tauopathies and α-synucleinopathies.

Alzheimer's disease

A diagnosis of Alzheimer's disease is made when a patient exhibits clinical evidence of progressive dementia and when a post-mortem examination of the brain reveals the characteristic neuropathology consisting of extracellular neuritic plaques and intracellular neurofibrillary lesions. Aβ, the 40–42-residue plaque component, is derived by proteolytic cleavage from the much larger amyloid precursor protein (APP) (Glenner & Wong 1984; Kang *et al.* 1987; Masters *et al.* 1995). Genetic evidence has shown that APP pathology has an important role in the aetiology and pathogenesis of at least a proportion of cases of Alzheimer's disease (Goate *et al.* 1991). The relationship between amyloid deposition and neurofibrillary lesions remains an important unresolved issue in our understanding of the pathogenesis of Alzheimer's disease.

Abundant amyloid deposits can be present in cognitively normal individuals and it is the presence of neurofibrillary lesions that correlates better with the presence of dementia (Arriagada *et al.* 1992). Until recently, a major question was whether this correlation implied a causal relation. It had repeatedly been suggested that neurofibrillary lesions might be nothing more than an epiphenomenon (Duff & Hardy 1995; Masters & Beyreuther 1998). The discovery of mutations in the tau gene in familial frontotemporal dementias, in conjunction with the presence of a filamentous tau pathology, has settled this question (see §3) (Hutton *et al.* 1998; Poorkaj *et al.* 1998; Spillantini *et al.* 1998*c*,*d*). It now seems clear that the formation of a filamentous tau pathology leads to nerve cell degeneration.

Table 21.1 *Intraneuronal filamentous inclusions in neurodegenerative diseases*

disease	*filamentous inclusion*	*main component*
Alzheimer's disease	neurofibrillary lesions	tau protein
Pick's disease	Pick bodies	tau protein
FTDP-17	neurofibrillary lesions, glialfibrillary lesions	tau protein
PSP	neurofibrillary lesions, glialfibrillary lesions	tau protein
CBD	neurofibrillary lesions, glialfibrillary lesions	tau protein
Parkinson's disease	Lewy bodies and neurites	α-synuclein
dementia with Lewy bodies	Lewy bodies and neurites	α-synuclein
multiple system atrophy	glial and neuronal inclusions	α-synuclein
Huntington's disease	intranuclear inclusions, dystrophic neurites	expanded glutamine repeats in huntingtin
SCA-1	intranuclear inclusions	expanded glutamine repeats in ataxin 1
SCA-3	intranuclear inclusions	expanded glutamine repeats in ataxin 3
SCA-7	intranuclear inclusions	expanded glutamine repeats in ataxin 7
DRPLA	intranuclear inclusions	expanded glutamine repeats in atrophin 1
SBMA	intranuclear inclusions	expanded glutamine repeats in androgen receptor

(Abbreviations: FTDP-17, frontotemporal dementia and Parkinsonism linked to chromosome 17; PSP, progressive supranuclear palsy; CBD, corticobasal degeneration; SCA, spinocerebellar ataxia; DRPLA, dentatorubral–pallidoluysian atrophy; SBMA, spinal and bulbar muscular atrophy.)

Natural history of neurofibrillary lesions

The neurofibrillary pathology forms within nerve cells of the cerebral cortex, the hippocampal formation and some subcortical nuclei. The nerve cells eventually degenerate and the insoluble neurofibrillary lesions are found in the extracellular space as ghost tangles (Alzheimer 1907, 1911). In the hippocampus, there exists an inverse correlation between the number of extracellular tangles and the number of surviving nerve cells (Bondareff *et al.* 1989; Cras *et al.* 1995; Fukutani *et al.* 1995), demonstrating that the nerve cells that degenerate developed neurofibrillary lesions. These lesions are found in nerve cell bodies and apical dendrites as neurofibrillary tangles, in distal dendrites as neuropil threads and in abnormal neurites that are often,

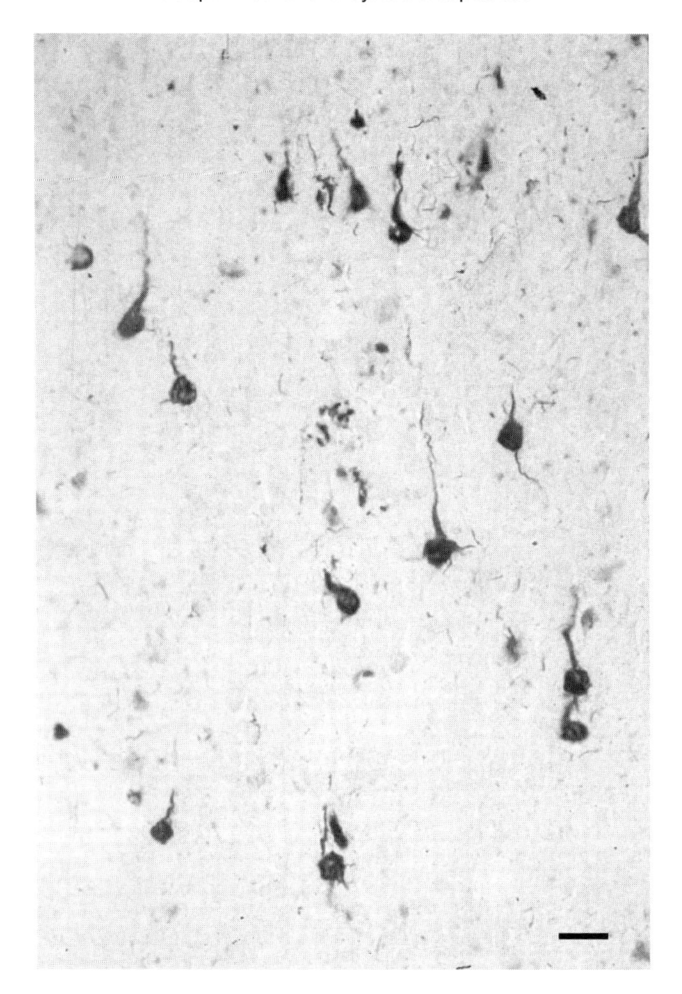

Figure 22.1 Neurofibrillary lesions in cerebral cortex from an Alzheimer's disease patient revealed with a phosphorylation-dependent anti-tau antibody. (See also colour plate section.)

but not always, associated with amyloid plaques (figure 21.1 (plate 14)). Ultrastructurally, the neurofibrillary pathology consists of paired helical filaments (PHFs) and the related straight filaments (SFs) (figure 21.2) (Kidd 1963; Crowther 1991). These filaments are made of the microtubule-associated protein tau, in a hyperphosphorylated state (Brion *et al.* 1985; Goedert *et al.* 1988; Kondo *et al.* 1988; Wischik *et al.* 1988; Lee *et al.* 1991).

A filamentous tau pathology indistinguishable from that of Alzheimer's disease is a frequent accompaniment of ageing. The difference with Alzheimer's disease lies in the much smaller number of affected nerve cells. The development of neurofibrillary lesions in ageing and Alzheimer's disease follows a stereotyped pattern with regard to affected cell types, cellular layers

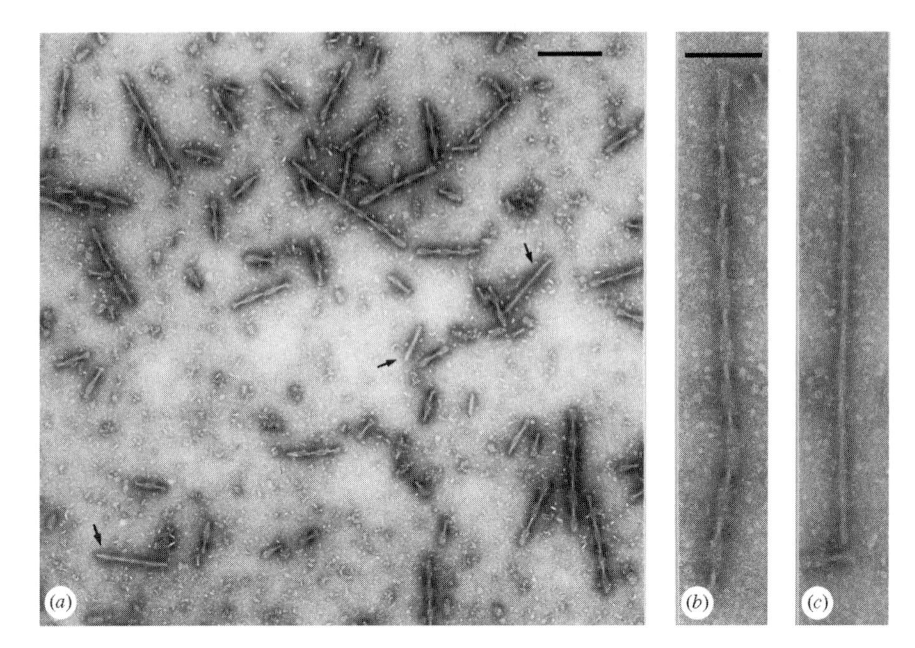

Figure 21.2 Electron micrographs of negatively stained abnormal filaments from the brain of a patient with Alzheimer's disease. (*a*) Low-power view showing predominantly paired helical filaments but with a few straight filaments (arrows); (*b*,*c*) high-power view of a paired helical filament (*b*) and a straight filament (*c*). Scale bars: (*a*) 200 nm; (*b*,*c*) 100 nm.

and brain regions, with little variation between individuals. This pattern has been used to define six neuropathological stages of Alzheimer's disease (Braak & Braak 1991, 1997). The very first nerve cells in the brain to develop neurofibrillary lesions are located in the pre-alpha layer of the transentorhinal region, thus defining stage I. Stage II shows a more severe involvement of this region, as well as a mild involvement of the pre-alpha layer of the entorhinal cortex. Patients with this pathology are unimpaired cognitively, indicating that stages I and II might represent clinically silent stages of Alzheimer's disease. Mild impairments of cognitive function become apparent in stages III and IV. Stage III is characterized by severe neurofibrillary lesions in the pre-alpha layers of both the entorhinal and transentorhinal regions. Stage III is also characterized by the appearance of the first extracellular tangles. In stage IV, the deep pre-alpha layer develops extensive neurofibrillary lesions. During stages III and IV, changes are also seen in layer I of Ammon's horn of the hippocampus and in a number of subcortical nuclei. The major feature of stages V and VI is the massive development of neurofibrillary lesions in isocortical association areas. They meet the criteria for the neuropathological diagnosis of Alzheimer's disease and are found in patients who were severely demented at the time of death.

As a function of age, most individuals develop stages I and II of neurofibrillary degeneration. Extensive studies by Braak & Braak (1997) have suggested a continuum between these initial changes and full-blown Alzheimer's disease. They have indicated that even small numbers of neurofibrillary lesions are pathological and might represent the early stages of Alzheimer's disease. Extracellular tangles were never observed in the absence of intracellular lesions, indicating that once initiated the neurodegenerative process follows its relentless course. These studies have also provided the important information that neurofibrillary lesions can develop in layer pre-alpha in the absence of extracellular Aβ deposits. They are inconsistent with the widely held view that the neurofibrillary pathology develops as the mere consequence of the neurotoxic action of Aβ deposits and suggest instead that nerve cells die from within.

Tau protein in normal brain

Tau is a microtubule-associated protein whose physiological functions are to promote microtubule assembly and to stabilize microtubules (Hirokawa 1994). In adult human brain, six isoforms of tau are expressed, which are produced by alternative mRNA splicing from a single gene located on the long arm of chromosome 17 (figure 21.3 (plate 15)) (Goedert *et al.* 1989*a*,*b*; Goedert & Jakes 1990; Andreadis *et al.* 1992). They differ by the presence of three or four tandem repeats of 31 or 32 residues each, located in the C-terminal region, in conjunction with inserts of 0, 29 or 58 residues located in the N-terminal region (figure 21.1 (plate 14)). There is also a larger tau isoform, with an additional 254-residue insert in the N-terminal region, which is mainly expressed in the peripheral nervous system (Couchie *et al.* 1992; Goedert *et al.* 1992*a*).

The repeat regions of tau and sequences flanking the repeats constitute microtubule-binding domains, with the functions of the N-terminal regions remaining uncertain (Gustke *et al.* 1994; Trinczek *et al.* 1995). Tau protein mRNA is expressed predominantly in nerve cells, with lower levels in some glial cells. Within nerve cells, tau protein is present mainly in axons (Binder *et al.* 1985). Tau does not seem to be an essential protein, because inactivation of its gene by homologous recombination leads to no overt phenotype, except a decrease in the number of microtubules in some small-calibre axons (Harada *et al.* 1994). Tau expression is developmentally regulated in that only the tau isoform with three repeats and no N-terminal inserts is present in foetal human brain (Goedert *et al.* 1989*a*,*b*). There exist true species differences in the expression of tau isoforms in adult brain. Thus, only three tau isoforms are expressed in rodent brain, each with four repeats and N-terminal inserts of 0, 29 or 58 residues (Götz *et al.* 1995). In contrast, all six tau isoforms are expressed in adult human brain, where tau isoforms with three repeats are slightly more abundant than isoforms with four repeats (Goedert & Jakes 1990). Tau is a phosphoprotein, and phosphorylation is also developmentally

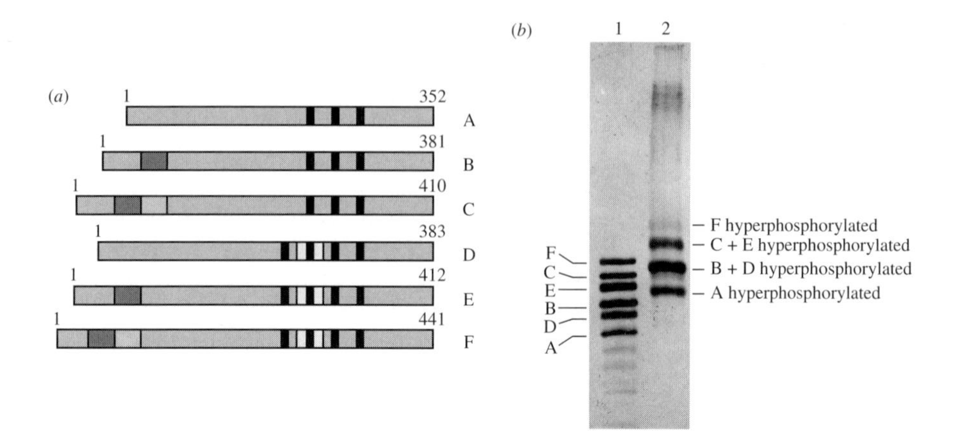

Figure 21.3 Isoforms of human tau. (*a*) Schematic representation of the six human brain tau isoforms (ranging from 352 to 441 residues). The region common to all isoforms is shown in blue, with the N-terminal inserts shown in red and green. The alternatively spliced repeat is in yellow. The three or four tandem repeats are indicated by black bars. Isoform 1 is expressed in foetal human brain, whereas all six isoforms (1–6) are expressed in adult human brain. (*b*) PHF-tau from Alzheimer's disease brain and recombinant human tau isoforms. Lane 1, mixture of recombinant human brain tau isoforms, with each isoform identified by a number; lane 2, the four PHF-tau bands of 60, 64, 68 and 72 kDa, with the tau isoform composition of each band identified by a number. After SDS–PAGE, the tau isoforms were revealed by immunoblotting with a phosphorylation-independent anti-tau antibody.

regulated. Thus, the shortest isoform is phosphorylated more during development than any of the six tau isoforms in adult brain (Kanemaru *et al.* 1992; Goedert *et al.* 1993).

Tau protein in Alzheimer's disease brain

In Alzheimer's disease brain, a proportion of tau protein is filamentous and of decreased solubility. Dispersed PHFs and SFs consist of three major tau bands of 60, 64 and 68 kDa and a minor band of 72 kDa (figure 21.3 (plate 15)) (Greenberg & Davies 1990; Lee *et al.* 1991). On dephosphorylation, six tau bands are seen that align with the six recombinant human brain tau isoforms (Goedert *et al.* 1992*b*). Thus, all the brain tau isoforms are present, each in a full-length form. Several approaches have helped to delineate which tau isoforms make up each PHF-tau band (figure 21.3 (plate 15)) (Goedert *et al.* 1992*b*; Mulot *et al.* 1994; Sergeant *et al.* 1997). The shortest and the longest tau isoforms constitute the 68 and 72 kDa bands, respectively. Each of the 64 and 68 kDa bands consists of two tau isoforms, one with three repeats and one with four repeats. PHF-tau from Alzheimer's disease brain therefore consists of all six tau isoforms, each in a hyperphosphorylated state.

Hyperphosphorylation and abnormal phosphorylation are major biochemical abnormalities of PHF-tau. They are early events in the development of the neurofibrillary lesions (Braak *et al.* 1994) and as a result tau is unable to bind to microtubules (Bramblett *et al.* 1993; Yoshida & Ihara 1993). Most phosphorylated sites are known (Morishima-Kawashima *et al.* 1995; Hanger *et al.* 1998). They consist of serine or threonine residues, many of which are followed by a proline in the tau sequence. A number of protein kinases have been implicated in the abnormal phosphorylation of tau, largely based on studies of tau phosphorylation *in vitro*. The latest additions to this growing list are stress-activated protein (SAP) kinases, especially SAP kinase 3 and SAP kinase 4 (Goedert *et al.* 1997). Protein phosphatase 2A is the major protein phosphatase activity in brain able to dephosphorylate tau phosphorylated by a number of protein kinases (Goedert *et al.* 1992*c*; Sontag *et al.* 1996).

Relatively little is known about which protein kinases phosphorylate tau in brain. This requires specific protein kinase inhibitors or the inactivation of individual protein kinase genes. The use of lithium chloride as a specific inhibitor of glycogen synthase kinase 3 has provided strong evidence that this protein kinase is involved in the phosphorylation of tau in normal brain (Hong *et al.* 1997; Munoz-Montado *et al.* 1997). However, the identity of the protein kinases and/or protein phosphatases that lead to the hyperphosphorylation of tau in the Alzheimer's disease brain remains to be established.

PHFs and SFs form from hyperphosphorylated full-length tau protein. After assembly, tau is proteolysed, mainly from the N-terminus, with extensively proteolysed PHFs and SFs only consisting of the microtubule-binding repeats (Wischik *et al.* 1988; Goedert *et al.* 1992*b*). Ubiquitination is a biochemical modification of PHF-tau that occurs after partial proteolysis from the N-terminus (Mori *et al.* 1987; Morishima-Kawashima *et al.* 1993). Four lysine residues in the microtubule-binding repeats of tau have been identified as the ubiquitin-conjugating sites (Morishima-Kawashima *et al.* 1993). Ubiquitination is an event that follows assembly into filaments and probably forms part of a largely unsuccessful attempt by the cellular machinery to degrade PHFs and SFs.

Synthetic tau filaments

Whether hyperphosphorylation and abnormal phosphorylation of tau are sufficient for PHF formation is unclear. Phosphorylated recombinant tau has consistently failed to assemble into PHF-like filaments in experiments *in vitro*. In contrast, incubation of recombinant tau with sulphated glycosaminoglycans such as heparin and heparan sulphate results in the bulk assembly of tau into Alzheimer-like filaments (figure 21.4) (Goedert *et al.* 1996; Pérez *et al.* 1996; Arrasate *et al.* 1997; Hasegawa *et al.* 1997; Friedhoff *et al.* 1998). Tau isoforms with three repeats assemble into twisted paired helical-like filaments, whereas tau isoforms with four repeats assemble into straight

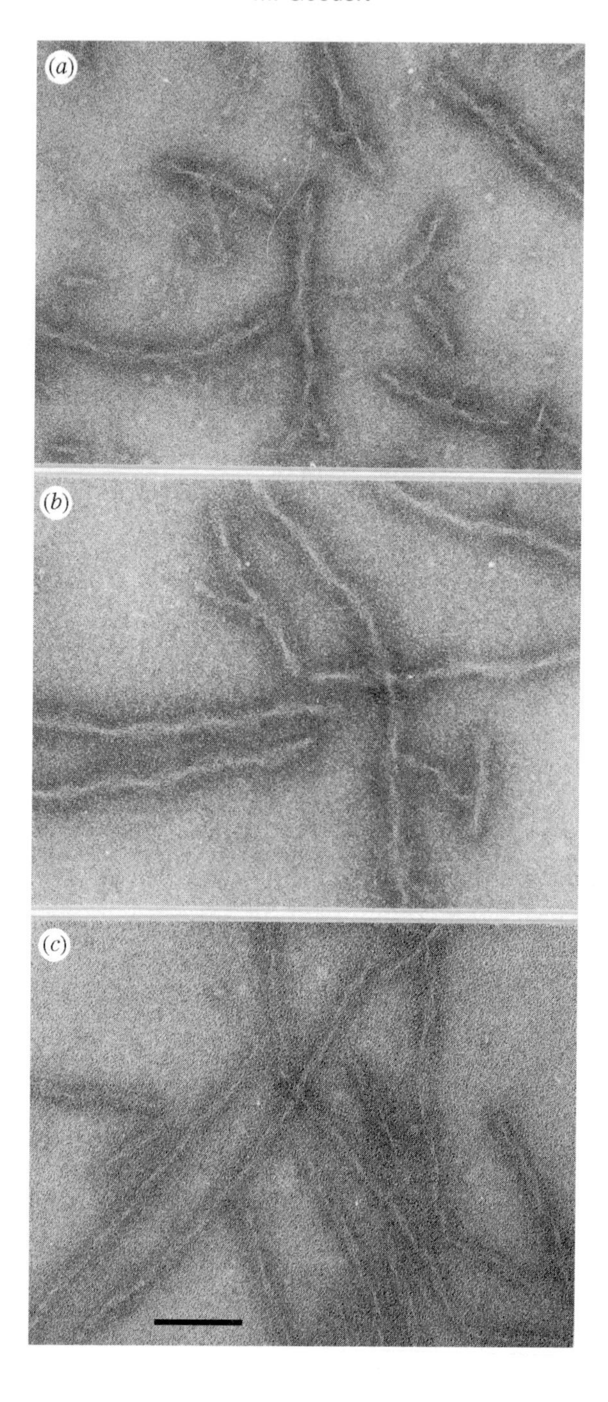
(a)
(b)
(c)

filaments. By immunoelectron microscopy, the paired helical-like filaments are decorated by antibodies directed against the N- and C-termini of tau, but not by an antibody directed against the microtubule-binding repeat region. These results, which indicate that in the filaments the repeat region of tau is inaccessible to the antibody, are identical to those previously obtained with PHFs from the brains of Alzheimer's disease patients (Goedert *et al.* 1992*b*, 1996). They establish that the microtubule-binding repeat region of tau is essential for sulphated glycosaminoglycan-induced filament formation. The dimensions of tau filaments formed in the presence of sulphated glycosaminoglycans are similar to those of filaments extracted from Alzheimer's disease brain, with a diameter of *ca.* 20 nm for twisted filaments and 15 nm for straight filaments, with a crossing-over spacing of *ca.* 80 nm for paired helical-like filaments, although their twist is in general less regular than in Alzheimer's disease filaments.

Sulphated glycosaminoglycans also stimulate the phosphorylation of tau by a number of protein kinases, prevent the binding of tau to taxol-stabilized microtubules and disassemble microtubules assembled from tau and tubulin (Hasegawa *et al.* 1997; Qi *et al.* 1998). Moreover, heparan sulphate has been detected in nerve cells in the early stages of neurofibrillary degeneration (Snow *et al.* 1989; Goedert *et al.* 1996). Sulphated glycosaminoglycans stimulate tau phosphorylation at lower concentrations than those required for filament formation. The pathological presence of heparan sulphate within the cytoplasm of some nerve cells, perhaps as a result of leakage from membrane-bound compartments, would lead first to the hyperphosphorylation of tau, resulting in its inability to bind to microtubules. At higher concentrations of heparan sulphate, tau would then assemble into PHFs and SFs. The formation of tau filaments is also observed after the incubation of recombinant tau with RNA, which has been shown to be sequestered in the neurofibrillary lesions of Alzheimer's disease (Kampers *et al.* 1996; Ginsberg *et al.* 1997, 1998; Hasegawa *et al.* 1997). Whether the presence of RNA is an early event remains to be determined.

Sulphated glycosaminoglycans and RNA share a repeat sugar backbone and negative charges in the form of sulphates or phosphates. Tau protein is thought to be an extended molecule with little secondary structure that becomes partly structured on binding to microtubules. Binding of sulphated glycosaminoglycans or RNA to tau might induce or stabilize a conformation of tau that brings the microtubule-binding repeats of individual tau molecules into close proximity, creating sites that favour polymerization into filaments.

Figure 21.4 Sulphated glycosaminoglycan-induced filament assembly. (*a*) Recombinant three-repeat htau37 (381-residue isoform of human tau) was incubated with heparin. (*b*) htau37 was incubated with heparan sulphate. (*c*) Recombinant four-repeat htau40 (441-residue isoform of human tau) was incubated with heparin. Note the presence of paired helical-like filaments in (*a*) and (*b*) and straight filaments in (*c*). Scale bar, 100 nm.

Experimental animal models

The work on synthetic tau filaments has provided the first robust methods by which to produce Alzheimer-like filaments from full-length tau. The same cannot yet be said of tau filaments in nerve cells. So far, there has been no demonstration of Alzheimer-like filaments in transgenic mice. Two studies have directly addressed this issue by expressing wild-type human tau in mouse brain (Götz *et al.* 1995; Brion *et al.* 1999; Goedert & Hasegawa 1999). It has been addressed indirectly in transgenic mouse models of Aβ deposition, which are based on the expression of mutated APP (Games *et al.* 1995; Hsiao *et al.* 1996; Sturchler-Pierrat *et al.* 1997). Although some staining for hyperphosphorylated tau has been described in nerve cell processes around Aβ deposits in transgenic mice expressing mutated APP, no somatodendritic staining of hyperphosphorylated tau was observed in these mice. Two of these mouse lines did not exhibit nerve cell loss, whereas a third showed a 17% decrease in the number of nerve cells in layer CA1 of the hippocampus (Irizarry *et al.* 1997*a,b*; Calhoun *et al.* 1998). However, it remains to be seen whether this cell loss is mechanistically related to the nerve cell loss observed in Alzheimer's disease hippocampus. Mutated APP is expressed at high levels in these mice and this could in itself result in the degeneration of some nerve cells.

Expression of human tau in transgenic mouse has used either the longest or the shortest brain isoforms (Götz *et al.* 1995; Brion *et al.* 1999). Both studies described broadly similar results, in that they showed strong somatodendritic and axonal staining for hyperphosphorylated tau in subpopulations of nerve cells. By electron microscopy, transgenic human tau was associated with microtubules in axons and dendrites but not in nerve cell bodies, where it was associated with ribosomes or distributed more diffusely (Brion *et al.* 1999). Overexpression of human tau in lamprey neurons has also been shown to lead to the presence of hyperphosphorylated human tau in the somatodendritic compartment (Hall *et al.* 1997). It therefore seems that an excess of tau over available binding sites on microtubules results in the accumulation of tau in nerve cell bodies.

Somatodendritic staining for hyperphosphorylated tau has been described as an early pathological change in human brain, where it is characteristic of the so-called 'pre-tangle' stage of Alzheimer's disease (Braak *et al.* 1994). In human brain, the pre-tangle pathology progresses to the filamentous tangle stage, which is followed by nerve cell degeneration and death. So far, tau filaments have not been observed in brains of mice transgenic for tau protein. There is no evidence to suggest the presence of nerve cell loss in these mice, indicating that the prolonged presence of hyperphosphorylated tau in the somatodendritic compartment of nerve cells is not sufficient to lead to nerve cell degeneration. The current transgenic mouse models therefore go only part of the way towards a filamentous tau pathology.

Other tauopathies

Filamentous tau protein deposits are the defining pathological characteristics of neurodegenerative diseases other than Alzheimer's disease (Spillantini & Goedert 1998). In these diseases, tau pathology is found in the absence of Aβ amyloid deposits. Pick's disease is the prototypical frontotemporal dementia, with Pick bodies as its central pathological characteristic (Alzheimer 1911). The latter consist of abnormal filaments that comprise hyperphosphorylated tau protein (table 21.1) (Rewcastle & Ball 1968; Pollock *et al.* 1986). Unlike PHFs and SFs from Alzheimer's disease brain, Pick body filaments contain only three tau isoforms, each with three microtubule-binding repeats (figure 21.5) (Sergeant *et al.* 1997; Delacourte *et al.* 1998). Over the past few years, familial

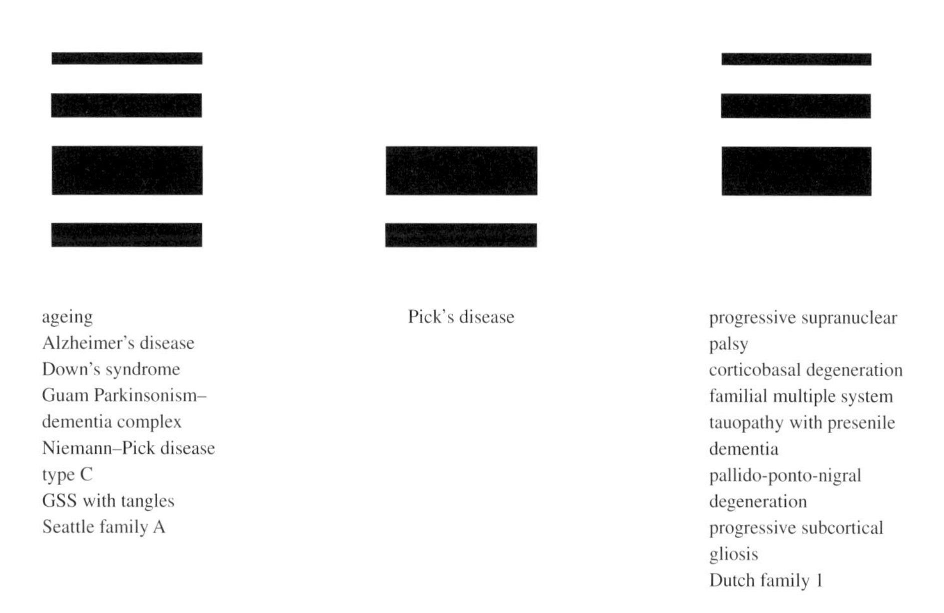

Figure 21.5 Schematic representation of tau bands from filamentous assemblies of different tauopathies. Type I includes Alzheimer's disease and a number of other dementing disorders and is characterized by tau bands of 60, 64, 68 and 72 kDa. It includes Seattle family A, an FTDP-17 with a V337M mutation in exon 12 of the tau gene. Type II consists of Pick's disease and is characterized by two major tau bands of 60 and 64 kDa and a minor tau band of 68 kDa. Type III comprises progressive supranuclear palsy, corticobasal degeneration, familial multiple system tauopathy with presenile dementia (MSTD), pallido-ponto-nigral degeneration (PPND), Dutch family 1 and familial progressive subcortical gliosis (PSG). Familial MSTD is an FTDP-17 with a mutation in the intron following exon 10 of the tau gene (at position +3), PPND is an FTDP-17 with a N279K mutation in exon 10, Dutch family 1 is an FTDP-17 with a P301L mutation in exon 10 of the tau gene and familial PSG is an FTDP-17 with a mutation in the intron following exon 10 of the tau gene (at position +16). Type III is characterized by two major tau bands of 64 and 68 kDa and a minor band of 72 kDa. Abbreviation: GSS, Gerstmann–Sträussler–Scheinker disease.

Table 21.2 *Tau mutations, isoforms and filaments in FTDP-17*

tau mutation	*soluble tau*	*filamentous tau*	*tau filaments*
P301L (exon 10)	normal ratio of 3- to 4-repeat isoforms (4-repeat isoforms mutated)	4-repeat isoforms; small amount of 3-repeat isoform	narrow twisted ribbons in neurons and glia
intron following exon 10, N279K (exon 10)	abnormal preponderance of 4- over 3-repeat isoforms	4-repeat isoforms	wide twisted ribbons in neurons and glia
V337M (exon 12), R406W (exon 13)	normal ratio of 3 to 4-repeat isoforms (all isoforms mutated)	all 6 isoforms	paired helical filaments and straight filaments in neurons

frontotemporal dementias, some with Parkinsonism, have been recognized as a previously unknown group of dementing disorders (Foster *et al.* 1997). Their unifying pathological characteristic is the presence of abundant filamentous tau deposits (Spillantini *et al.* 1998*c*). In some of these families, tau deposits are found in both nerve cells and glial cells, whereas in others, only nerve cells are affected. Ultrastructurally, depending on the families, tau filaments are either identical to those from Alzheimer's disease brain or show twisted ribbon-like morphologies (table 21.2) (Spillantini *et al.* 1996, 1997*b*, 1998*e*; Reed *et al.* 1997, 1998). Biochemically, these filaments fall into at least three separate groups: they consist either of all six brain tau isoforms (as in Seattle family A), or they consist predominantly (as in Dutch family 1 and in pallido-ponto-nigral degeneration) or exclusively (as in familial multiple system tauopathy with presenile dementia (MSTD) and in familial progressive subcortical gliosis (PSG)) of only three tau isoforms, each with four microtubule-binding repeats (table 21.2, figure 21.5) (Spillantini *et al.* 1996, 1997*b*, 1998*e*; Reed *et al.* 1998; Goedert *et al.* 1999). A biochemical tau pattern similar to that of familial MSTD and familial PSG is present in progressive supranuclear palsy and corticobasal degeneration, two largely sporadic tauopathies (table 21.1, figure 21.5)

Figure 21.6 Mutations in the tau gene in frontotemporal dementia and Parkinsonism linked to chromosome 17 (FTDP-17). (*a*) Schematic diagram of the six tau isoforms (A–F) that are expressed in adult human brain. Alternatively spliced exons are shown in red (exon 2), green (exon 3) and yellow (exon 10); black bars indicate the microtubule-binding repeats. Seven missense mutations and one deletion mutation in the coding region are shown. They affect all six tau isoforms, with the exception of N279K, ΔK280, P301L, P301S and S305N, which affect only tau isoforms with four microtubule-binding repeats. Amino acid numbering corresponds to the 441-residue isoform of human brain tau. (*b*) Predicted stem–loop in the pre-mRNA at the exon 10–5' intron boundary. The probable destabilizing effects of the S305N mutation and the four intronic mutations are indicated. Exon sequences are shown in upper-case and intron sequences in lower-case letters. (See also colour plate section.)

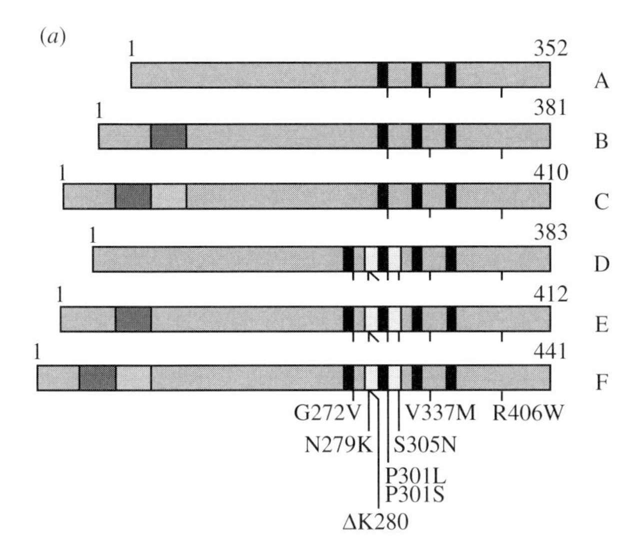
(a)
1
352
A
1
381
B
1
410
C
1
383
D
1
412
E
1
441
F
G272V
V337M
R406W
N279K
S305N
P301L
P301S
ΔK280

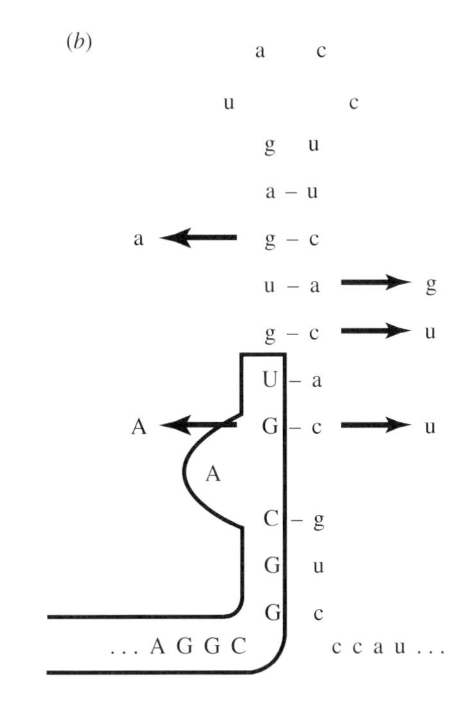
(b)
a c
u c
g u
a – u
a
g – c
u – a
g
g – c
u
U – a
A
G – c
u
A
C – g
G u
G c
. . . A G G C
c c a u . . .

(Flament *et al.* 1991; Ksiezak-Reding *et al.* 1994). Intriguingly, an intronic polymorphism in the tau gene has been reported to be a risk factor for progressive supranuclear palsy (Conrad *et al.* 1997).

Tau mutations in FTDP-17

Aside from having a filamentous tau pathology in common, the familial frontotemporal dementias also share genetic linkage to chromosome 17q21–22, the same region as that containing the tau gene (Wilhelmsen *et al.* 1994). They have therefore been grouped together under the heading of 'frontotemporal dementia and Parkinsonism linked to chromosome 17' (FTDP-17) (Foster *et al.* 1997; Spillantini *et al.* 1998*c*).

Over the past two years, the first mutations in the tau gene have been discovered in a number of these families (figure 21.6 (plate 16)) (Hutton *et al.* 1998; Poorkaj *et al.* 1998; Spillantini *et al.* 1998*d*). The speed with which mutations are being discovered suggests that a defective tau gene is a major cause of inherited dementing disease. The mutations are either missense or deletion mutations in the microtubule-binding repeat region and the C-terminal region, or intronic mutations located close to the splice-donor site of the intron following exon 10 (figure 21.6 (plate 16)) (Goedert *et al.* 1998*b*; Hardy *et al.* 1998). Missense mutations have been found in exons 9, 10, 12 and 13 of the tau gene. They change glycine-272 to valine (G272V), asparagine-279 to lysine (N279K), proline-301 to either leucine (P301L) or serine (P301S), serine-305 to asparagine (S305N), valine-337 to methionine (V337M) and arginine-406 to tryptophan (R406W) (figure 21.6 (plate 16)) (Clark *et al.* 1998; Dumanchin *et al.* 1998; Hutton *et al.* 1998; Poorkaj *et al.* 1998; Spillantini *et al.* 1998*d*; Bugiani *et al.* 1999; Iijima *et al.* 1999; Mirra *et al.* 1999). A mutation that deletes lysine-280 (ΔK280) has been found in exon 10 (figure 21.6 (plate 16)) (Rizzu *et al.* 1999). So far, five different mutations have been described in the 31-residue alternatively spliced exon 10. They affect only four-repeat tau isoforms. In contrast, the other four missense mutations are present in all six brain tau isoforms. Four different intronic mutations have been found close to the splice-donor site of the intron following exon 10 (at positions +3, +13, +14 and +16, with the first nucleotide of the invariant splice-donor site sequence taken as +1) (figure 21.6 (plate 16)) (Hutton *et al.* 1998; Spillantini *et al.* 1998*d*; Goedert *et al.* 1999). These mutations disrupt a predicted stem–loop structure located at the exon 10–5' intron boundary. The S305N missense mutation, which is located in the last residue of exon 10, also disrupts the predicted stem–loop (Iijima *et al.* 1999).

The presence of a stem–loop at the exon 10–5' intron boundary of the tau gene has been inferred from predictions of secondary structure (Hutton *et al.* 1998; Spillantini *et al.* 1998*d*). Its existence has been put on a firm footing with the determination of the three-dimensional structure by NMR spectroscopy of a 25-mer oligonucleotide extending from positions –5 to +19 (Varani *et al.* 1999), showing that the stem–loop forms a stable, folded structure. The stem consists of a lower and an upper part, separated by a bulged adenine at position –2, with the loop consisting of six nucleotides. The structure of the tau exon 10 regulatory element (Varani *et al.* 1999) differs in several respects from the predicted structures (Hutton *et al.* 1998; Spillantini *et al.* 1998*d*). The known intronic mutations are located in the upper part of the stem, which is destabilized as a result, as judged by a marked decrease in melting temperatures. Binding of the aminoglycoside antibiotic neomycin to the mutated stem–loops results in their stabilization, suggesting a possible therapeutical avenue for tauopathies resulting from the increased splicing of exon 10 (Varani *et al.* 1999).

Recombinant tau proteins carrying missense mutations or the deletion mutation have a decreased ability to promote microtubule assembly, which is more marked for three-repeat than for four-repeat isoforms (Hasegawa *et al.* 1998; Bugiani *et al.* 1999; Rizzu *et al.* 1999). Of the mutations tested, the P301L and ΔK280 mutations in exon 10 had the largest effects (Hasegawa *et al.* 1998; Rizzu *et al.* 1999). The likely primary effect of these mutations is thus a decreased ability of mutated tau to interact with microtubules, which amounts to a partial loss of function.

The net effect of the intronic mutations is increased splicing in of exon 10, leading to a change in the ratio of three-repeat to four-repeat tau isoforms and resulting in an overproduction of four-repeat isoforms (Hutton *et al.* 1998; Spillantini *et al.* 1998*d*; Goedert *et al.* 1999). Earlier work had suggested that three-repeat and four-repeat tau isoforms might bind to different sites on microtubules (Goode & Feinstein 1994). The overproduction of tau isoforms with four repeats might result in an excess of tau over available binding sites on microtubules, equivalent to a partial loss of function of the unbound excess tau. Increased splicing in of exon 10 also seems to be the primary mechanism by which the missense mutations N279K and S305N lead to dementia (Hasegawa *et al.* 1999). Recombinant tau proteins with these mutations do not show a decreased ability to promote microtubule assembly, unlike tau proteins with other missense mutations. However, by exon trapping, both mutations lead to increased splicing in of exon 10, exactly as in the intronic mutations (Hasegawa *et al.* 1999). This is explained by the fact that the N279K mutation creates an exon-splice enhancer sequence, whereas the S305N mutation destabilizes the predicted stem–loop at the exon10–5' intron boundary (Clark *et al.* 1998; Hong *et al.* 1998; Iijima *et al.* 1999). Thus, the known mutations in the tau gene either produce a decreased ability to promote microtubule assembly or lead to increased splicing in of exon 10, resulting in the overproduction of four-repeat isoforms (Goedert *et al.* 1998*b*).

Pathogenesis of FTDP-17

The pathway leading from a mutation in the tau gene to neurodegeneration is unknown. A partial loss of function of tau resulting from the mutations could lead to the destabilization of microtubules, with deleterious consequences for cellular processes, such as rapid axonal transport. However, in the intronic mutations, where four-repeat tau is overproduced, this seems

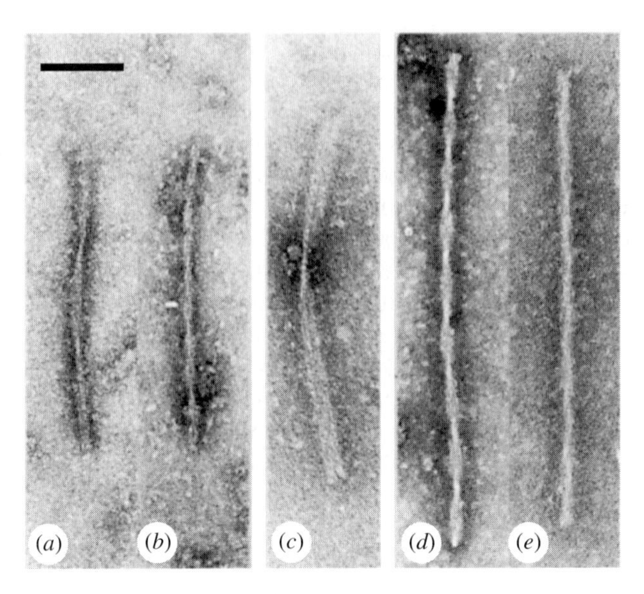

Figure 21.7 Tau filaments in FTDP-17. Dutch family 1 (with the P301L mutation in exon 10) is characterized by the presence of narrow twisted ribbons (*a*) and occasional rope-like filaments (*b*). The tau pathology is both neuronal and glial. Familial multiple system tauopathy with presenile dementia (with the +3 intronic mutation) is characterized by wide twisted ribbons (*c*), which might be formed by two copies of the narrow twisted ribbons joined across the central axis. The tau pathology is both neuronal and glial. Seattle family A (with the V337M mutation in exon 12) is characterized by the presence of paired helical (*d*) and straight (*e*) filaments. The tau pathology is largely neuronal. Scale bar, 100 nm. (Reproduced from Goedert *et al.* (1998*b*).)

unlikely. Moreover, mutations in exon 10 will affect only 20–25% of tau molecules, with 75–80% of tau being normal (Goedert *et al.* 1998*b*).

It is possible, however, that a correct ratio of wild-type three-repeat to four-repeat tau is essential for the normal function of tau in human brain. An alternative hypothesis is that a partial loss of function of tau is necessary for setting in motion the mechanisms that ultimately lead to filament assembly. Besides leading to a partial loss of function phenotype, tau mutations might have additional effects on phosphorylation and filament assembly.

Where studied, pathological tau from FTDP-17 brain is hyperphosphorylated. As the known mutations in tau do not create additional phosphorylation sites (with the possible exception of the P301S mutation), hyperphosphorylation of tau must be an event downstream of the primary effects of the mutations and might be a consequence of the partial loss of function. It probably reinforces the effects of the mutations, because it is well established that hyperphosphorylated tau is unable to bind to microtubules (Bramblett *et al.* 1993; Yoshida & Ihara 1993). The steps lying between hyperphosphorylation of tau and assembly into filaments are at present unknown.

The emerging picture is that of a remarkably direct correspondence between the locations of tau mutations, the cellular pathology, and the isoform compositions and morphologies of tau filaments (table 21.2, figure 21.7). Mutations in exon 10 lead to a neuronal and glial tau pathology, with the narrow twisted ribbon-like filaments being made predominantly of four-repeat tau isoforms. Mutations in the intron following exon 10 lead to a neuronal and glial tau pathology, with the wide twisted ribbon-like filaments being made of only four-repeat tau isoforms. This contrasts with missense mutations located outside exon 10 that lead to a mostly neuronal pathology, with the Alzheimer-type PHFs and SFs being made of all six tau isoforms (Goedert *et al.* 1998*b*).

It is unclear why some mutations lead to a neuronal and glial tau pathology, whereas others result in a largely neuronal pathology. In normal human brain, tau protein is confined mostly to nerve cells, where it is concentrated in axons (Binder *et al.* 1985). Although it is not known which isoforms account for the low levels of tau in glial cells, the cellular pathology of FTDP-17 is compatible with nerve cells expressing all six tau isoforms and glial cells expressing predominantly four-repeat isoforms. For mutations located outside exon 10, the ordered assembly of tau into filaments might be driven by three-repeat isoforms, leading to the formation of PHFs and SFs in nerve cells. Four-repeat tau isoforms drive filament assembly when tau mutations are located in exon 10 or in the intron following exon 10, resulting in the formation of twisted ribbon-like filaments in both nerve cells and glial cells.

It is perhaps not suprising that mutations that affect predominantly four-repeat isoforms give rise to filaments with different morphologies from those resulting from mutations that affect all six tau isoforms. It is well established that the repeat region of tau forms the densely packed core of PHFs and SFs, with the N- and C-terminal parts of the molecule forming a proteolytically sensitive coat (Wischik *et al.* 1988; Goedert *et al.* 1992*b*). Moreover, the morphology of filaments assembled *in vitro* in the presence of sulphated glycosaminoglycans depends on the number of repeats in the tau isoform used (Goedert *et al.* 1996). Thus, mutations in the repeat region or a change in the relative amounts of three- and four-repeat isoforms could well influence filament morphology.

The most important feature of the tau assemblies might be their extended filamentous nature and the deleterious effects that this has on intracellular processes, rather than the detailed morphology of the different filaments. The new work has firmly established that the events leading to a filamentous tau pathology or the mere presence of tau filaments are sufficient for the degeneration of affected nerve cells and glial cells and the onset of dementia.

Lewy body diseases

Parkinson's disease

Parkinson's disease is a movement disorder characterized by tremor, rigidity and bradykinesia. Neuropathologically, it is defined by nerve cell loss in the substantia nigra and several other regions of the nervous system and by the presence of Lewy bodies and Lewy neurites. Under the light microscope, brainstem Lewy bodies appear as round, intracytoplasmic inclusions, 5–25 μm in diameter, with a dense eosinophilic core and a clearer surrounding corona (Lewy 1912). Ultrastructurally, they are composed of a core of filamentous and granular material that is surrounded by radially oriented filaments 10–20 nm in diameter (Duffy and Tennyson 1965; Forno 1996). Lewy neurites constitute an important component of the pathology of Parkinson's disease. They correspond to abnormal neurites that have the same immunohistochemical staining profile as Lewy bodies and consist ultrastructurally of abnormal filaments similar to those found in Lewy bodies.

Despite much work, the biochemical nature of the Lewy body filament remained unknown until recently. On the basis of immunohistochemical findings, some had reached the conclusion that neurofilaments constitute the major filament component (Anderton 1997). However, this type of work does not distinguish between intrinsic Lewy body components and normal cellular constituents that merely become trapped in the filaments that make up the Lewy body. A similar problem plagued the field of Alzheimer's disease over part of the 1980s. It was solved with the purification and analysis of PHFs and SFs

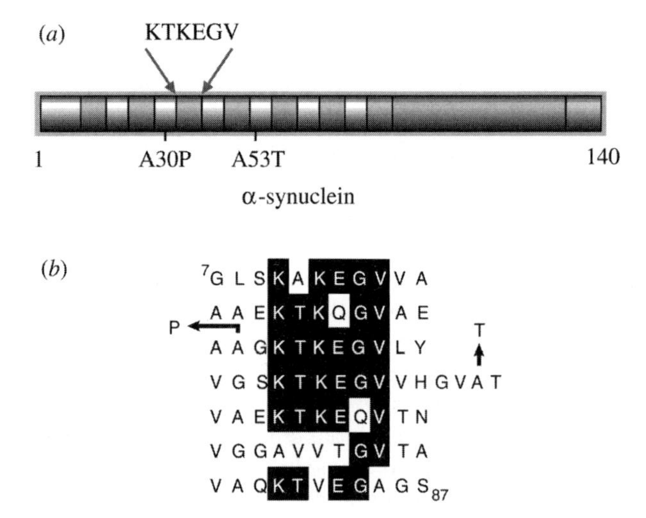

Figure 21.8 Mutations in the α-synuclein gene in familial Parkinson's disease. (*a*) Schematic diagram of human α-synuclein. The seven repeats with the consensus sequence KTKEGV are shown as green bars. The hydrophobic region is shown in blue and the negatively charged C-terminus in red. The two known missense mutations are indicated. (*b*) Repeats in human α-synuclein. Residues 7–87 of the 140-residue protein are shown. Amino acid identities between at least five of the seven repeats are indicated by black bars. The Ala→Pro mutation at residue 30 between repeats two and three and the Ala→Thr mutation at residue 53 between repeats four and five are shown. (Reproduced from Goedert & Spillantini (1998).) (See also colour plate section.)

(Goedert *et al.* 1988; Kondo *et al.* 1988; Wischik *et al.* 1988). A similar approach with Lewy bodies has met with only partial success, mainly because Lewy bodies and Lewy neurites are less abundant than neurofibrillary lesions (Iwatsubo *et al.* 1996). This was the situation until the middle of 1997, when genetics came to the rescue, taking us straight to the very core of the Lewy body filament.

Most cases of Parkinson's disease are sporadic, without an obvious family history. However, a small percentage is familial and inherited in an autosomal-dominant manner. Two separate missense mutations have been discovered in the α-synuclein gene in kindreds with early-onset familial Parkinson's disease (figure 21.8 (plate 17)) (Polymeropoulos *et al.* 1997; Krüger *et al.* 1998). The first mutation, which changes residue 53 in α-synuclein from alanine to threonine (A53T), was identified in a large Italian–American kindred and three smaller Greek pedigrees. The second mutation, which changes alanine-30 to proline (A30P), was found in a German pedigree with early-onset Parkinson's disease.

α-Synuclein is a 140-residue protein of unknown function that is abundantly expressed in brain, where it is located in presynaptic nerve terminals, with little staining of nerve cell bodies and dendrites (Uéda *et al.* 1993; Jakes *et al.* 1994). Two related proteins, called β-synuclein and γ-synuclein (or BCSG1), have also been described in brain (Nakajo *et al.* 1992; Jakes *et al.* 1994; Ji *et al.* 1997). The N-terminal half of each synuclein is taken up by imperfect repeats, with the consensus sequence KTKEGV (single-letter codes). These repeats are followed by a hydrophobic middle region and a negatively charged C-terminal region (figure 21.8 (plate 17)). Both the A30P and A53T mutations lie in the repeat region of α-synuclein. It seems likely that α-synuclein binds through its repeats to other cellular components. Recent work has shown that it does bind through the repeats to synthetic vesicles and to vesicle preparations from rat brain (Davidson *et al.* 1998; Jensen *et al.* 1998). Interestingly, the A30P mutation was found to be devoid of significant vesicle-binding activity (Jensen *et al.* 1998).

Although the A53T mutation in α-synuclein accounts for only a small percentage of familial cases of Parkinson's disease, its identification was quickly followed by the discovery that α-synuclein is the major component of Lewy bodies and Lewy neurites in all cases of Parkinson's disease (table 21.1, figure 21.9 (plate 18)) (Spillantini *et al.* 1997*a*). Full-length, or close to full-length, α-synuclein has been found in Lewy bodies and Lewy neurites, with both the core and the corona of the Lewy body being stained. Staining for α-synuclein has been found to be more extensive than staining for ubiquitin, which was until then the most sensitive marker for Lewy bodies and Lewy neurites (Kuzuhara *et al.* 1988; Spillantini *et al.* 1998*a*). The Lewy body pathology does not stain for β-synuclein or γ-synuclein. Thus, of the three brain synucleins, only α-synuclein is of relevance in the context of Parkinson's disease. The original finding that α-synuclein is present in Lewy bodies and Lewy neurites (Spillantini *et al.* 1997*a*) was rapidly confirmed and extended (Wakabayashi *et al.* 1997, 1998*b*; Arima *et al.* 1998*b*; Baba *et al.* 1998; Irizarry *et al.* 1998; Lippa *et al.* 1998; Mezey *et al.* 1998*a*,*b*; Takeda *et al.* 1998*a*,*b*).

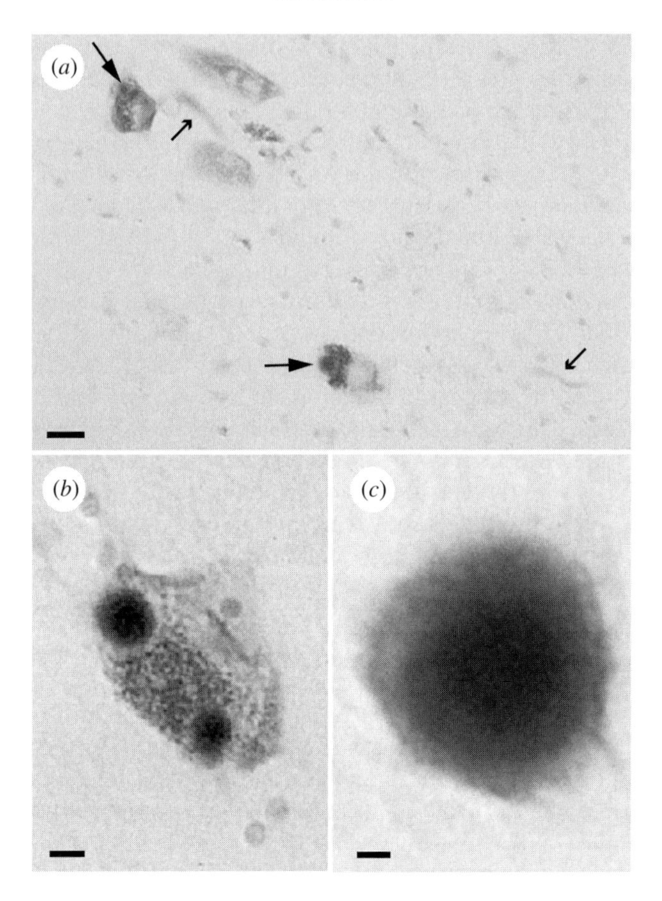

Figure 21.9 Substantia nigra from patients with Parkinson's disease immunostained for α-synuclein. (*a*) Two pigmented nerve cells, each containing an α-synuclein-positive Lewy body (thin arrows). Lewy neurites (thick arrows) are also immunopositive. Scale bar, 20 μm. (*b*) Pigmented nerve cell with two α-synuclein-positive Lewy bodies. Scale bar, 8 μm. (*c*) α-Synuclein-positive extracellular Lewy body. Scale bar, 4 μm. (Reproduced from Spillantini *et al.* (1997*a*).) (See also colour plate section.)

The A30P and A53T mutations might promote the aggregation of α-synuclein into filaments, resulting in the formation of Lewy bodies and Lewy neurites. Alternatively, they might interfere with a normal property of α-synuclein that could in turn indirectly facilitate assembly into filaments. In either case, the net effect would be akin to a gain of toxic function. In idiopathic Parkinson's disease, as yet unknown modifications in α-synuclein or interactions with other components might lead to aggregation into filaments. Lewy bodies and Lewy neurites are space-occupying lesions that fill most of the cytoplasm of affected nerve cells. This might, in turn, lead to the entrapment of normal cellular components, possibly explaining the variable staining of Lewy bodies and Lewy neurites for neurofilaments and other proteins.

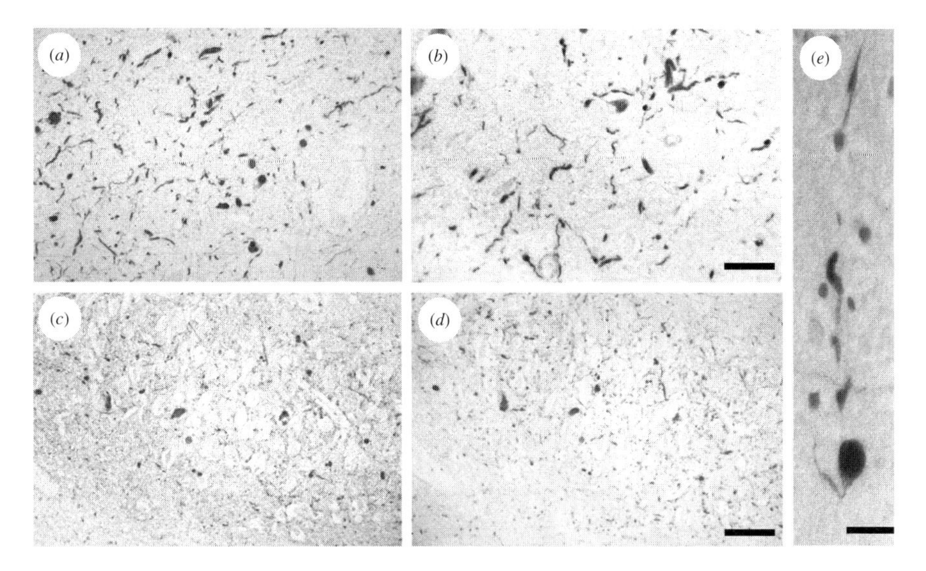

Figure 21.10 Brain tissue from patients with dementia with Lewy bodies immunostained for α-synuclein. (*a*,*b*) α-Synuclein-positive Lewy bodies and Lewy neurites in substantia nigra stained with antibodies recognizing the N-terminal (*a*) or the C-terminal (*b*) region of α-synuclein. Scale bar in (*b*), 100 μm (for (*a*) and (*b*)). (*c*,*d*) α-Synuclein-positive Lewy neurites in serial sections of hippocampus stained with antibodies recognizing the N-terminal (*c*) or the C-terminal (*d*) region of α-synuclein. (*e*) α-Synuclein-positive intraneuritic Lewy body in a Lewy neurite in substantia nigra stained with an antibody recognizing the C-terminal region of α-synuclein. Scale bar, 40 μm. (Reproduced from Spillantini *et al.* (1998*a*).) (See also colour plate section.)

Over time, the presence of Lewy bodies and Lewy neurites is likely to lead to nerve cell degeneration (Goedert 1997; Goedert & Spillantini 1998).

Dementia with Lewy bodies

Lewy bodies and Lewy neurites also constitute the defining neuropathological characteristics of dementia with Lewy bodies, a common late-life dementia that exists in a pure form or overlaps with the neuropathological characteristics of Alzheimer's disease, especially Aβ deposits. Some studies have suggested that dementia with Lewy bodies is the second most common cause of dementia, after Alzheimer's disease.

Unlike Parkinson's disease, it is characterized by the presence of numerous Lewy bodies and Lewy neurites in cerebral cortex (Kosaka 1978). As in Parkinson's disease, the Lewy body pathology is also present in the substantia nigra and other subcortical regions. Lewy bodies and Lewy neurites from dementia with Lewy bodies are strongly immunoreactive for α-synuclein, exactly as in the pathological features of Parkinson's disease (table 21.1, figure 21.10 (plate 19)) (Spillantini *et al.* 1997*a*). It suggests, but does not

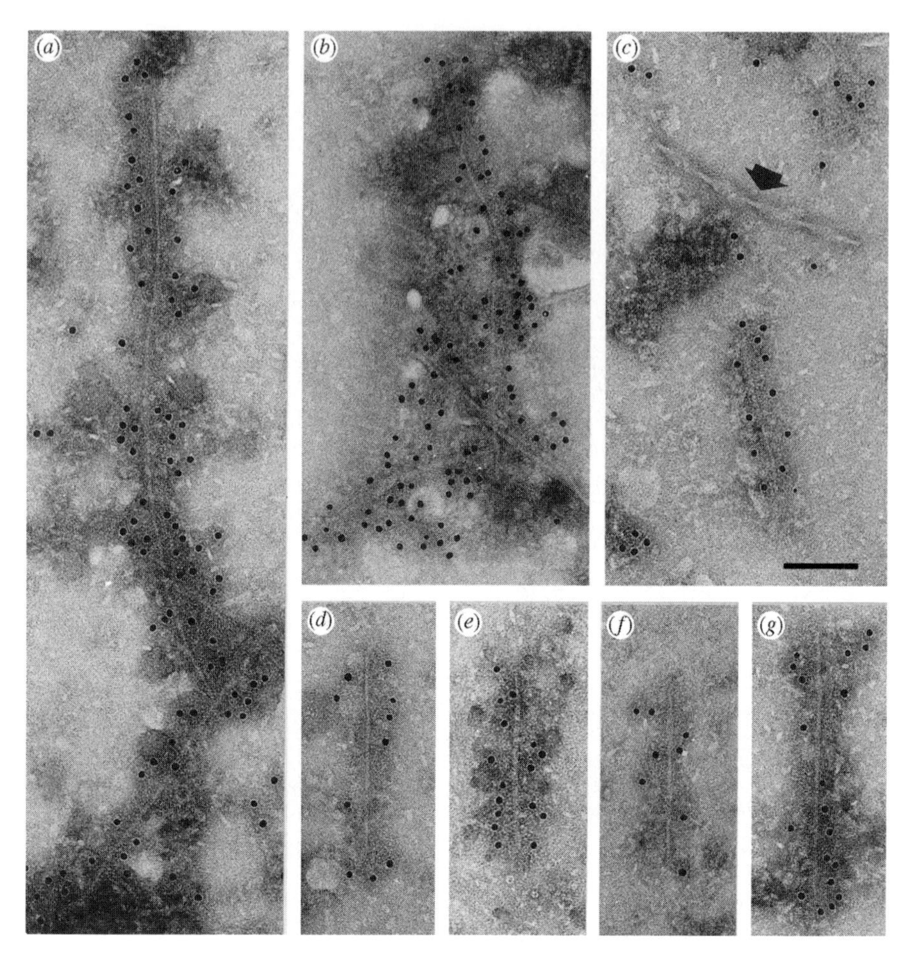

Figure 21.11 (*a,b*) Filaments from cingulate cortex of patients with dementia with Lewy bodies immunolabelled for α-synuclein. Small clumps of α-synuclein filaments. (*c*) A labelled α-synuclein filament and an unlabelled paired helical filament (arrow). (*d–g*) The labelled filaments have various morphologies, including 5 nm filament (*d*), 10 nm filament with dark stain penetrating centre line (*e*), twisted filament showing alternating width (*f*) and 10 nm filament with slender 5 nm extensions at ends ((*g*), also (*c*)). The 10 nm gold particles attached to the secondary antibody appear as black dots. Scale bar, 100 nm (in (*c*)). (Reproduced from Spillantini *et al.* (1998*a*).)

Figure 21.12 Immunolabelling of synthetic α-synuclein(1–120) filaments and filaments extracted from diseased brains, labelled with an antibody against the C-terminal region of α-synuclein (*a–e*) or labelled with an antibody directed against the N-terminal region of α-synuclein (*f–i*). (*a–c*) α-Synuclein(1–120) filaments showing a labelled clump and individual filaments. (*d,e*) Filaments extracted from brains with Lewy body dementia (*d*) or multiple system atrophy (*e*). (*f–i*) End-labelled filaments, showing α-synuclein(1–120) filaments (*f–h*) and a filament from a brain with multiple system atrophy (*i*). The 10 nm gold particles attached to the secondary antibody appear as black dots. Scale bar, 100 nm. (Reproduced from Crowther *et al.* (1998).)

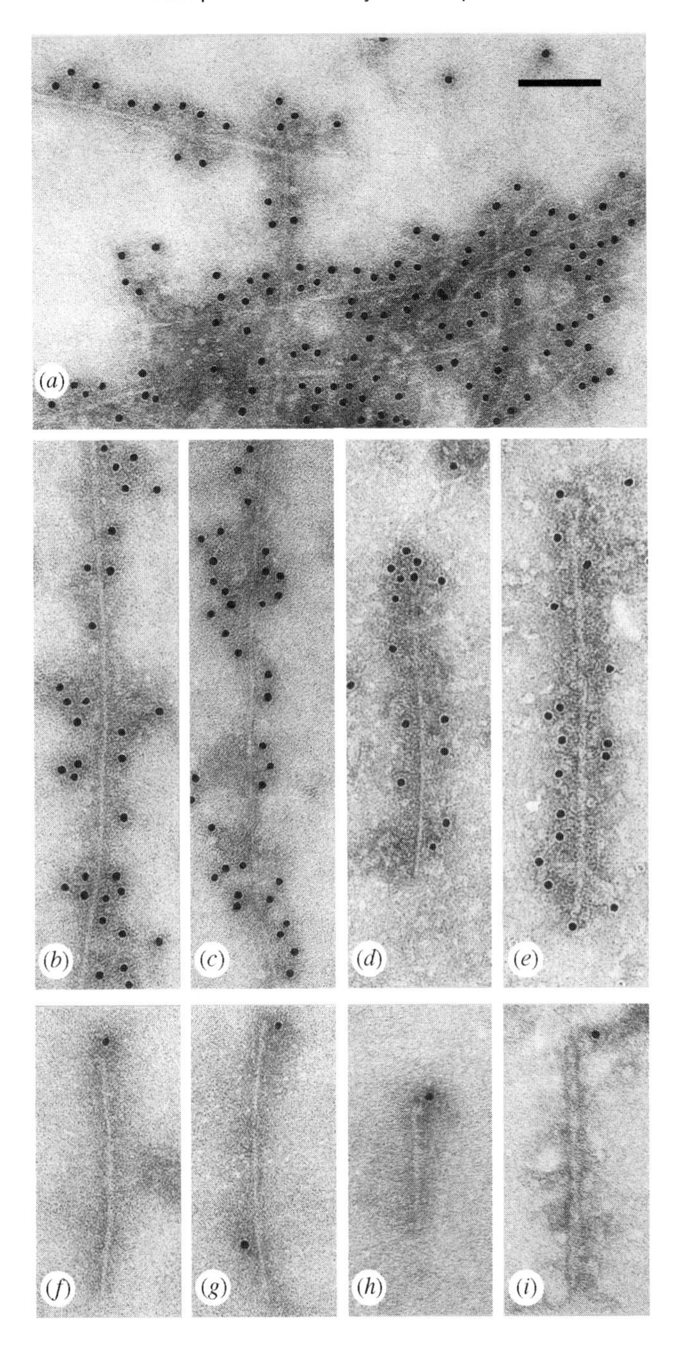
(a)
(b)
(c)
(d)
(e)
(f)
(g)
(h)
(i)

prove, that α-synuclein is the major component of the abnormal filaments that make up Lewy bodies and Lewy neurites. The pathological changes are particularly numerous in cingulate cortex, facilitating the extraction of filaments. Isolated filaments were strongly labelled for α-synuclein along their entire lengths, demonstrating that they contain α-synuclein as a major component (figure 21.11) (Spillantini *et al.* 1998*a*). Filament morphologies and staining characteristics with several antibodies have led to the suggestion that α-synuclein molecules might run parallel to the filament axis and that the filaments are polar structures (figure 21.12). Moreover, under the electron microscope, some filaments and granular material in partly purified Lewy bodies seem to be labelled by α-synuclein antibodies (Baba *et al.* 1998). Immunoelectron microscopy has shown decoration of Lewy body filaments in tissue sections from brain of individuals with dementia with Lewy bodies.

As in Parkinson's disease, the presence of abnormal filamentous α-synuclein inclusions in nerve cells is probably the cause of nerve cell degeneration in dementia with Lewy bodies. Although most Lewy bodies are confined to the cell soma, this is not their only location. Thus, the intraneuritic Lewy body, which consists of a large, circular Lewy body within a neurite, is a most striking pathological feature of Parkinson's disease and dementia with Lewy bodies (figure 21.10*e* (plate 19)). Its presence is bound to lead to an interruption of axonal transport, akin to a nerve ligation, with obvious deleterious consequences for the whole nerve cell. Pathological structures such as this serve to illustrate the implausibility of the view expressed by some in the context of glutamine repeat diseases, namely that filamentous intraneuronal inclusions might be neutral or even have a protective effect (Saudou *et al.* 1998; Sisodia 1998).

Synthetic α-synuclein filaments

The discovery of α-synuclein filaments in Parkinson's disease and dementia with Lewy bodies has led to attempts aimed at producing synthetic α-synuclein filaments. A first study has reported that removal of the C-terminal 20–30 residues of α-synuclein leads to spontaneous assembly into filaments within 24–48 h at 37°C, with morphologies and staining characteristics indistinguishable from those of Lewy body filaments (figure 21.12) (Crowther *et al.* 1998). This indicates that the packing of α-synuclein molecules in the filaments *in vitro* is very similar to that of filaments extracted from brain. A large proportion of α-synuclein extracted from partly purified Lewy bodies has been found to be truncated (Baba *et al.* 1998). In conjunction with the results of studies *in vitro*, this suggests that proteolytic degradation might have a role in the assembly of α-synuclein in Lewy body diseases. A second study on synthetic α-synuclein filaments has reported assembly from full-length protein with the A53T mutation, after incubations ranging from three weeks to two months at 37°C (Conway *et al.* 1998). It remains to be seen whether synthetic filaments can be produced from mutated full-length α-synuclein under certain conditions or whether C-terminal truncation is an obligatory step for assembly.

Whatever exact mechanisms underlie assembly, these findings help to establish firmly that Lewy body filaments are made of α-synuclein. Together with the light microscopic and electron microscopic studies of Lewy bodies and Lewy neurites, they refute the notion that Lewy body filaments are made of neurofilaments. They also provide first assays for the testing of compounds aimed at preventing the assembly of α-synuclein into Lewy body-like filaments.

Multiple system atrophy

Multiple system atrophy is a neurodegenerative disorder that comprises cases of olivopontocerebellar atrophy, striatonigral degeneration and Shy–Drager syndrome (Graham & Oppenheimer 1969). Clinically, it is characterized by a combination of cerebellar, extrapyramidal and autonomic symptoms.

Neuropathologically, glial cytoplasmic inclusions (GCIs), which consist of filamentous aggregates, are the defining feature of multiple system atrophy (Papp *et al.* 1989). They are found mostly in the cytoplasm and, to a lesser extent, in the nucleus of oligodendrocytes. Inclusions are also observed in the cytoplasm and nucleus of some nerve cells, as well as in neuropil threads. They consist of straight and twisted filaments, with reported diameters of 10–30 nm (Kato & Nakamura 1990). At the light microscopic level, GCIs are immunoreactive for ubiquitin and, to a lesser extent, for cytoskeletal proteins such as tau and tubulin. However, until recently, the biochemical composition of GCI filaments was unknown.

This has changed with the discovery that GCIs are strongly immunoreactive for α-synuclein and that filaments isolated from the brains of patients with multiple system atrophy are strongly labelled by α-synuclein antibodies (table 21.1) (Arima *et al.* 1998*a*; Gai *et al.* 1998; Mezey *et al.* 1998*a*; Spillantini *et al.* 1998*d*; Tu *et al.* 1998; Wakabayashi *et al.* 1998*a*,*b*). The filament morphologies and their staining characteristics were found to be very similar to those of filaments extracted from cingulate cortex of patients with dementia with Lewy bodies (figure 21.12) (Spillantini *et al.* 1998*d*). As for the latter, staining for α-synuclein was far more extensive than staining for ubiquitin, until then the most sensitive immunohistochemical marker of GCIs (Spillantini *et al.* 1998*d*). This work refutes the view that filaments from brain with multiple system atrophy are made of tau protein or other cytoskeletal components. It indicates that α-synuclein is the major component of the GCI filaments and reveals an unexpected molecular link between multiple system atrophy and the Lewy body disorders Parkinson's disease and dementia with Lewy bodies.

Conclusion

The discovery that tau protein and α-synuclein account for the filamentous neuronal and glial inclusions of most late-onset neurodegenerative diseases has provided a unifying theme to our understanding of these disorders. The presence of mutations in the tau gene in FTDP-17 and in the α-synuclein gene in familial Parkinson's disease has underscored the crucial importance of tau and α-synuclein for the neurodegenerative process. Tau and α-synuclein are soluble proteins in normal brain. Understanding their abnormal assembly into filaments is thus central to the study of these diseases. Filament assembly seems to be an energetically unfavourable, nucleation-dependent process that requires a critical concentration of tau or α-synuclein (Goedert *et al.* 1996; Crowther *et al.* 1998). The concentration dependence of assembly might be part of the reason why some cells are much more prone to developing pathology than others. Many cells might have levels of tau or α-synuclein below the critical concentration. Other cells might have effective mechanisms for preventing the formation of nuclei or might be able to degrade them once they have formed. Insufficient protective mechanisms and tau or α-synuclein concentrations above the critical concentration might underlie the selective degeneration of nerve cells and glial cells, which constitutes a central characteristic of most neurodegenerative diseases and is responsible for the distinctive clinical phenotype of each disease (Goedert *et al.* 1998*b*).

Although tau and α-synuclein share no sequence similarities, they have some properties in common. They are both natively unfolded proteins, without much secondary structure, as reflected in the fact that they are both heat-stable (Jakes *et al.* 1994; Schweers *et al.* 1994; Weinreb *et al.* 1996). Like tau, α-synuclein contains repeats through which it can bind to other cellular components and become structured in the process. It seems that the primary effect of the tau and α-synuclein mutations is to decrease their ability to interact effectively with their respective binding partners (Hasegawa *et al.* 1998; Jensen *et al.* 1998). This partial loss of function might be necessary for setting in motion the mechanisms that lead to filament formation and the subsequent demise of affected nerve cells and glial cells. Ordered assembly into filaments as the gain of toxic function is an emerging theme in the study of neurodegenerative diseases.

References

Alzheimer, A. 1907 Über eine eigenartige Erkrankung der Hirnrinde. *Allg. Z. Psychiat. Psych. Gerichtl. Med.* **64**, 146–148.

Alzheimer, A. 1911 Über eigenartige Krankheitsfälle des späteren Alters. *Z. Ges. Neurol. Psychiat.* **4**, 356–385.

Anderton, B. H. 1997 Changes in the ageing brain in health and disease. *Philos. Trans. R. Soc. Lond.* B **352**, 1781–1792.

Andreadis, A., Brown, M. W. & Kosik, K. S. 1992 Structure and novel exons of the human tau gene. *Biochemistry* **31**, 10626–10633.

Arima, K., Uéda, K., Sunohara, N., Arakawa, K., Hirai, S., Nakamura, M., Tonozuka-Uehara, H. & Kawai, M. 1998a NACP/α-synuclein immunoreactivity in fibrillary components of neuronal and oligodendroglial cytoplasmic inclusions in the pontine nuclei in multiple system atrophy. *Acta Neuropathol.* **96**, 439–444.

Arima, K., Uéda, K., Sunohara, N., Hirai, S., Izumiyama, Y., Tonozuka-Uehara, H. & Kawai, M. 1998*b* Immunoelectron-microscopic demonstration of NACP/α-synuclein epitopes on the filamentous component of Lewy bodies in Parkinson's disease and in dementia with Lewy bodies. *Brain Res.* **808**, 93–100.

Arrasate, M., Pérez, M., Valpuesta, J. M. & Avila, J. 1997 Role of glycosaminoglycans in determining the helicity of paired helical filaments. *Am. J. Pathol.* **151**, 1115–1122.

Arriagada, P. V., Growdon, J. H., Hedley-White, E. T. & Hyman, B. T. 1992 Neurofibrillary tangles but not senile plaques parallel duration and severity of Alzheimer's disease. *Neurology* **42**, 631–638.

Baba, M., Nakajo, S., Tu, P.-H., Tomita, T., Nakaya, K., Lee, V. M.-Y., Trojanowski, J. Q. & Iwatsubo, T. 1998 Aggregation of α-synuclein in Lewy bodies of sporadic Parkinson's disease and dementia with Lewy bodies. *Am. J. Pathol.* **152**, 879–884.

Binder, L. I., Frankfurter, A. & Rebhun, L. I. 1985 The distribution of tau in the mammalian nervous system. *J. Cell Biol.* **101**, 1371–1378.

Bondareff, W., Mountjoy, C. Q., Roth, M. & Hauser, D. L. 1989 Neurofibrillary degeneration and neuronal loss in Alzheimer's disease. *Neurobiol. Aging* **10**, 709–715.

Braak, H. & Braak, E. 1991 Neuropathological stageing of Alzheimer-related changes. *Acta Neuropathol.* **82**, 239–259.

Braak, H. & Braak, E. 1997 Frequency of stages of Alzheimer-related lesions in different age categories. *Neurobiol. Aging* **18**, 351–357.

Braak, E., Braak, H. & Mandelkow, E. M. 1994 A sequence of cytoskeleton changes related to the formation of neurofibrillary tangles and neuropil threads. *Acta Neuropathol.* **87**, 554–567.

Bramblett, G. T., Goedert, M., Jakes, R., Merrick, S. E., Trojanowski, J. Q. & Lee, V. M.-Y. 1993 Abnormal tau phosphorylation at Ser396 in Alzheimer's disease recapitulates development and contributes to reduced microtubule binding. *Neuron* **10**, 1089–1099.

Brion, J. P., Passareiro, H., Nunez, J. & Flament-Durand, J. 1985 Mise en évidence immunologique de la protéine tau au niveau des lésions de dégénérescence neurofibrillaire de la maladie d'Alzheimer. *Arch. Biol.* **95**, 229–235.

Brion, J. P., Tremp, G. & Octave, J. N. 1999 Transgenic expression of the shortest human tau affects its compartmentalization and its phosphorylation as in the pretangle stage of Alzheimer's disease. *Am. J. Pathol.* **154**, 255–270.

Bugiani, O. (and 15 others) 1999 Frontotemporal dementia and corticobasal degeneration in a family with a Pro301Ser mutation in tau. *J. Neuropathol. Exp. Neurol.* (In the press.)

Calhoun, M. E., Wiederholt, K. H., Abramowski, D., Phinney, A. L., Probst, A., Sturchler-Pierrat, C., Staufenbiel, M., Sommer, B. & Jucker, M. 1998 Neuron loss in APP transgenic mice. *Nature* **395**, 755–756.

Clark, L. N. (and 18 others) 1998 Pathogenic implications of mutations in the tau gene in pallido-ponto-nigral degeneration and related neurodegenerative disorders linked to chromosome 17. *Proc. Natl Acad. Sci. USA* **95**, 13103–13107.

Conrad, C. (and 12 others) 1997 Genetic evidence for the involvement of tau in progressive supranuclear palsy. *Ann. Neurol.* **41**, 277–181.

Conway, K. A., Harper, J. D. & Lansbury, P. T. 1998 Accelerated *in vitro* fibril formation by a mutant α-synuclein linked to early-onset Parkinson disease. *Nature Med.* **4**, 1318–1320.

Couchie, D., Mavilia, C., Georgieff, I. S., Liem, R. K. H., Shelanski, M. L. & Nunez, J. 1992 Primary structure of high molecular weight tau present in the peripheral nervous system. *Proc. Natl Acad. Sci. USA* **89**, 4378–4381.

Cras, P., Smith, M. A., Richey, P. L., Siedlak, S. L., Mulvihill, P. & Perry, G. 1995 Extracellular neurofibrillary tangles reflect neuronal loss and provide further evidence of extensive protein cross-linking in Alzheimer disease. *Acta Neuropathol.* **89**, 291–295.

Crowther, R. A. 1991 Straight and paired helical filaments in Alzheimer disease have a common structural unit. *Proc. Natl Acad. Sci. USA* **88**, 2288–2292.
Crowther, R. A., Jakes, R., Spillantini, M. G. & Goedert, M. 1998 Synthetic filaments assembled from C-terminally truncated α-synuclein. *FEBS Lett.* **436**, 309–312.
Davidson, W. S., Jonas, A., Clayton, D. F. & George, J. M. 1998 Stabilization of α-synuclein secondary structure upon binding to synthetic membranes. *J. Biol. Chem.* **273**, 9443–9449.
Delacourte, A., Sergeant, N., Wattez, A., Gauvreau, D. & Robitaille, Y. 1998 Vulnerable neuronal subsets in Alzheimer's and Pick's diseases are distinguished by their tau isoform distribution and phosphorylation. *Ann. Neurol.* **43**, 193–204.
Duff, K. & Hardy, J. 1995 Alzheimer's disease. Mouse model made. *Nature* **373**, 476–477.
Duffy, P. E. & Tennyson, V. M. 1965 Phase and electron microscopic observations of Lewy bodies and melanin granules in the substantia nigra and locus coeruleus in Parkinson's disease. *J. Neuropathol. Exp. Neurol.* **24**, 398–414.
Dumanchin, C. (and 12 others) 1998 Segregation of a missense mutation in the microtubule-associated protein tau gene with familial frontotemporal dementia and parkinsonism. *Hum. Mol. Genet.* **7**, 1825–1829.
Flament, S., Delacourte, A., Verny, M., Hauw, J. J. & Javoy-Agid, F. 1991 Abnormal tau proteins in progressive supranuclear palsy. Similarities and differences with the neurofibrillary degeneration of the Alzheimer type. *Acta Neuropathol.* **81**, 591–596.
Forno, L. S. 1996 Neuropathology of Parkinson's disease. *J. Neuropathol. Exp. Neurol.* **55**, 259–272.
Foster, N. L. (and 40 others) 1997 Frontotemporal dementia and Parkinsonism linked to chromosome 17: a consensus statement. *Ann. Neurol.* **41**, 706–715.
Friedhoff, P., Schneider, A., Mandelkow, E. M. & Mandelkow, E. 1998 Rapid assembly of Alzheimer-like paired helical filaments from microtubule-associated protein tau monitored by fluorescence in solution. *Biochemistry* **37**, 10223–10230.
Fukutani, Y., Kobayashi, K., Nakamura, I., Watanabe, K., Isaki, K. & Cairns, N. J. 1995 Neurons, intracellular and extracellular neurofibrillary tangles in subdivisions of the hippocampal cortex in normal ageing and Alzheimer's disease. *Neurosci. Lett.* **200**, 57–60.
Gai, W. P., Power, J. H. T., Blumbergs, P. C. & Blessing, W. W. 1998 Multiple system atrophy: a new α-synuclein disease? *Lancet* **352**, 547–548.
Games, D. (and 23 others) 1995 Alzheimer-type neuropathology in transgenic mice overexpressing V717F β-amyloid precursor protein. *Nature* **373**, 523–527.
Glenner, G. G. & Wong, C. W. 1984 Alzheimer's disease: initial report of the purification and characterization of a novel cerebrovascular amyloid protein. *Biochem. Biophys. Res. Commun.* **120**, 885–890.
Ginsberg, S. D., Crino, P. B., Lee, V. M. Y., Eberwine, J. H. & Trojanowski, J. Q. 1997 Sequestration of RNA in Alzheimer's disease neurofibrillary tangles and senile plaques. *Ann. Neurol.* **41**, 200–209.
Ginsberg, S. D., Galvin, J. E., Chiu, T. S., Lee, V. M. Y, Masliah, E. & Trojanowski, J. Q. 1998 RNA sequestration to pathological lesions of neurodegenerative diseases. *Acta Neuropathol.* **96**, 487–494.
Goate, A. (and 20 others) 1991 Segregation of a missense mutation in the amyloid precursor protein gene with familial Alzheimer's disease. *Nature* **349**, 704–706.
Goedert, M. 1997 Familial Parkinson's disease. The awakening of α-synuclein. *Nature* **388**, 232–233.
Goedert, M. & Hasegawa, M. 1999 The tauopathies. Towards an experimental animal model. *Am. J. Pathol.* **154**, 1–6.
Goedert, M. & Jakes, R. 1990 Expression of separate isoforms of human tau protein: correlation with the tau pattern in brain and effects on tubulin polymerization. *EMBO J.* **9**, 42250–4230.

Goedert, M. & Spillantini, M. G. 1998 Lewy body diseases and multiple system atrophy as α-synucleinopathies. *Mol. Psychiatr.* **3**, 462–465.
Goedert, M., Wischik, C. M., Crowther, R. A., Walker, J. E. & Klug, A. 1988 Cloning and sequencing of the cDNA encoding a core protein of the paired helical filament of Alzheimer disease. *Proc. Natl Acad. Sci. USA* **85**, 4051–4055.
Goedert, M., Spillantini, M. G., Potier, M. C., Ulrich, J. & Crowther, R. A. 1989*a* Cloning and sequencing of the cDNA encoding an isoform of microtubule-associated protein tau containing four tandem repeats: differential expression of tau protein mRNAs in human brain. *EMBO J.* **8**, 393–399.
Goedert, M., Spillantini, M. G., Jakes, R., Rutherford, D. & Crowther, R. A. 1989*b* Multiple isoforms of human microtubule-associated protein tau: sequences and localization in neurofibrillary tangles of Alzheimer's disease. *Neuron* **3**, 519–526.
Goedert, M., Spillantini, M. G. & Crowther, R. A. 1992*a* Cloning of a big tau microtubule-associated protein characteristic of the peripheral nervous system. *Proc. Natl Acad. Sci. USA* **89**, 1983–1987.
Goedert, M., Spillantini, M. G., Cairns, N. J. & Crowther, R. A. 1992*b* Tau proteins of Alzheimer paired helical filaments: abnormal phosphorylation of all six brain isoforms. *Neuron* **8**, 159–168.
Goedert, M., Cohen, E. S., Jakes, R. & Cohen, P. 1992*c* p42 MAP kinase phosphorylation sites in microtubule-associated protein tau are dephosphorylated by protein phosphatase $2A_1$. *FEBS Lett.* **312**, 95–99.
Goedert, M., Jakes, R., Crowther, R. A., Six, J., Lübke, U., Vandermeeren, M., Cras, P., Trojanowski, J .Q. & Lee, V. M.-Y. 1993 The abnormal phosphorylation of tau protein at Ser-202 in Alzheimer disease recapitulates phosphorylation during development. *Proc. Natl Acad. Sci. USA* **90**, 5066–5070.
Goedert, M., Jakes, R., Spillantini, M. G., Hasegawa, M., Smith, M. J. & Crowther, R. A. 1996 Assembly of microtubule-associated protein tau into Alzheimer-like filaments induced by sulphated glycosaminoglycans. *Nature* **383**, 550–553.
Goedert, M., Hasegawa, M., Jakes, R., Lawler, S., Cuenda, A. & Cohen, P. 1997 Phosphorylation of microtubule-associated protein tau by stress-activated protein kinases. *FEBS Lett.* **409**, 57–62.
Goedert, M., Spillantini, M. G. & Davies, S. W. 1998*a* Filamentous nerve cell deposits in neurodegenerative diseases. *Curr. Opin. Neurobiol.* **8**, 619–632.
Goedert, M., Crowther, R. A. & Spillantini, M. G. 1998*b* Tau mutations cause frontotemporal dementias. *Neuron* **21**, 955–958.
Goedert, M. (and 11 others) 1999 Tau gene mutation in familial progressive subcortical gliosis. *Nature Med.* 5, 454–457.
Goode, B. L. & Feinstein, S. C. 1994 Identification of a novel microtubule binding and assembly domain in the developmentally regulated inter-repeat region of tau. *J. Cell Biol.* **124**, 769–782.
Götz, J., Probst, A., Spillantini, M. G., Schäfer, T., Jakes, R., Bürki, K. & Goedert, M. 1995 Somatodendritic localisation and hyperphosphorylation of tau protein in transgenic mice expressing the longest human brain tau isoform. *EMBO J.* **14**, 1304–1313
Graham, J. C. & Oppenheimer, D. R. 1969 Orthostatic hypotension and nicotine sensitivity in a case of multiple system atrophy. *J. Neurol. Neurosurg. Psychiatr.* **32**, 28–34.
Greenberg, S. G. & Davies, P. 1990 A preparation of Alzheimer paired helical filaments that displays distinct tau proteins by polyacrylamide gel electrophoresis. *Proc. Natl Acad. Sci. USA* **87**, 5827–5831.
Gustke, N., Trincczek, B., Biernat, J., Mandelkow, E. M. & Mandelkow, E. 1994 Domains of tau protein and interaction with microtubules. *Biochemistry* **33**, 9511–9522.
Hall, G. F., Yao, J. & Lee, G. 1997 Human tau becomes phosphorylated and forms filamentous deposits when overexpressed in lamprey central neurons *in situ*. *Proc. Natl Acad. Sci. USA* **94**, 4733–4738.

Hanger, D. P., Betts, J. C., Loviny, T. L. F., Blackstock, W. P. & Anderton, B. H. 1998 New phosphorylation sites identified in hyperphosphorylated tau (paired helical filament-tau) from Alzheimer's disease brain using nanoelectrospray mass spectrometry. *J. Neurochem.* **71**, 2465–2476.

Harada, A., Oguchi, K., Okabe, S., Kuno, J., Terada, S., Ohshima, T., Sato-Yoshitake, R., Takei, Y., Noda, T. & Hirokawa, N. 1994 Altered microtubule organization in small-calibre axons of mice lacking tau protein. *Nature* **369**, 488–491.

Hardy, J. & Gwinn Hardy, K. 1998 Genetic classification of primary neurodegenerative disease. *Science* **282**, 1075–1079.

Hardy, J., Duff, K., Gwinn Hardy, K., Perez-Tur, J. & Hutton, M. 1998 Genetic dissection of Alzheimer's disease and related dementias: amyloid and its relationship to tau. *Nature Neurosci.* **1**, 355–358.

Hasegawa, M., Crowther, R. A., Jakes, R. & Goedert, M. 1997 Alzheimer-like changes in microtubule-associated protein tau induced by sulfated glycosaminoglycans. Inhibition of microtubule binding, stimulation of phosphorylation, and filament assembly depend on the degree of sulfation. *J. Biol. Chem.* **272**, 33118–33124.

Hasegawa, M., Smith, M. J. & Goedert, M. 1998 Tau proteins with FTDP-17 mutations have a reduced ability to promote microtubule assembly. *FEBS Lett.* **437**, 207–210.

Hasegawa, M., Smith, M. J., Iijima, M., Tabira, T. & Goedert, M. 1999 FTDP-17 mutations N279K and S305N in tau produce increased splicing of exon 10. *FEBS Lett.* **443**, 93–96.

Hirokawa, N. 1994 Microtubule organization and dynamics dependent on microtubule-associated proteins. *Curr. Opin. Cell Biol.* **6**, 74–81.

Hong, M., Chen, D. C. R., Klein, P. S. & Lee, V. M.-Y. 1997 Lithium reduces tau phosphorylation by inhibition of glycogen synthase kinase-3. *J. Biol. Chem.* **272**, 25326–25332.

Hong, M. (and 14 others) 1998 Mutation-specific functional impairments in distinct tau isoforms and hereditary FTDP-17. *Science* **282**, 1914–1917.

Hsiao, K., Chapman, P., Nilsen, S., Eckman, C., Harigaya, Y., Younkin, S., Yang, F. & Cole, G. 1996 Correlative memory deficits, Aβ elevation, and amyloid plaques in transgenic mice. *Science* **274**, 99–102.

Hutton, M. (and 50 others) 1998 Association of missense and 5'-splice-site mutations in tau with the inherited dementia FTDP-17. *Nature* **393**, 702–705.

Iijima, M. (and 12 others) 1999 A distinct familial presenile dementia with a novel missense mutation in the tau gene. *NeuroReport* **10**, 497–501.

Irizarry, M. C., Soriano, F., McNamara, M., Page, K. J., Schenk, D., Games, D. & Hyman, B. T. 1997*a* Aβ deposition is associated with neuropil changes, but not with overt neuronal loss in the human amyloid precursor protein V717F (PDAPP) transgenic mouse. *J. Neurosci.* **17**, 7053–7059.

Irizarry, M. C., McNamara, M., Fedorchak, K., Hsiao, K. & Hyman, B. T. 1997*b* APP_{Sw} transgenic mice develop age-related Aβ deposits and neuropil abnormalities, but no neuronal loss in CA1. *J. Neuropathol. Exp. Neurol.* **56**, 965–973.

Irizarry, M. C., Growdon, W., Gomez-Isla, T., Newell, K., George, J. M., Clayton, D. F. & Hyman, B. T. 1998 Nigral and cortical Lewy bodies and dystrophic nigral neurites in Parkinson's disease and cortical Lewy body disease contain α-synuclein immunoreactivity. *J. Neuropathol. Exp. Neurol.* **57**, 334–337.

Iwatsubo, T., Yamaguchi, H., Fujimoro, M., Yokosawa, H., Ihara, Y., Trojanowski, J. Q. & Lee, V. M.-Y. 1996 Purification and characterization of Lewy bodies from the brains of patients with diffuse Lewy body disease. *Am. J. Pathol.* **148**, 1517–1529.

Jakes, R., Spillantini, M. G. & Goedert, M. 1994 Identification of two distinct synucleins from human brain. *FEBS Lett.* **345**, 27–32.

Jensen, P. H., Nielsen, M. H., Jakes, R., Dotti, C. G. & Goedert, M. 1998 Binding of α-synuclein to rat brain vesicles is abolished by familial Parkinson's disease mutation. *J. Biol. Chem.* **273**, 26292–26294.

Ji, H., Liu, Y. E., Jia, T., Wang, M., Liu, J., Xiao, G., Joseph, B. K., Rosen, C. & Shi, Y. E. 1998 Identification of a breast cancer-specific gene, *BCSG1*, by direct differential cDNA sequencing. *Cancer Res.* **57**, 759–764.

Kampers, T., Friedhoff, P., Biernat, J., Mandelkow, E. M. & Mandelkow, E. 1996 RNA stimulates aggregation of microtubule-associated protein tau into Alzheimer-like paired helical filaments. *FEBS Lett.* **399**, 344–349.

Kanemura, K., Takio, K., Miura, R., Titani, K. & Ihara, Y. 1992 Fetal-type phosphorylation of the tau in paired helical filaments *J. Neurochem.* **58**, 1667–1675.

Kang, J., Lemaire, H. G., Unterbeck, A., Salbaum, J. M., Masters, C. L., Grzeschik, K. H., Multhaup, G., Beyreuther, K. & Müller-Hill, B. 1987 The precursor of Alzheimer's disease amyloid A4 protein resembles a cell-surface receptor. *Nature* **325**, 733–736.

Kato, S. & Nakamura, H. 1990 Cytoplasmic argyrophilic inclusions in neurons of pontine nuclei in patients with olivopontocerebellar atrophy: immunohistochemical and ultrastructural studies. *Acta Neuropathol.* **79**, 584–594.

Kidd, M. 1963 Paired helical filaments in electron microscopy of Alzheimer's disease. *Nature* **197**, 192–193.

Kondo, J., Honda, T., Mori, H., Hamada, Y., Miura, R., Ogawara, H. & Ihara, Y. 1988 The carboxyl third of tau is tightly bound to paired helical filaments. *Neuron* **1**, 827–837.

Kosaka, K. 1978 Lewy bodies in cerebral cortex. Report of three cases. *Acta Neuropathol.* **42**, 127–134.

Krüger, R., Kuhn, W., Müller, T., Woitalla, D., Graeber, M., Kösel, S., Przuntek, H., Epplen, J. T., Schöls, L. & Riess, O. 1998 Ala30Pro mutation in the gene encoding α-synuclein in Parkinson's disease. *Nature Genet.* **18**, 106–108.

Ksiezak-Reding, H., Morgan, K., Mattiace, L. A., Davies, P., Liu, W. K., Yen, S.-H., Weidenheim, K. & Dickson, D. W. 1994 Ultrastructure and biochemical composition of paired helical filaments in corticobasal degeneration. *Am. J. Pathol.* **145**, 1496–1508.

Kuzuhara, S., Mori, H., Izumiyama, N., Yoshimura, M. & Ihara, Y. 1988 Lewy bodies are ubiquitinated. A light and electron microscopic immunocytochemical study. *Acta Neuropathol.* **75**, 345–353.

Lee, V. M.-Y., Balin, B. J., Otvos, L. & Trojanowski, J. Q. 1991 A68—a major subunit of paired helical filaments and derivatized forms of normal tau. *Science* **251**, 675–678.

Lewy, F. 1912 Paralysis agitans. In *Handbuch der Neurologie*, vol. 3 (ed. M. Lewandowski), pp. 920–933. Berlin: Springer Verlag.

Lippa, C. F. (and 16 others) 1998 Lewy bodies contain altered α-synuclein in brains of many familial Alzheimer's disease patients with mutations in presenilin and amyloid precursor protein genes. *Am. J. Pathol.* **153**, 1365–1370.

Masters, C. L. & Beyreuther, K. 1998 Alzheimer's disease. *Br. Med. J.* **316**, 446–448.

Masters, C. L., Simms, G., Weinman, N. A., Multhaup, G., McDonald, B. L. & Beyreuther, K. 1995 Amyloid plaque core protein in Alzheimer disease and Down syndrome. *Proc. Natl Acad. Sci. USA* **82**, 4245–4249.

Mezey, E., Dehejia, A., Harta, G., Papp, M. I., Polymeropoulos, M. H. & Brownstein, M. J. 1998*a* Alpha synuclein in neurodegenerative disorders: murderer or accomplice? *Nature Med.* **4**, 755–757.

Mezey, E., Dehejia, A. M., Harta, G., Suchy, S. F., Nussbaum, R. L., Brownstein, M. J. & Polymeropoulos, M. H. 1998*b* Alpha synuclein is present in Lewy bodies in sporadic Parkinson's disease. *Mol. Psychiatr.* **3**, 493–499.

Mirra, S. S. (and 13 others) 1999 Tau pathology in a family with dementia and a P301L mutation in tau. *J. Neuropathol. Exp. Neurol.* **58**, 335–345.

Mori, H., Kondo, J. & Ihara, Y. 1987 Ubiquitin is a component of paired helical filaments in Alzheimer's disease. *Science* **235**, 1641–1644.

Morishima-Kawashima, M., Hasegawa, M., Takio, K., Suzuki, M., Titani, K. & Ihara, Y. 1993 Ubiquitin is conjugated with N-terminally processed tau in paired helical filaments. *Neuron* **10**, 1151–1160.

Morishima-Kawashima, M., Hasegawa, M., Takio, K., Suzuki, M., Yoshida, H., Titani, K. & Ihara, Y. 1995 Proline-directed and non-proline-directed phosphorylation of PHF-tau. *J. Biol. Chem.* **270**, 823–829.

Mulot, S. F. C., Hughes, K., Woodgett, J. R., Anderton, B. H. & Hanger, D. P. 1994 PHF-tau from Alzheimer's brain comprises four species on SDS–PAGE which can be mimicked by *in vitro* phosphorylation of human tau by glycogen synthase kinase-3β. *FEBS Lett.* **349**, 359–364.

Munoz-Montano, J.R., Moreno, F.J., Avila, J. & Diaz-Nido, J. 1997 Lithium inhibits Alzheimer's disease-like tau protein phosphorylation in neurons. *FEBS Lett.* **411**, 183–188.

Nakajo, S., Tsukada, K., Omata, K., Nakamura, Y. & Nakaya, K. 1993 A new brain-specific 14-kDa protein is a phosphoprotein. Its complete amino acid sequence and evidence for phosphorylation. *Eur. J. Biochem.* **217**, 1057–1063.

Papp, M. I., Kahn, J. E. & Lantos, P. L. 1989 Glial cytoplasmic inclusions in the CNS of patients with multiple system atrophy. *J. Neurol. Sci.* **94**, 79–100.

Pérez, M., Valpuesta, J. M., Medina, M., Montejo de Garcini, E. & Avila, J. 1996 Polymerization of tau into filaments in the presence of heparin: the minimal sequence requirement for tau–tau interaction. *J. Neurochem.* **67**, 1183–1190.

Pollock, N. J., Mirra, S. S., Binder, L. I., Hansen, L. A. & Wood, J. G. 1986 Filamentous aggregates in Pick's disease, progressive supranuclear palsy, and Alzheimer's disease share antigenic determinants with microtubule-associated protein, tau. *Lancet* **ii**, 1211.

Polymeropoulos, M. H. (and 19 others) 1997 Mutation in the α-synuclein gene identified in families with Parkinson's disease. *Science* **276**, 2045–2047.

Poorkaj, P., Bird, T. D., Wijsman, E., Nemens, E., Garruto, R. M., Anderson, L., Andreadis, A., Wiederholt, W. C., Raskind, M. & Schellenberg, G. D. 1998 Tau is a candidate gene for chromosome 17 frontotemporal dementia. *Ann. Neurol.* **43**, 815–825.

Qi, Z., Zhu, X., Goedert, M., Fujita, D. J. & Wang, J. H. 1998 Effect of heparin on phosphorylation site specificity of neuronal Cdc2-like kinase. *FEBS Lett.* **423**, 227–230.

Reed, L. A. (and 10 others) 1997 Autosomal dominant dementia with widespread neurofibrillary tangles. *Ann. Neurol.* **42**, 564–572.

Reed, L. A., Schmidt, M. L., Wszolek, Z. K., Balin, B. J., Soontornniyomkij, V., Lee, V. M.-Y., Trojanowski, J. Q. & Schelper, R. L. 1998 The neuropathology of a chromosome 17-linked autosomal dominant Parkinsonism and dementia (pallido-ponto-nigral degeneration). *J. Neuropathol. Exp. Neurol.* **57**, 588–601.

Rewcastle, N. B. & Ball, M. J. 1968 Electron microscopic structure of the inclusion bodies in Pick's disease. *Neurology* **18**, 1205–1213.

Rizzu, P. (and 12 others) 1999 High prevalence of mutations in the microtubule-associated protein tau in a population study of frontotemporal dementia in the Netherlands. *Am. J. Hum. Genet.* **64**, 414–421.

Saudou, F., Finkbeiner, S., Devys, D. & Greenberg, M. E. 1998 Huntingtin acts in the nucleus to induce apoptosis but death does not correlate with the formation of intranuclear inclusions. *Cell* **95**, 55–66.

Schweers, O., Schönbrunn-Hanebeck, E., Marx, A. & Mandelkow, E. 1994 Structural studies of tau protein and Alzheimer paired helical filaments show no evidence for beta-structure. *J. Biol. Chem.* **269**, 24290–24297.

Sergeant, N., David, J. P., Lefranc, D., Vermersch, P., Wattez, A. & Delacourte, A. 1997*a* Different distribution of phosphorylated tau protein isoforms in Alzheimer's and Pick's diseases. *FEBS Lett.* **412**, 578–582.

Sergeant, N., David, J. P., Goedert, M., Jakes, R., Vermersch, P., Buée, L., Lefranc, D., Wattez, A. & Delacourte, A. 1997*b* Two-dimensional characterization of paired helical filament-tau from Alzheimer's disease: demonstration of an additional 74-kDa component and age-related biochemical modifications. *J. Neurochem.* **69**, 834–844.

Sisodia, S. S. 1998 Nuclear inclusions in glutamine repeat disorders: are they pernicious, coincidental, or beneficial? *Cell* **95**, 1–4.

Snow, A. D., Mar, S., Nochlin, D. & Wight, T. N. 1989 Cationic dyes reveal proteoglycans structurally integrated within the characteristic lesions of Alzheimer's disease. *Acta Neuropathol.* **78**, 113–123.

Sontag, E., Nunbhadki-Craig, V., Bloom, G. S. & Mumby, M. C. 1996 Regulation of the phosphorylation state and microtubule-binding activity of tau by protein phosphatase 2A. *Neuron* **17**, 1201–1207.

Spillantini, M. G. & Goedert, M. 1998 Tau protein pathology in neurodegenerative diseases. *Trends Neurosci.* **21**, 428–433.

Spillantini, M. G., Crowther, R. A. & Goedert, M. 1996 Comparison of the neurofibrillary pathology in Alzheimer's disease and familial presenile dementia with tangles. *Acta Neuropath.* **92**, 42–48.

Spillantini, M. G., Schmidt, M. L., Lee, V. M.-Y., Trojanowski, J. Q., Jakes, R. & Goedert, M. 1997*a* α-Synuclein in Lewy bodies. *Nature* **388**, 839–840.

Spillantini, M. G., Goedert, M., Crowther, R. A., Murrell, J. R., Farlow, M. J. & Ghetti, B. 1997*b* Familial multiple system tauopathy with presenile dementia: a disease with abundant neuronal and glial tau filaments. *Proc. Natl Acad. Sci. USA* **94**, 4113–4118.

Spillantini, M. G., Crowther, R. A., Jakes, R., Hasegawa, M. & Goedert, M. 1998*a* α-Synuclein in filamentous inclusions of Lewy bodies from Parkinson's disease and dementia with Lewy bodies. *Proc. Natl Acad. Sci. USA* **95**, 6469–6473.

Spillantini, M. G., Crowther, R. A., Jakes, R., Cairns, N. J., Lantos, P. L. & Goedert, M. 1998*b* Filamentous α-synuclein inclusions link multiple system atrophy with Parkinson's disease and dementia with Lewy bodies. *Neurosci. Lett.* **251**, 205–208.

Spillantini, M. G., Bird, T. D. & Ghetti, B. 1998*c* Frontotemporal dementia and Parkinsonism linked to chromosome 17: a new group of tauopathies. *Brain Pathol.* **8**, 387–402.

Spillantini, M. G., Murrell, J. R., Goedert, M., Farlow, M. R., Klug, A. & Ghetti, B. 1998*d* Mutation in the tau gene in familial multiple system tauopathy with presenile dementia. *Proc. Natl Acad. Sci. USA* **95**, 7737–7741.

Spillantini, M. G., Crowther, R. A., Kamphorst, W., Heutink, P. & Van Swieten, J. C. 1998*e* Tau pathology in two Dutch families with mutations in the microtubule-binding region of tau. *Am. J. Pathol.* **153**, 1359–1363.

Sturchler-Pierrat, C. (and 15 others) 1997 Two amyloid precursor protein transgenic mouse models with Alzheimer disease-like pathology. *Proc. Natl Acad. Sci. USA* **94**, 13287–13292.

Takeda, A., Mallory, M., Sundsmo, M., Honer, W., Hansen, L. & Masliah, E. 1998*a* Abnormal accumulation of NACP/α-synuclein in neurodegenerative disorders. *Am. J. Pathol.* **152**, 367–372.

Takeda, A., Hashimoto, M., Mallory, M., Sundsmo, M., Hansen, L., Sisk, A. & Masliah, E. 1998*b* Abnormal distribution of the non-Aβ component of Alzheimer's disease amyloid precursor/α-synuclein in Lewy body disease as revealed by proteinase K and formic acid pretreatment. *Lab. Invest.* **78**, 1169–1177.

Trinczek, B., Biernat, J., Baumann, K., Mandelkow, E. M. & Mandelkow, E. 1995 Domains of tau protein, differential phosphorylation, and dynamic instability of microtubules. *Mol. Biol. Cell* **6**, 1887–1902.

Tu, P. H., Galvin, J. E., Baba, M., Giasson, B., Tomita, T., Leight, S., Nakajo, S., Iwatsubo, T., Trojanowski, J. Q. & Lee, V. M.-Y. 1998 Glial cytoplasmic inclusions in white matter oligodendrocytes of multiple system atrophy brains contain insoluble α-synuclein. *Ann. Neurol.* **44**, 415–422.

Uéda, K. (and 10 others) 1993 Molecular cloning of cDNA encoding an unrecognized component of amyloid in Alzheimer disease. *Proc. Natl Acad. Sci. USA* **90**, 11282–11286.

Varani, L., Hasegawa, M., Spillantini, M. G., Smith, M. J., Murrell, J. R., Ghetti, B., Klug, A., Goedert, M. & Varani, G. 1999 Structure of tau exon 10 splicing regulatory element RNA and disruption by FTDP-17 mutations. *Proc. Natl Acad. Sci. USA* **96**. (In the press.)

Wakabayashi, K., Matsumoto, K., Takayama, K., Yoshimoto & Takahashi, H. 1997 NACP, a presynaptic protein, immunoreactivity in Lewy bodies in Parkinson's disease. *Neurosci. Lett.* **239**, 45–48.

Wakabayashi, K., Yoshimoto, M., Tsuji, S. & Takahashi, H. 1998*a* α-Synuclein immunoreactivity in glial cytoplasmic inclusions in multiple system atrophy. *Neurosci. Lett.* **249**, 180–182.

Wakabayashi, K., Hayashi, S., Kakita, A., Yamada, M., Toyoshima, Y., Yoshimoto, M. & Takahashi, H. 1998*b* Accumulation of α-synuclein/NACP is a cytopathological feature common to Lewy body disease and multiple system atrophy. *Acta Neuropathol.* **96**, 445–452.

Weinreb, P. H., Zhen, W., Poon, A. W., Conway, K. A. & Lansbury, P. T. 1996 NACP, a protein implicated in Alzheimer's disease and learning, is natively unfolded. *Biochemistry* **35**, 13710–13715.

Wilhelmsen, K. C., Lynch, T., Pavlou, E., Higgins, M. & Nygaard, T. G. 1994 Localization of disinhibition–dementia–Parkinsonism–amyotrophy complex to 17q21–22. *Am. J. Hum. Genet.* **55**, 1159–1165.

Wischik, C. M., Novak, M., Thogersen, H. C., Edwards, P. C., Runswick, M. J., Jakes, R., Walker, J. E., Milstein, C., Roth, M. & Klug, A. 1988 Isolation of a fragment of tau derived from the core of the paired helical filament of Alzheimer disease. *Proc. Natl Acad. Sci. USA* **85**, 4506–4510.

Yoshida, H. & Ihara, Y. (1993) Tau in paired helical filaments is functionally distinct from fetal tau: assembly incompetence of paired helical filament tau. *J. Neurochem.* **61**, 1183–1186.

Index